SLIPPER ORCHIDS

of THE TROPICAL AMERICAS

Frontispiece. The oldest painting of a *Phragmipedium* is that of *Phragmipedium caudatum* painted in 1786 near Pillao, Peru by Jose Brunete for Ruiz and Pavon.

SLIPPER ORCHIDS of THE TROPICAL AMERICAS

Phillip Cribb and Christopher Purver

Plates by **Carol Woodin**

with additional illustrations by

Stig Dalström, Walter Hood Fitch, Matilda Smith,

Everard Im Thurn and Robert Schomburgk

Line drawings by

Judi Stone, Paul Harwood and Carol Woodin

Natural History Publications (Borneo)
Kota Kinabalu

The Orchid Digest Corporation
Laguna Niguel

in association with

Royal Botanic Gardens Kew

2017

Published by

Natural History Publications (Borneo) Sdn. Bhd. (216807-X)
A913, 9th Floor, Wisma Merdeka Phase 1,
P.O. Box 15566,
88864 Kota Kinabalu, Sabah, Malaysia.
Tel: +6088-233098 Fax: +6088-240768
e-mail: info@nhpborneo.com
Website: www.nhpborneo.com

and

The Orchid Digest Corporation
P.O. Box 6966,
Laguna Niguel,
CA 92607-6966, USA.
Email: odeditor@cox.net

in association with

Royal Botanic Gardens, Kew
Richmond,
Surrey TW9 3AB, United Kingdom.

Slipper Orchids of the Tropical Americas
by Phillip Cribb and Christopher Purver

ISBN 978-983-812-175-0

First published March 2017.

Frontispiece: *Phragmipedium caudatum*. (Courtesy of Real Jardin Botanico, Madrid).

Printed in Taiwan.

CONTENTS

LIST OF COLOUR PLATES

PREFACE

My first botanical expedition from December 1970 until April 1971 took me to Peru and Bolivia and introduced me to the amazingly rich orchid flora of the Andes. It also provided me with my first view of a *Phragmipedium*, namely *P. caudatum*, an epiphytic species spotted in the montane forests near Machu Picchu. I little imagined that some 45 years later I would be involved in preparing an account of the genus for The Royal Botanic Gardens, Kew, this being the third volume in a revision of all of the slipper orchids, an isolated subfamily of the orchids with less than 200 species. However, what they lack in numbers they more than make up for in their fascinating evolution, morphology, biology and horticultural interest.

Three genera found in the American tropics from Mexico south to Bolivia and east to Brazil are included in this volume. They are, respectively, *Selenipedium*, a small genus of bamboo-like terrestrials with long erect stems, many well-spaced, pleated leaves and small flowers; *Mexipedium*, a monotypic genus from southern Mexico with a lithophytic rhizomatous habit, short stems, a fan of hard leaves and small white flowers; and *Phragmipedium*, the largest genus with over 20 species including terrestrials, lithophytes or epiphytes with short stems, fans of hard leaves and small to large and sometimes colourful flowers. A fourth genus, *Cypripedium*, with elongate stems, several pleated leaves and large yellow flowers, has three species within the range of genera covered here but they were included in an earlier monograph in this series (Cribb, 1997).

Phragmipediums were introduced into cultivation in the 19th century and enjoyed a brief period of popularity in the late Victorian era. *Mexipedium* appeared in the 1990s in cultivations and is grown as a curiosity. Selenipediums do not adapt well to cultivation and their large size and small flowers are not of horticultural interest.

The discovery of the scarlet-flowered *Phragmipedium besseae* in Peru in the early 1980s marked a turning point in horticultural interest in the genus *Phragmipedium*. Prior to this event, the hard-leaved tropical American slipper orchids had been of marginal interest. Orchid nurseries often grew one or two species and a few of the hybrids that had been bred in the late 19th century but hybridisation had reached a plateau and few nurseries saw a future in further developing them as a popular breeding group. After all, most of the species and hybrids had rather dull coloured flowers in shades of green, brown and purple. The only species that bucked the trend was *P. schlimii*, a species with small white to rose-pink flowers that had been used to produce several attractive hybrids, such as *P.* Sedenii and *P.* Cardinale with pink and white flowers.

Phragmipedium besseae caused a sensation when it was first introduced into cultivation in 1982. A plant shown for award at the Royal Horticultural Society is one of the few novelties that has produced an audible gasp of amazement from the committee of experienced judges. It was quickly hybridised with every other species and hybrid that could be found. The Eric Young Orchid Foundation in Jersey was in the forefront of this breeding programme. Its first flowered hybrids were extremely encouraging, the red pigment being dominant in the hybrids. Eric Young employed the cytogeneticist Don Wimber to experiment with chromosome doubling both species and hybrids. His efforts with *Phragmipedium besseae* hybrids were little short of sensational, producing vigorous plants that were easy to grow and produced larger more deeply coloured flowers that lasted much longer than those of their diploid siblings.

A new generation began to grow these sensational slipper orchids which coincidentally began to win awards at nearly every show where they were exhibited. The discovery of *P. besseae* also stimulated further exploration to see if other species had been overlooked, particularly in the Andes of Peru, Ecuador and Colombia. The conclusion of this renewed interest was the discovery in 2001 of *Phragmipedium kovachii*, the most distinctive species in the genus with extraordinarily large royal purple flowers borne on tall red-hairy flower stalks. Its discovery made the front pages of the national press in the Americas, Europe, Asia and Australia. It reappeared shortly afterwards as a 'cause célèbre' because its introduction into cultivation was illegal, leading to a court case in the USA that was widely covered by both the press and television.

Needless to say, the publicity stimulated the popularity of phragmipediums, awareness of the vulnerability of horticulturally exciting new species and the need to protect them in the wild. Fortunately, an enlightened attitude by the Peruvian government allowed two Peruvian nurseries to propagate plants for export. New hybrids of *P. kovachii* soon appeared, the first less interesting than the species, but the second generation are showing far greater promise.

Eric Young asked me many years ago to sort out the nomenclature of *Phragmipedium* but my interests at the time were mainly in African and Asian orchids. However, given the increasing number of novelties that have been discovered and the fine range of hybrids that have been bred since his request, I have maintained an interest and worked closely with Chris Purver, now manager of the Eric Young Foundation. This book is the result of our continuing collaboration.

Phillip Cribb
Royal Botanic Gardens, Kew
October 2015

ACKNOWLEDGEMENTS

We would like to thank Carol Woodin for her superb watercolour illustrations, Judi Stone for the bulk of the line drawings and Paul Harwood for additional line drawings. They have worked to the highest standard, often with limited material and we are delighted with the results. Other watercolours come from the archives of illustrations held at Kew and from Stig Dalström. Christopher Mills, Martyn Rix, Julia Buckley, Trishya Long and Lynn Parker selected the historical plates from the *Botanical Magazine*, John Day's scrapbooks and other illustrations held in the Kew archives and herbarium; Trish Long scanned them. They are used with permission of the Director and Trustees of the Royal Botanic Gardens, Kew.

The following have kindly lent materials for illustration: Jerry Fischer, Lou Hegedus, Margie and Tom Kohler, Joseph Kunisch, Terry Partin, Pepe Portilla, Robert and Lynn Wellenstein, and Harry and Rosemarie Zelenko. Dr José Luis Fernández Alonso and Esther García Guillén of the Jardín Botaníco del Madrid gave permission to reproduce the paintings from the Ruiz and Pavon expedition to Peru and Chile. Henry Oakeley provided the digital images of these plates.

Information, herbarium, spirit material or photographs have been kindly provided by: Chuck Acker, Roberto Angulo, Luis Baquero, João Batista Fernandes da Silva, N. Baumbach, Bridget Beattie, Nicky Biggs, Ian Chalmers, Stig Dalström, the late Galfrith (Stalky) Dunsterville, Jerry Fischer, Eric Hagsater, Paul Harcourt Davies, Ray Harley, Lou Hegedus, J. Hermans, E. Kaes, Ch'ien Lee, J. Levy, Eve Lucas, Alfredo Manrique, Isobel and the late David Miller, Henry Oakeley, I. Olmedo, Eduardo Alberto Pérez Garcia, Bridget Riley, Clark Riley, the late Isaias Rolando, Gustavo Romero, Gerardo Salazar, Rolando Solano Gómez, Barbara Tischerman, Francisco Tobar Perez, Cassio van den Berg, Dick Warren, Oliver Whaley, Mark Whitten and Harry Zelenko. The copyright of images that have been used remains with the photographers.

One of the authors (PC) would like to thank his companions on several enjoyable field trips, particularly Sam Crothers, Eric Hagsater, the late Ed Greenwood, Harold Koopowitz, Steve Hampson, the late Miguel Soto Arenas, Gerardo Salazar, Eduardo Alberto Pérez Garcia, Alex and Pepe Portilla, and Oliver Whaley.

Jing Wei (Brian) Yap is thanked for kindly running the DNA sequences for the cladograms published here that clarify the phylogetic issues raised by previous authors. Hassan Rankou and Mike Fay of the Jodrell Laboratory, Kew, generously provided the thorough conservation assessments. Aaron Davis and Tim Wilkinson are kindly thanked for helping with producing the distribution maps. Samuel Sprunger, Lucas Loew and Vic Kettle have helped with scanning and cleaning the images.

The following herbaria have been visited in the preparation of this monograph: Ames (AMES), Mexican Orchid Association (AMO), Basel (RENZ), Edinburgh (E), Florence (F), Geneva (G), Kew (K), Missouri (MO), Natural History Museum (BM), New York (NY), Paris (P), Marie Selby (SEL), Quito (QCA), and Vienna (W & W-RCHB).

We are particularly grateful to David Simpson, Bill Baker, and Gina Fullerlove and other staff of the Royal Botanic Gardens, Kew, and the Trustees and staff of The Eric Young Orchid Foundation in Jersey for their support while preparing this book.

Finally, we would like to thank Datuk Chan Chew Lun, Natural History Publications (Borneo), Kota Kinabalu, who is the designer and publisher of this monograph. Without his support, talents and care, this volume would be much the poorer. The publication of this work has been made possible through the generosity of Professor Harold Koopowitz and the Orchid Digest Foundation who arranged sponsorship. Our sponsors are also gratefully acknowledged.

Photographic acknowledgements and copyright

Chuck Acker (CA), Roberto Angulo (RA), João Batista Fernandes da Silva (JB), N. Baumbach (NB), Bridget Beattie (BB), Nicky Biggs (NB), Stig Dalström (SD), the late Galfrith (Stalky) Dunsterville (GD), Eric Young Orchid Foundation (EYOF), Jerry Fischer (JF), Eric Hagsater (EH), Paul Harcourt Davies (PD), Lou Hegedus (LH), E. Kaes (EK), Ch'ien Lee (CL), J. Levy (JL), Henry Oakeley (HO), I. Olmedo (IO), Eduardo Alberto Pérez Garcia (EP), Pepe Portilla (PP), Gustavo Romero (GR), Royal Botanic Gardens, Kew (Kew), Rolando Solano Gómez (RS), Swiss Orchid Foundation (SFO), Francisco Tobar Perez (FT), Cassio van den Berg (CB), Dick Warren (RW), Oliver Whaley (OW).

This book has been made possible due to the generosity of the sponsors listed below:

LAWRENCE & LOIS DAUELSBERG
STEVE HAMPSON
RICHARD HESS
HILO ORCHID FARM (JAMES FONG)
HAROLD KOOPOWITZ
KRULL-SMITH (FRANK SMITH)
LEHUA ORCHIDS (GRAHAM WOOD)
FRED MISSBACH
MICHAEL PUGH & STEVE GOLLIS
SLIPPER ORCHID ALLIANCE
SANDRA TILLISCH SVOBODA
PAUL & ANNE TUSKES
TAN JIEW HOE

Plate 1. *Phragmipedium boissierianum.* Watercolour by Francisco Pulgar prepared on the Ruiz and Pavon expedition in Peru in 1787. (Courtesy of Real Jardin Botanico, Madrid)

INTRODUCTION

The Spanish King Carlos III commissioned the Spanish botanists Hipolito Ruiz Lopez (1754–1816) and José Pavon Jimenez (1754–1840) to undertake the botanical exploration of Peru and Chile between 1777 and 1788 and they were the first to discover and collect tropical American slipper orchids (Pupulin, 2012) The results of their expedition, over 3000 collections of herbarium specimens and 2500 watercolour illustrations, formed the basis of their proposed ten-volume *Florae Peruviana et Chilensis* (1798–1802) of which only the first four volumes appeared in their life-times. The illustrations, made for them by the artists José Brunete Casto Dubua (1746–1787), Isidro Gálvez Galo (1754–1809), Francisco Pulgar (fl. 1784–1788), José Gabriel Rivera (died 1815) and Francisco Cortés Alcocer (1770–1841) are deposited in the archives of the Jardín Botaníco de Madrid. Pupulin (2012) has recently published, for the first time, the complete collection of their watercolour paintings of orchids.

Phragmipedium caudatum was the first slipper orchid to be collected by them in 1786 from Pillao, Peru followed by *P. boissierianum* in 1787. Amongst the illustrations are watercolours of both *P. caudatum* (Frontispiece) by Brunete and *P. boissierianum* (Plate 1) by Pulgar. Unfortunately, neither species was described at the time by Ruiz and Pavon.

John Lindley saw a flower of each that had been brought from the Ruiz and Pavon collection, still in Lima, by Andrew Matthews (1801–1841) for Sir William Hooker. Lindley described one, as *Cypripedium caudatum*, but the second flower proved to be too badly preserved for description (Lindley, 1840). At the same time he also described *Cypripedium lindleyanum* based on a specimen collected and named by Robert Schomburgk that had been collected by him on Mount Roraima in Guyana. Lindley commented at the time "I unwillingly consent, at the particular instance of Mr Schomburgk, to allow this plant to bear my name, who (sic) have no title to the compliment". Schomburgk's original watercolour illustration (Plate 2) is lodged in Lindley's herbarium at Kew.

The range of *Phragmipedium* was extended to Brazil by José Mariano Vellozo (1831) when he described *Cypripedium vittatum* (now *Phragmipedium vittatum* (Vell.) Rolfe).

The genus *Phragmipedium* has had a chequered nomenclatural history. John Lindley (1799–1865) coined the earliest name for the genus in 1846 when he established *Uropedium* for *U. lindenii*, the only species to lack a slipper-shaped lip (Plate 3). For many years, until 1975, the *International Code of Botanical Nomenclature* (ICBN) allowed names based on teratological or monstrous plants to be discounted for purposes of priority. Hence, the later name *Phragmipedium*, established in 1896 by Robert Rolfe for Lindley's *Cypripedium caudatum*, became the established name for this genus of hard-leaved tropical American slipper orchids. More recently, this rule was overturned and *Phragmipedium* has now been conserved over *Uropedium* (in *Taxon* 64/104. 9 Feb 1996).

Well before that, in 1854, H.G. Reichenbach established the genus *Selenipedium* for *S. palmifolium*, a terrestrial plant with slender elongate erect stems bearing many lanceolate pleated leaves along their length and with short subterminal inflorescences of small, yellow, slipper-shaped flowers (Plate 4). He placed six other species, now in *Phragmipedium*, in the genus. In the same year, he and Jean Linden described *Selenipedium* (now *Phragmipedium*) *schlimii*, a hard-leaved species with a very short stem and pink flowers,

Plate 2. *Phragmipedium lindleyanum*. Drawn by Robert Schomburgk in Guyana between 1835 and 1839 and now in the Lindley herbarium at Kew. (Lindley herbarium, Kew).

Plate 3. *Phragmipedium lindenii* (as *Uropedium lindenii*). From H.G. Reichenbach's *Xenia Orchidacea* 1: t. 15 (1858).

Plate 4. *Selenipedium chica* and *S. palmifolium* from H.G. Reichenbach's *Xenia Orchidacea* 1: t.2 (1854).

Plate 5. *Phragmipedium kovachii*. Drawn in habitat in Peru by Stig Dalström.

Plate 6. *Phragmipedium besseae*. Drawn by Stig Dalström for *Curtis's Botanical Magazine*, n.s. t. 135 (1989) from a cultivated plant.

based upon a collection made for Linden by his nephew, Louis Schlim, while collecting orchids near Ocaña in Colombia (Plate 7). This was the first species to be described that did not have greenish, yellowish or brown flowers and was later significant as a parent in early *Phragmipedium* breeding, introducing pink and white into the early hybrids. Until his death in 1885, Reichenbach continued to include all other tropical American slipper orchids in *Selenipedium* rather than in *Cypripedium*, adding another ten hard-leaved species to the genus.

Selenipedium chica, a second species with elongate stems, pleated leaves and small flowers, found in Panama was described by Reichenbach (1854) alongside *S. palmifolium*. A third species was discovered in Brazil by João Barbosa Rodrigues when he collected and described *S. isabelianum* (Diario do Grão Pará, 285 (8 Dec. 1874)).

Robert Rolfe (1896) was the first to recognise *Phragmipedium* as a distinct genus from *Selenipedium*, based upon its short stem, fan of hard conduplicate leaves and trilocular ovary, and included nine species in the genus. Pfitzer (1898) changed the spelling to *Phragmopedilum*, but this is treated as an orthographic variant and has not been taken up by later authors.

The novelties added to the genus for over 80 years after its establishment were broadly similar to *P. caudatum* and *P. longifolium* in floral morphology and colouration. In 1981, however, *P. besseae*, a remarkable species with orange-red or red flowers, was discovered in Peru and was described by Calaway Dodson and Janet Kuhn (Plate 6). Its successful introduction into cultivation and hybridising programmes initiated a resurgence of interest in *Phragmipedium* breeding. The resulting progeny, often inheriting the red and orange-red tones of *P. besseae*, have been winning awards around the world ever since.

Castaño *et al.* (1984) extended the range of the genus to southern Mexico with the description of *P. exstaminodium*, a long-petalled species in which the staminode was much reduced.

Since then, the discovery in 2001 of *P. kovachii*, another Peruvian species with magnificent imperial purple flowers of a size never before seen in tropical hard-leaved slipper orchids, has created further interest in the genus (Plate 5). Its early hybrids, while less impressive than the species, have the potential to further increase the popularity of the genus in horticulture.

Leslie Garay (1979) revised *Phragmipedium* and recognised five sections and 21 species. His species delimitation relied largely on floral morphology with particular emphasis on petal shape, lip lobing and staminode structure. John Atwood (1988) was the first to test the monophyly of the slipper orchid genera. He confirmed the monophyly of *Phragmipedium* as sister to the South-east Asian genus *Paphiopedilum* but was unable to satisfactorily resolve *Selenipedium* and *Cypripedium* other than that the latter was derived from the former. Lucille McCook (1989 a & b, 1990, 1998) revised *Phragmipedium* for her doctoral thesis, reducing the number of sections to four by combining sections *Phragmipedium* and *Himantopetalum*, the latter being basal in the larger clade that includes species of section *Phragmipedium*. She recognised 15 species in the genus, including the recently described *P. xerophyticum*. Her treatment was later published in summary with a detailed explanation of her broad species concept (McCook, 1998). The present work is greatly indebted to her treatment of the genus.

Plate 7. *Phragmipedium schlimii*. Drawn by Walter Hood Fitch for *Curtis's Botanical Magazine*, t. 6514 (1867).

Olaf Gruss (1996) published a series of papers on the genus in the journal *Die Orchidee* which were later published as a Supplement to the journal. He recognised five sections and 16 species in his treatment, adopting a broad species concept based largely on floral morphology. Gruss (2003) published an updated classification, recognising 20 species and 33 taxa in his treatment.

Guido Braem (2004) established a new section *Schluckebieria* for *P. kovachii*, distinguishing it by its large distinctive purple flowers, the form and position of both the lip and staminode, and the lack of translucent windows at the back of the pouch. Braem (2011) raised three sections to subgeneric rank: *Schluckebieria*, *Micropetalum* and *Platypetalum*; and established a fourth, namely, *Longifolium*. He recognised two sections within the last, *Lorifolia* and *Viridiflorae*. His recognition of *Schluckebieria* at subgeneric rank was only validated by him in a later publication (Braem, 2015). His treatment is considered here as taxonomic inflation and unnecessary in such a small genus.

Gruss (2014) has recently published in German an account of *Phragmipedium* bringing together his many articles from orchid journals published over the previous 23 years. He recognised 26 species and six varieties in the genus and largely followed Garay's 1979 infrageneric classification of the genus.

The discovery and description of the dwarf, hard-leaved, small-flowered slipper orchid in southern Mexico produced considerably more interest in the scientific than in the horticultural community because its flowers combined morphological features of both *Phragmipedium* and the Asiatic *Paphiopedilum*. Miguel Soto, Gerardo Salazar and Eric Hagsater described it in 1990 as *Phragmipedium xerophyticum* based on a collection from Oaxaca in Mexico where it grows on rocks. It has a xerophytic habit, and is a small plant with tough leathery leaves and small white flowers on a successively flowering and branching inflorescence. It was originally included by Soto *et al.* (1990) in *Phragmipedium* because of its possession of sepals with valvate aestivation, the absence of sinuous epidermal cells in the perianth, the fusion of the side lobes of its lip and the ventral synsepal being larger than the dorsal sepal. Victor Albert and Mark Chase (1992) agreed that a branched inflorescence is found in *Phragmipedium* but is absent from *Paphiopedilum* but noted that only the first criterion listed by Soto *et al.* was consistent in distinguishing *Phragmipedium* (including *P. xerophyticum*) from *Paphiopedilum*. With the benefit of data from DNA analysis which showed that *P. xerophyticum* was sister to the clade comprising the rest of *Phragmipedium*, Albert and Chase established the genus *Mexipedium* to accommodate this strange species, and their view has been accepted by later workers. Based upon the observation that *Phragmipedium*, *Mexipedium* and *Paphiopedilum* form a larger clade that is monophyletic (Albert, 1994), Albert and Pettersson (1994) then sunk the first two into *Paphiopedilum*, the earliest name. This treatment has not been taken up by other botanists or horticulturists, mainly because of the distinctiveness of the chromosome number and breeding behaviour of these genera.

MORPHOLOGY

The morphological differences found in slipper orchids are the basis for the generic delimitation of the five recognised genera, four of which occur in the tropical Americas. Only *Paphiopedilum*, found in South and South-East Asia across to New Guinea and the Solomon Islands, is not found in the tropical Americas. *Paphiopedilum* plants resemble those of *Phragmipedium* but the flowers have an unilocular (one-celled) ovary with parietal placentation and sepals that are imbricate not valvate, as found in all the other genera (Rolfe, 1896).

Selenipedium is the most distinctive of the slipper orchid genera, having long bamboo-like stems, many plicate, relatively thin-textured leaves and inflorescences in which the flowers bloom successively (Fig. 1). The flowers are small compared with the plant and the flowers of most other slipper orchids and are persistent. The dorsal sepal and synsepal are similar, the petals are linear-tapering and not twisted and the lip has an incurved margin with no claw visible. The flowers are concolorous or sometimes the lip is distinctively coloured. The lateral lip lobes are free, not united as in *Phragmipedium*. The column bears two lateral anthers and a decurved apical staminode that is usually ovate, trullate, deltoid or linear. The seeds of *Selenipedium* are unique amongst slipper orchids, being subglobose rather than fusiform

Cypripedium, like *Selenipedium*, has plicate leaves but they are fewer in number and borne on a shorter fleshy stem and a persistent perianth. The three species found in Mexico and Central America have a stem bearing a few hairy pleated leaves and a terminal inflorescence bearing a few large flowers in succession. The flowers have a dorsal sepal and synsepal and a concolorous inflated lip with an incurved margin. At the back of the lip large transparent windows can be found, presumed to assist pollinators in exiting the flower.

Plants of *Phragmipedium* (Fig. 3) and the monotypic *Mexipedium* (Fig. 2) resemble closely those of the tropical Asian sister genus *Paphiopedilum*. They have several hard conduplicate leaves arranged distichously in a fan on a short stem and produce an inflorescence of several to many flowers. Their resemblance is close enough to have led some authors to consider them to belong to the same genus (Albert & Pettersson, 1994). Indeed, the hard-leaved slipper orchids belong to a monophyletic clade sister to the temperate genus *Cypripedium*. Most modern authors continue to treat the hard-leaved slipper orchids in three genera, an approach that is supported here. How then does *Phragmipedium* differ from its allies? It differs in several features from the Asiatic genus *Paphiopedilum*, notably in its trilocular (three-celled) ovary with axile placentation and distinct chromosome number. *Mexipedium* differs from *Phragmipedium* in having an elongate creeping rhizome, tiny white flowers and a unilocular ovary.

A detailed account of the morphology of *Phragmipedium* is provided by McCook (1988).

Fig. 1. *Selenipedium palmifolium.* **A**, habit; **B**, base of plant and roots; **C**, dorsal sepal; **D**, synsepal; **E**, petal; **F**, lip, longitudinal section; **G**, **H**, column, side and front views; **I**, ovary, transverse section. All drawn by Judi Stone: from *Maas & Westra* 3962, *Trinidad Forest Department* 2728 & *Philcox* 6120 (Kew Spirit Collection no. 37409). Single bar = 1 mm; double bar = 10 mm.

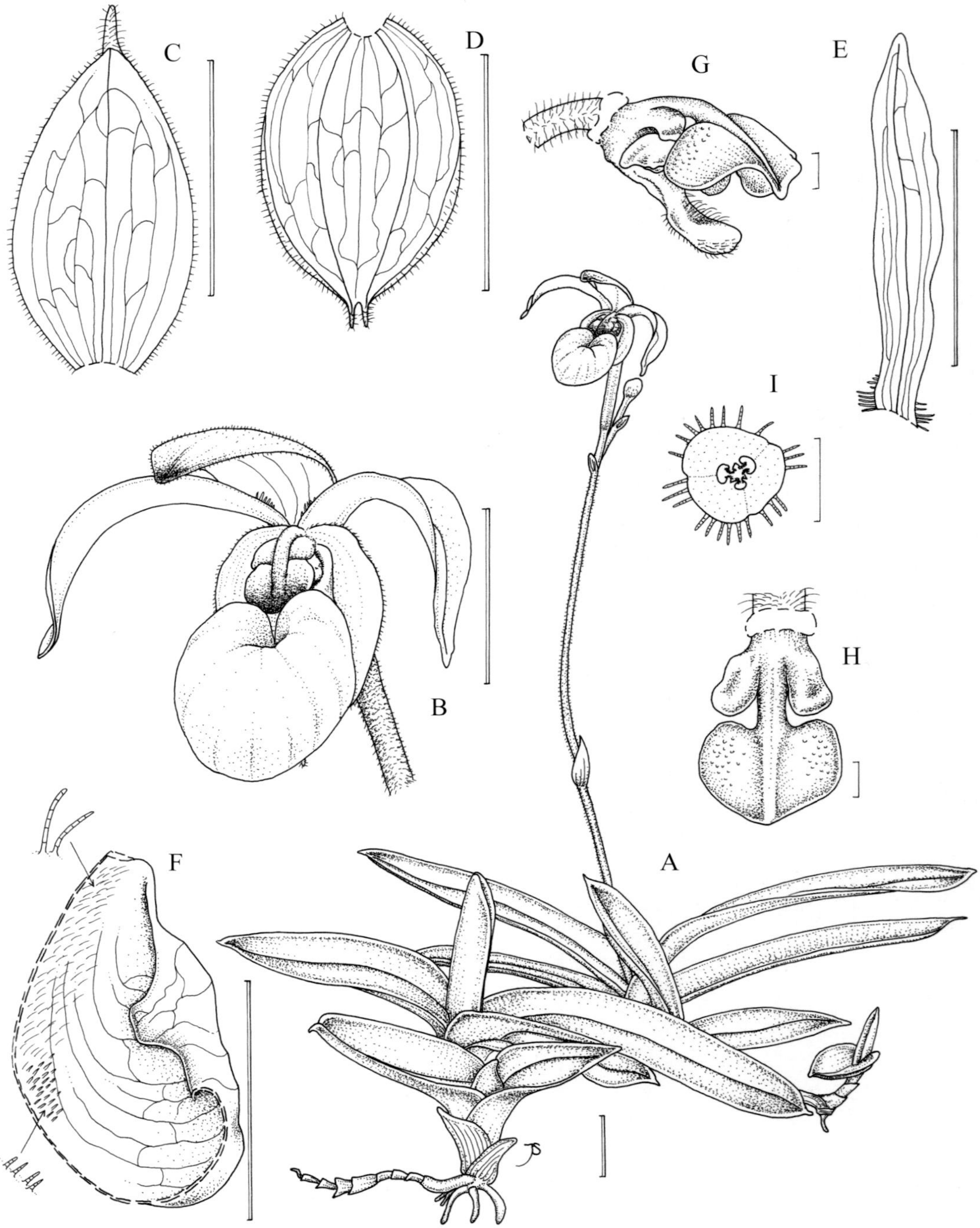

Fig. 2. ***Mexipedium xerophyticum.*** **A**, habit; **B**, flower; **C**, dorsal sepal; **D**, synsepal; **E**, petal; **F**, lip, longitudinal section; **G** & **H**, column, oblique and top views; **I**, ovary, transverse section. Drawn by Judi Stone after Soto *et al.* (1990) and Kew Spirit Collection 50896. Single bar = 1 mm; double bar = 10 mm.

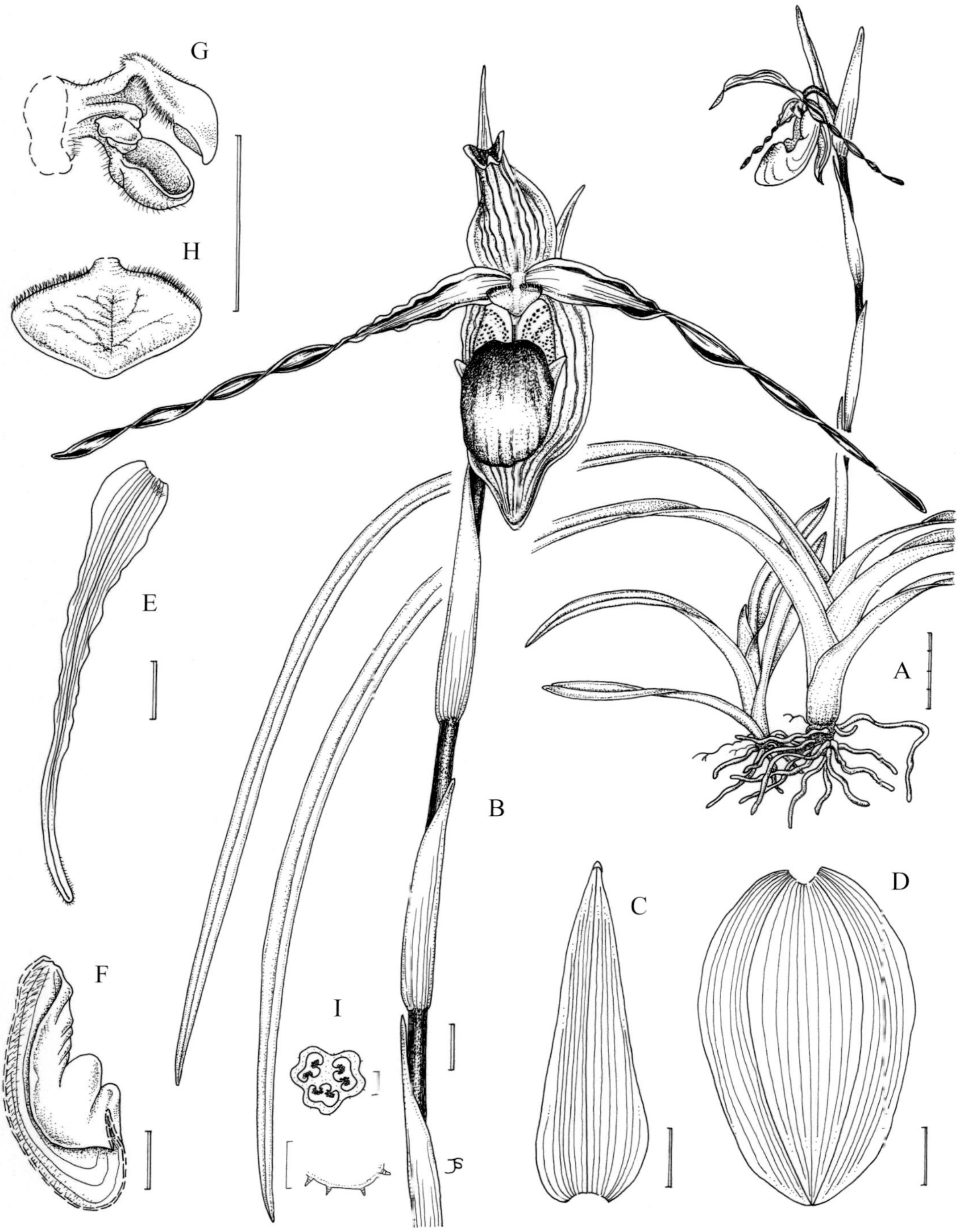

Fig. 3. ***Phragmipedium longifolium.*** **A**, habit; **B**, inflorescence and flower; **C**, dorsal sepal; **D**, synsepal; **E**, petal; **F**, lip, longitudinal section; **G**, **H**, column, side and front views; **I**, ovary, transverse section. Drawn by Judi Stone from *Hugh-Jones* 162 and Kew Spirit Collection (cult. Kunitsch)' with 'Hugh-Jone. Single bar = 1 mm; double bar = 10 mm.

Habit

Selenipedium and the tropical American *Cypripedium* species bear several pleated leaves on an elongate stem. *Selenipedium* species are tall clump-forming plants growing from a short rhizome. Their stems, often mistaken for bamboo or a *Sobralia*, are leafy along the upper two-thirds of their length and bear terminal inflorescences. Cypripediums are similar but generally shorter with a fleshy stem, although *C. irapeanum* can exceed a metre or more in height.

In contrast, all *Phragmipedium* species and *Mexipedium xerophyticum* possess a short stem and conduplicate green leaves. The leaf type falls into two broad categories: the majority of the species have leathery linear-ligulate leaves but species of *Phragmipedium* section Micropetalum have broader leaves of a thinner texture. Most species have leaves with either acute tips but the long-petalled species of section *Phragmipedium* have rounded tips. A number of *Phragmipedium* species have leaves with thin yellow margins.

Inflorescence

All slipper orchids bear terminal inflorescences. Most species have a racemose inflorescence, often reduced to a single flower (but not in the tropical American slipper orchids), but branching inflorescences can be found in a few species, e.g. *Mexipedium xerophyticum*, *Phragmipedium dalessandroi*, *P. caricinum* and *P. lindleyanum* which regularly have branched paniculate inflorescences but can also produce unbranched ones. A number of other species will also occasionally produce branched inflorescences. Successive flowering, considered the pleisiomorphic state, is normal in the tropical American slipper orchids. Elsewhere simultaneous flowering is common.

Several features of the inflorescence are taxonomically useful. The peduncle can be sheathed or unsheathed with sterile bracts which resemble small leaves. The species of *Phragmipedium* with elongate petals lack such sheaths. The peduncle and rachis can be glabrous or pubescent. The fertile bracts can be shorter or longer than the ovary. In some cases the bracts can be brightly coloured red-purple.

Flower

Flower number can vary a lot, but all of the tropical American species have two or more flowers. Flowers of some species enlarge after anthesis, but flowering at full size is considered the pleisiomorphic condition. The buds can be glabrous or be covered with multi-cellular hairs, the latter being found in *Selenipedium*, *Cypripedium* and in species of *Phragmipedium* sect. *Platypetalum*. Species of *Selenipedium* and *Cypripedium* retain their flowers after senescence but *Phragmipedium* and *Mexipedium* species have early deciduous flowers which drop from the tip of the ovary at senescence.

Selenipedium and *Phragmipedium* have trilocular ovaries, but *Cypripedium* and *Mexipedium* have unilocular ones, a distinction first identified by Reichenbach in 1854. Ovaries can be pubescent, glandular or hairy.

All of the tropical American slipper orchids have a calyx of two parts: a dorsal sepal and a ventral synsepal, the latter comprising the united two lateral sepals of other orchids. Both are usually convex and elliptic to ovate in shape.

The corolla of slipper orchids comprises two lateral petals and a distinctly shaped, urn- or slipper-like ventral pouch. The petals can spread at right angles or at various angles below the horizontal. Petals are linear-tapering and incurved in *Selenipedium*, more or less elliptic and spreading in the Mexican and Central American *Cypripedium* species, linear-oblong and incurved in *Mexipedium* and elliptic, linear-tapering or spatulate and spreading to pendent in *Phragmipedium*. *Selenipedium* has glabrous petals, the others are variously pubescent, particularly at the inner base and often on the margins.

The lip is the most distinctive feature of slipper orchids, one of the defining feature of the subfamily. All species have a slipper-shaped or urn-shaped lip except *Phragmipedium lindenii* and *Selenipedium chironianum* where the lips are linear-tapering, petal-like and pendent. They are not surprisingly self-pollinated.

The simplest lips are found in *Selenipedium* and *Cypripedium* where the entire margin is incurved around a more or less circular to elliptic orifice. All species are furnished with hairs within on the ventral surface of the lip, allowing any trapped insect to climb up the hairs and exit the lip on either side of the base of the column where an opening exists. The orange, yellow, brown or lilac lip in *Selenipedium* usually has two purple marks on each side of the orifice in its basal half or in the middle. The lip of the three *Cypripedium* species is yellow, the same colour as the other floral segments. *Mexipedium* has a white flower and also has a lip with a simple incurved margin. Lip shape in more variable in *Phragmipedium* where the lip is three-lobed, the side lobes incurved and more or less adnate and fused along their margins. The incurved side lobes can be tuberculate, spotted or ribbed on the claw face. The midlobe is deeply saccate or urceolate, much larger and has either an incurved or a fused margin. Translucent windows occur in the back of the lip in the cypripediums and in *Phragmipedium* sect. *Micropetalum.*

The column of slipper orchids is a short pubescent stalk and somewhat triangular in cross-section. It bears two lateral stamens, each somewhat bilocular with pollen borne in masses. A sterile anther or staminode of varying shape is borne at the apex of the column. Staminode shape has traditionally been used in species delimitation but has almost certainly been over-emphasised in its significance. Staminodes can be linear, transversely rhombic or broadly ovoid, elliptic or oblong and variously coloured and decorated with purple and yellow. Only in *P. humboldtii* var. *exstaminodium* is the staminode reduced to a short knob that does not cover the entrance to the lip. Unsurprisingly, it is self-pollinating.

The stigma is borne ventrally on the column, hidden from view by the basal lobes of the lip. It is usually shortly stalked, and more or less triangular with a noticeable inverted Y-shaped scar dividing the stigma into three parts. Staminodes are often papillate.

ANATOMY

A detailed account of the anatomy of *Selenipedium*, *Mexipedium* and *Phragmipedium* is provided by Pridgeon in Pridgeon *et al.* (1999). Some of the more pertinent details are described below.

Selenipedium

The roots of *Selenipedium* have a uniseriate (one-layer thick) velamen that lacks root hairs. A multiseriate layer of sclerenchyma cells lies immediately inside the uniseriate endodermis. The sclerenchyma strengthens the roots. Simply perforate vessels and metaxylem tracheids are both present in the roots.

The rhizome is fleshy and bears large starch grains. Its epidermis is sloughed and lacks hairs and stomata. Vessels are absent but tracheids are present. The stem possesses a thick cuticle, stomata and both non-glandular and glandular hairs. A thick schlerenchyma is present as a sheath around the vascular bundle which lacks vessels.

The thin-textured plicate leaves and bracts have an epidermis with straight-walled cells of small volume. Stomata are present on both surfaces, the guard cells being the smallest among slipper orchids. Glandular and non-glandular hairs are found on both surfaces. The vascular strands are sheathed in schlerenchyma, providing support for the leaves.

Mexipedium

Mexipedium xerophyticum has, as its name suggests, xerophytic features, stout roots, long stolons, a short stem and succulent leaves. It is likely that its anatomy is similar to that of *Phragmipedium*. However, as far as we know, its anatomy has not been examined in detail.

Phragmipedium

The roots of *Phragmipedium* are enclosed in a velamen of four to seven layers. Unicellular hairs are present on the outermost layer of the roots. The exodermis comprises long cells with U-shaped thickenings. Tilosomes have been reported for *P. boissierianum* and *P. lindenii*. The vascular bundle is surrounded by a multiseriate pericycle of sclerotic parenchyma cells. Vessels with simple perforation plates and metaxylem tracheids are present in the vascular bundle.

The stem and rhizome have a uniseriate epidermis lacking both stomata and hairs. The vascular cylinder contains several vascular bundles and has a sclerotic sheath. Vessels are absent and tracheids have annular, spiral and scalariform thickenings.

The leaves have abaxial stomata. Micropapillae and sculpturing have been reported as adorning the epidermis of several species. The vascular strand is surrounded by a multiseriate pericycle of sclerotic parenchyma cells. Vessels have simple perforations and metaxylem tracheids with scalariform pitting.

CYTOLOGY

The cytology of the slipper orchids has been well studied, mainly by scientists in Japan. These studies have proved instructive in slipper orchids, particularly in the genus *Paphiopedilum* (Karasawa, 1979). *Phragmipedium* and *Paphiopedilum*, the two most important genera in horticulture, differ in basic chromosome number, a factor that explains the poor results when species of the two genera are crossed, as evidenced by the low viability and poor performance of the few resulting seedlings. In plant breeding the late Don Wimber, working for the Eric Young Orchid Foundation in Jersey, doubled the chromosome number of several *Phragmipedium* species using colchicine. Hybridisation using these chromosome doubled plants as parents has resulted in exciting results that have revolutionised interest in the genus.

Karasawa (1980, 1988) published detailed accounts of chromosome counts and karyo-morphological profiles for several species of *Phragmipedium* and a single species of *Selenipedium* (Karasawa, 2005).

A chromosome count of 2n = 20 was recorded from cultivated material of *Selenipedium aequinoctiale* Garay, this number being also found in *Cypripedium* and some *Phragmipedium* species and derived from a basic number of 2n = 18 by centric fission of chromosomes. However, Karasawa (2005) reported that *Selenipedium* has a distinctive cytological profile. The total chromosome length of 40 μm was shorter than that found in other slipper orchid genera. He also reported that the mitotic prophase was similar to that of *Phragmipedium*, whereas the early condensed segments of *Cypripedium* and *Paphiopedilum* at mitotic prophase were diffused continuously in both arms.

The earliest recorded chromosome counts for *Phragmipedium* are those for *P. caudatum* of 2n = 32, for *P. longifolium* var. *hartwegii* of 2n = 20, both reported by Duncan (1959), and *P. boissierianum* of 2n = 18 by Karasawa and Tanaka (1976).

The more detailed cytological investigations of *Phragmipedium* by Karasawa (1980, 1988) analysed 16 taxa of fourteen species, one variety and one artificial hybrid. His chromosome counts and more recently published ones are listed in Table 1 below.

Table 1. Chromosome counts in *Phragmipedium* and *Selenipedium*.

Species	**2n**	**Source**
Selenipedium aequinoctiale	20	Karasawa, 2005
Phragmipedium kovachii	32	Fischer, pers. comm.
P. schlimii	20	Karasawa, 1980
P. besseae	24. Also 25, 26, 36 from colchicine-treated plants	Karasawa, 1988; Moon & Cribb, 1998

Species	2n	Source
P. fischeri	36	Fischer, pers. comm..
P. dalessandroi	28	Wimber, pers. comm.
P. lindleyanum	22	Karasawa, 1980
P. sargentianum	22	Karasawa, 1980
P. caudatum	28, 32	Karasawa, 1980; Duncan, 1959
P. wallisii (= *P. warszewiczianum*)	28	Karasawa, 1980
P. lindenii	28	Karasawa, 1980
P. ecuadorense (= *P. pearcei*)	22	Karasawa, 1980
P. pearcei	20	Karasawa, 1980
P. boissierianum	18	Karasawa & Tanaka, 1976; Karasawa, 1980
P. vittatum	18	Karasawa, 1980
P. longifolium var. *hartwegii*	20	Duncan, 1959, Karasawa, 1980
P. longifolium	23	Karasawa, 1980
P. longifolium var. *gracile*	21	Karasawa, 1980
P. longifolium var. *roezlii*	22	Karasawa, 1980

He concluded that the basic chromosome number for the genus was 2n = 18, as in *Selenipedium*. This number, found in both *P. vittatum* and *P. boissierianum*, comprised two large metacentric chromomes, eight medium-sized metacentric ones and eight small metacentric ones. Chromosome numbers of 2n = 20, 22, 26, 28 and 30, found in species of sections *Micropetalum*, *Phragmipedium* and *Lorifolia*, showed a decrease in metacentric chromosomes and a corresponding increase in telocentric ones. He suggested that this was due to centric fission of the metacentric chromosomes.

Species in section *Platypetalum* had a chromosome number of 2n = 22, suggesting derivation from 2n = 17 due to the loss of telocentric chromosomes following centric fission.

In *P. pearcei* he found different chromosome numbers (2n = 20, 21, 22) in three different clones that he examined. He postulated that the clones were, in fact, hybrids of *P. pearcei* and other species since the karyotypes of their chromosomes were heterogeneous.

Three varieties of *P. longifolium*, var. *longifolium*, var. *gracile* and var. *roezlii* (as *P. roezlii*), were examined and found to have a chromosome number of 2n = 20 but differed in having different number of chromosome fragments.

POLLINATION

Few detailed observations have been made of the pollination of *Phragmipedium* in the wild. Some observations have been made of the removal of pollinia but not of their deposition on the stigma. Pemberton (2013) provides a summary of the known and surmised pollination syndromes of tropical American slipper orchids.

Dodson (1966) reported the first observation of pollinium removal of *Phragmipedium longifolium* in Ecuador by a halictid bee (*Chlerogella* sp.) and a hoverfly (*Syrphus* sp.). He noted that the insects investigated the spotting on the infolded margins of the lip above its mouth. On Chiriqui in Panama, McCook observed the hoverfly *Syrphus shorae* (Fluke) emerging from the lip of *Phragmipedium humboldtii* (as *P. caudatum*) with a pollinium smeared on its thorax. Koopowitz (2008) examined populations of *P. boissierianum* and *P. longifolium* in Ecuador and confirmed that the former was self-pollinating (autogamous), whereas the flowers of the latter could be either out-crossing or self-pollinating.

Pemberton (2011) observed a number of *Phragmipedium* species near Puyo in Ecuador. In a population of *P. pearcei* on a steep bank by a mountain stream at 2950 m, he noted the removal of a pollinium by a female hoverfly, *Ocyptamus antiphales* (Walker), on two occasions. He surmised that the flies mistook the spots on the infolded lip margins as the eggs of aphids, because the fly's larvae feed on aphids. He also observed, near Lita in northern Ecuador, a visit by a hoverfly and a halictid bee to flowers of *P. longifolium*, but neither were pollinators since the bee escaped the flower through the mouth of the lip and the fly emerged from one of the basal orifices of the lip without any residue of a pollinium, being too small to pick it up. He confirmed that *P. boissierianum* (as *P. reticulatum*) is self-pollinated, noting that a very high fruit set occurred without any pollinators visiting the flowers. Microscopic observation confirmed that the lateral margins of the stigma are enlarged and in contact with the anthers and that pollinia are in contact with the stigmatic surface to ensure autogamy. He also examined the flowers of *P. lindenii* which lacks a lip

Fig. 4 (left). *Meriania maxima* (Melastomataceae), a showy tree growing in the forests of the Ecuadorian Andes, pollinators of it might mistake the flowers of *P. kovachii* for it and also pollinate them by mistake. (PC) **Fig. 5** (right). *Meriania superba* (Melastomataceae), another tree with flowers similar to those of *P. kovachii* in colouration. (PC)

and bears three rather than the typical two anthers of other *Phragmipedium* species. The central anther lies beneath the stigma and appressed to it, thereby ensuring self-pollination.

In February 2011, during a lecture at the Machu Picchu Orchid Conservation Conference, he also reported that a photograph taken by Robert Dressler near Machu Picchu in Peru showed a female hoverfly emerging from one of the basal orifices of the lip of *P. caudatum*. However, the fly was dead and thus not the pollinator of the orchid.

Phragmipedium schlimii, P. besseae and their allies might be expected to be bee-pollinated, paralleling the situation in Asian species of *Paphiopedilum* sect. *Parvisepalum* and sect. *Concolor* with brightly coloured or white flowers (Cribb, 1998, Bänziger *et al.,* 2012). However, Pemberton (2013) reported the pollination of the orange-scarlet flowers of *P. besseae* by unidentified tiphiid wasps. These are parasitic on subterranean scarab (beetle) larvae. According to Koopowitz (2008), *P. schlimii* is occasionally autogamous.

The pollinator of *P. kovachii* has not been confirmed but its flower colour and pattern in the centre of its flower resemble closely those of sympatric tree and shrub species of Melastomataceae, such as *Meriania maxima* Markg. (Fig. 4) that have large purple flowers with white or yellow marks in the centre. It seems highly likely to be pollinated by large bees.

DISTRIBUTION

Slipper orchids have a wide distribution in the Americas (Map 1). *Cypripedium* ranges from Alaska south to Central America (Cribb, 1997; Frosch & Cribb, 2012). Its range only overlaps that of *Phragmipedium* at the southern end of its range. The genus *Selenipedium* is cryptic, resembling bamboos and other tall grass-like orchids, such as *Sobralia*. As such, it is undoubtedly under-collected and under-recorded. It is predominantly a tropical South American genus but reaches as far north as Panama and Trinidad and as far south as Ecuador and the eastern seaboard of Brazil (Map 2). Its distribution is very patchy and it is not common. *Selenipedium chica* is restricted to Panama, *S. palmifolium* is the most widespread species found in Trinidad and the Guyanan shield from Venezuela, Guiana, Surinam, French Guyana and adjacent Brazil. It is nowhere common and would benefit from a study of morphological variability since the herbarium material is difficult to resurrect. Plants from French Guyana on the internet purporting to be this species differ in flower colour and staminode shape from the type material. The recently described *S. chironianum* is a peloric form that lacks a saccate lip, a condition found also in other slipper orchid genera. A second species, *S. steyermarkii*, is found in Venezuela but is a narrow endemic. *Selenipedium isabelianum* and *S. vanillocarpum* are both Brazilian endemics: the former with a few scattered populations in eastern Brazil from Para to Bahia states; the latter known only from the type collection from Goias. Needless to say, more detailed studies would help ascertain the true status of both taxa. Two or three taxa have been reported from north-western South America in Colombia and Ecuador. *Selenipedium aequinoctiale* was described from the western flanks of the Ecuadorian Andes. Plants found in coastal Colombia near Buenaventura may belong here but examination of living and preserved material is needed to confirm its status. Finally, *S. dodsonii* has recently been described from the lower eastern flanks of the Andes of northern Ecuador.

Mexipedium xerophyticum, the sole species in the genus, must be one of the rarest and most endangered of all orchids. It is known only from a very limited area in Oaxaca State in peninsular Mexico (Map 12).

Phragmipedium, the third genus covered in this account, is the largest and most widespread, found from southern Mexico south to Brazil and Bolivia (Map 13). The epiphytic *P. humboldtii* var. *exstaminodium* is the only species that has been found in Chiapas State in Mexico where it just creeps across the Guatemalan border where it also occurs. The typical variety has a disjunct distribution in Guatemala, Honduras and Panama. *Phragmipedium longifolium* is the most widespread species, found in Costa Rica and Panama and also scattered in tropical South America as far south as Ecuador and Brazil.

Several species occur scattered along the northern Andes: *P. schlimii* in Colombia, *P. boissierianum* in Ecuador and Peru, *P. warszewiczianum* in Colombia, *P. lindenii* in western Venezuela, Colombia and Ecuador, *P. caudatum* in Peru, and *P. caricinum* in Bolivia. Several are much more limited in their distribution, notably: *P. andreettae*, *P. anguloi* and *P. fischeri*, all from south-western Colombia and/or north-eastern Ecuador; *P. hirtzii* in north-eastern Ecuador; *P. besseae* in northern Peru and southern Ecuador, *P. dalessandroi* in southern Ecuador, and *P. kovachii* in northern Peru.

Away from the Andes, *P. klotzschianum* and *P. lindleyanum* are found on the Guyanan shield from Venezuela and the Guyanas. The recently described *P. guianense* is a narrow endemic from French Guyana. Three species occur in Brazil. *Phragmipedium longifolium* has been found there just once and its status as a native species is debatable. *Phragmipedium sargentianum* occurs in eastern tropical Brazil. Finally, *P. vittatum* is another endemic but found from Matto Grosso State across to just north of Rio de Janeiro. None of the Brazilian species are common and existing populations are small in extent but can contain large numbers of plants.

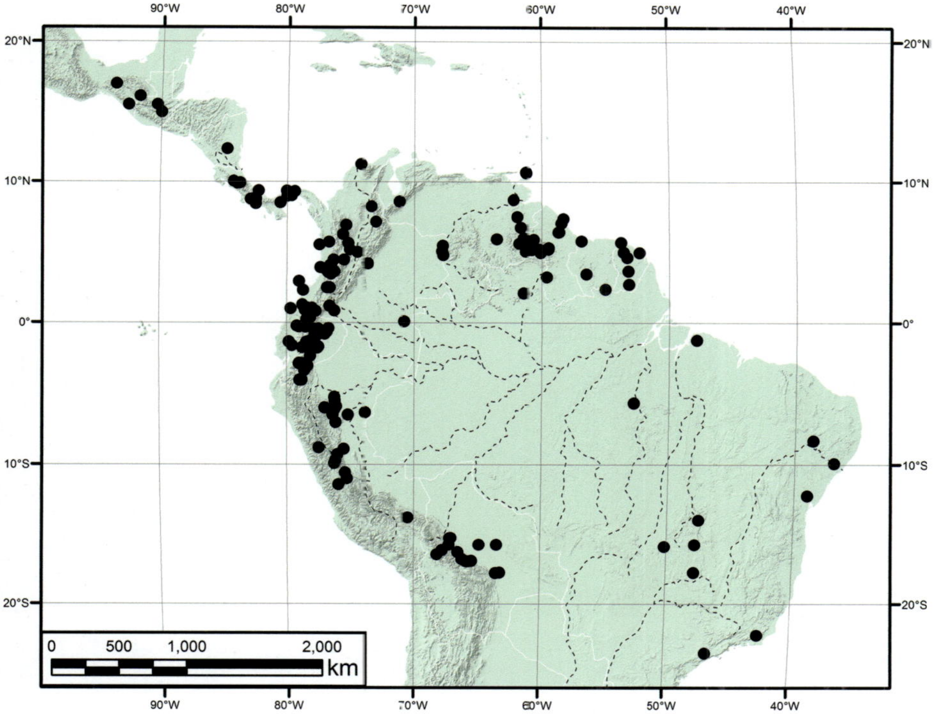

Map 1. Distribution of slipper orchids in the tropical Americas.

ECOLOGY

Slipper orchids are found in a wide variety of habitats in the tropical Americas but few are widespread in their distribution, several being narrow endemics. Few species are locally common but, where there is limited competition from other plants, species can form sizeable populations. The reason is not hard to find; they, like many other orchids, are often primary colonisers of bare areas of rock, soil or tree bark. Many colonies are very limited in extent, sometimes confined to one to a few plants. This makes them extremely vulnerable to habitat disturbance, deforestation and collection (see Conservation chapter).

The majority of species are terrestrial but habitats vary from roadside banks to seasonally flooded grassland by river margins. A few species are truly epiphytic but a larger number grow happily as lithophytes on old volcanic lava flows, cliffs, rocks and boulders in rivers. It can be difficult to distinguish between the lithophytes and terrestrials.

Selenipedium is distributed over a large area from Panama and the northern Andes and across to eastern Brazil but it is nowhere common. In fact, the total number of collections housed in the world's herbaria does not reach 100 samples. It has a relict distribution, indicative of an ancient group that was formerly more widespread. All *Selenipedium* species are terrestrials, growing in forest, on forest margins or amongst scrub, grasses and grass-like orchids. Their resemblance to some species of bamboo and terrestrial orchid, especially *Sobralia* and *Elleanthus*, makes them difficult to see and they are almost certainly overlooked even when in flower. *Selenipedium chica* is found in the tropical forests of lowlands of Panama, plants reaching 5 m in height. *Selenipedium aequinoctiale* grows on the west-facing slopes of the Andes in northern Ecuador at elevations between 300 and 600 m. Its best-known and most visited locality is a steep roadside bank, where it grows in secondary vegetation of scrub, ferns and *Sobralia* orchids in full sun (Figs. 6 & 7). Dunsterville and Dunsterville (1971) give an account of the habitat of *S. steyermarkii* which is only slightly shorter than *S. chica* and is found in a limited area of the Venezuela–Guyana border region from Auyantepui to Cerro Venamo. It grows in scrub and in more open places on the flanks of these mountains from 800 to 1800 m elevation in sandy soil, a mixture of sand, weathered from sandstone, and some leaf-mould. The most widespread species, *S. palmifolium*, has been recorded growing in wet low evergreen forest bordering

Fig. 6 (left). *Selenipedium aequinoctiale* growing on a disturbed roadside bank in northern Ecuador.
Fig. 7 (right). Habitat of *Selenipedium aequinoctiale,* near Lita, NW Ecuador. (PC)

savannas, in bush-land on sandy loam, and in marshy areas at between 200 and 400 m elevation.

The monotypic *Mexipedium* grows on rocks and cliffs in southern Mexico. A detailed account of its habitat has been provided by Perez-Garcia (2011). It grows on vertical walls of karst limestone outcrops surrounded by tropical rainforest at 320 m elevation in the isthmus of Tehuantepec, Oaxaca, Mexico (Fig. 8). The plants are in the shade and do not receive direct sunlight except in the rainy season when its xerophytic morphological adaptations come into play. It is a stoloniferous species spreading over the vertical rock-face so that the fan-like leafy stems are well separated. Woody plants growing on these outcrops include *Bursera simaruba*, *Plumeria rubra* and *Pseudobombax ellipticum*, while the herbaceous flora includes *Agave*, *Acanthocereus*, *Begonia*, *Catopsis*, *Peperomia*, *Philodendron*, *Pitcairnia*, *Tillandsia*, *Selaginella*, many ferns and the orchids *Bletia* sp., *Cyrtopodium macrobulbon* and *Encyclia* cf. *parviflora*. This region contains numerous other endemic species.

The largest genus, *Phragmipedium*, also occupies the greatest variety of habitats. The majority of species are lithophytic, growing in a thin layer of soil over a rocky substrate or are terrestrial. *Phragmipedium longifolium*, the most widespread species, is commonly found growing on roadside banks (Figs. 113 & 114). *Phragmipedium boissierianum* is also frequently found in similar situations and on low cliffs by the roadside (Fig. 11). Both species appear to be primary colonisers of these sites which quickly become covered in grasses, shrubs and other herbs, especially *Sobralia* orchids. The Brazilian *Phragmipedium vittatum* is restricted to marshes in elevated areas of the 'Cerrado' of the south-eastern and central-western regions of Brazil (Menezes, 1989). It is only found in peaty habitats, where it may be locally abundant (Fowlie, 1973, Menezes, 2004). It can also grow on wet rock faces with the roots attached to the rock some 8 cm below the surface of the turf (Fig. 124).

A number of species grow in montane rain forests near fast-flowing streams and rivers. *Phragmipedium schlimii*, *P. fischeri* (Fig. 83), *P. andreettae* and *P. anguloi* all grow on shady vertical banks of montane streams and rivers where the atmosphere is humid. *Phragmipedium besseae* (Fig. 16), *P. dalessandroi* and *P. kovachii* (Figs. 58 & 59) are all found on the eastern flanks of the Andes where they grow on vertical banks and rocky cliffs above rivers. These places are humid and somewhat shaded but often quite high above the river or stream. Several species, notably *Phragmipedium caricinum*, *P. hirtzii* (Fig. 10) and *P. pearcei* (Fig. 9) that grow on rocks in fast-flowing Andean streams, have a particularly rigorous environment to survive, being periodically inundated by water during flood conditions. *Phragmipedium klotzschianum* can also grow on rocks in rivers but is also found in periodically inundated grassland by rivers (Figs. 12 & 13).

Fig. 8. *Mexipedium xerophyticum,* growing on rocks by the type locality in Oaxaca, Mexico (EP).

Fig. 9 (top left). *Phragmipedium pearcei* growing on a rock in a mountain stream on the eastern flank of the Andes in central Ecuador. (IO). **Fig. 10** (top right). Habitat of *Phragmipedium hirtzii* in northern Ecuador (PC). **Fig. 11** (middle left). The habitat of *Phragmipedium boissierianum* in northern Peru. **Fig. 12** (middle right). *Phragmipedium klotzschianum* growing in a hollow in a rock by a river in Venezuela. (GR) **Fig. 13** (bottom left). The marshy riverside habitat of *Phragmipedium klotzschianum* in Venezuela. (GR) **Fig. 14** (bottom right). Habitat of *Phragmipedium lindenii,* growing on old lava flows on the lower slopes of the Tungurahua volcano in Ecuador (PC).

Phragmipedium lindleyanum grows at 1000 m elevation in semi-shade on the ground and as a low epiphyte in scrubby vegetation below bare black rock facing north-east in water seepages with *Clusia rosea* (Gorinsky, 1973)(Fig. 98). Its var. *kaieteurum* is found on rocks in the spray zone of waterfalls, notably Kaieteur Falls in Venezuela. In São Caetano do Norte in Pernambuco, *Phragmipedium sargentianum* grows in marshy seepages in forest at 850–950 m elevation on steep east-facing ridges (Ghillany, 1972).

Phragmipedium lindenii, the only species that lacks a slipper-shaped lip, commonly grows on old lava flows where it can form extensive colonies, particularly around the foot of the Tunguragua volcano in Ecuador (Fig. 14).

A few species, notably *P. caudatum* (Fig. 15), *P. guianense* and *P. humboldtii* are truly epiphytic, growing in partial shade on the trunks and in the crotches of large branches of trees in rain forest.

Fig. 15 (left). *Phragmipedium caudatum* growing in forest, Monopampa, near Muna, Huanuco Province, Peru. 2300 metres asl. (HO). **Fig. 16** (right). *Phragmipedium besseae* growing on ledges above a river in S Ecuador. (PC).

CONSERVATION

The slipper orchids are amongst the most popular of all orchid groups in cultivation but many species are naturally rare and occur in scattered and often small populations which can range from a few to thousands of plants. Many species are primary colonisers of bare ground, especially roadside banks, rocks and lava flows. As vegetation grows up to compete with them, the orchids decline and eventually disappear. Thus, declining numbers of plants can be a natural phenomenon. This renders them particularly vulnerable to habitat change, e.g. competition, drought and landslides, destruction, e.g. by fire or logging, and collection for horticulture or traditional medicine. The threat to these distinctive orchids has been recognised for many years and all of the species of the Asiatic genus *Paphiopedilum* and the tropical American *Phragmipedium* are afforded protection on Appendix 1 of the Convention on International Trade of Endangered Species of Fauna and Flora (CITES). This effectively prevents international trade in wild-collected plants and controls trade in artificially propagated ones. The other genera are listed on Appendix 2 of the convention where trade is permitted but also controlled. Unfortunately, slipper orchids continue to disappear because this convention does not control collection or destruction within a country. In these cases, national and local legislation can effectively protect the plants, but often fails to do so.

In the tropical Americas, some species of *Phragmipedium* and *Mexipedium xerophyticum* are still collected for horticulture and are particularly vulnerable. The horticultural trade

Fig. 17. Wild-collected orchids including plants of *Phragmipedium kovachii* on sale in a roadside stall in northern Peru. (PC)

Fig. 18. A roadside stall selling *Phragmipedium kovachii* plants close to the type locality of the orchid on the Moyobamba road in northern Peru. (PC)

in the showy species of *Phragmipedium* received a massive boost with the discovery in Peru of *P. besseae* in 1981 and *P. kovachii* in 2001. They introduced new colours into breeding programmes and the hybrids are spectacular and popular with growers, leading to the large-scale depredation of known habitats. Fortunately, not all species are rare in nature but, since all are readily available as nursery-raised plants, there is no need to collect more plants from nature.

Orchids can produce thousands of minute, wind-dispersed seeds within a single capsule, but only a few will land on a suitable substrate. The seeds require appropriate nutrients, moisture and, crucially, a symbiotic species of fungus, which, acting in combination, will enable germination. However, the number of seedlings produced will usually be only a fraction of the number of released seeds.

Different sites provide different resources and harbour different levels of threat, and so individuals of the same species occurring at two locations will stand very different chances of completing their life cycle. Most slipper orchid species appear to be highly dependent on certain habitat factors and are able to survive in only a narrowly defined habitat, some species being known from only one or a few places. But the presence of ideal habitat does not mean that it will necessarily host its dependent orchid, either temporarily or permanently. Dispersal of seeds across the mosaic of habitat types to another site suitable for colonisation is a highly random occurrence. And, since the diversity of the gene pool is maintained by subsequent cross-pollination, the chance of pollen being carried by a pollinating insect between one site and the other can be low.

Orchid distribution and abundance

Orchids differ markedly in their distribution, in the number of individuals of the species present in a population (abundance) and how those individuals are arranged in the area. It is possible for the same number of individuals of a species to be either clustered into a single population or scattered in a number of smaller subpopulations across the area. The implications of either scenario for conservation are quite different. For example, a specific threat or random disturbance could have catastrophic consequences for the single population, whereas losses at one location could be buffered by the persistence of individuals at other locations in the case of multiple scattered subpopulations. Conversely, widely separated subpopulations may be restricted in the extent to which they can engage in gene flow in the form of pollen transfer and/or seed dispersal, with a concomitant reduction in genetic diversity through inbreeding at both sites. Quantifying a species' abundance and distribution therefore becomes a useful way to gauge its extinction risk, and accurate knowledge of these parameters is of crucial relevance to conservation biology, land management, protection and planning.

Orchid occurrence

The data for conservation assessments come from verified dried herbarium and spirit-preserved specimens that have been annotated and provide a verifiable means of confirming

Fig. 19. Close-up of wild plants of *Phragmipedium kovachii* potted up in a private nursery in southern Ecuador. (PC)

the occurrence of a plant species at a particular site at a particular point in time. A list of all collections examined is provided in an Appendix. Photographs are also fertile sources of *in situ* records. The authors and a number of other colleagues have also provided records from field work in recent years where collection has not been an option.

Threats

Both natural and anthropogenic events can negatively impact the distribution and abundance of species. The main direct threats to tropical American slipper orchids are habitat loss through degradation and development, and collection for horticulture. However, natural events, including periods of drought and intense rainstorms leading to flash floods, can also destroy populations. Plants growing in exposed places or in stream beds are vulnerable and circumstances may occasionally be sufficiently severe to extirpate populations. Temperature variations over the course of a single growing season may also affect plant survival. Apart from species that are known from one or a small number of sites, protecting plants against catastrophes of this sort may be impractical.

Loss of pollinators can lead to a decline in plant reproductive success. Fortunately most slipper orchids are long-lived perennials and can survive by vegetative reproduction for many years.

In contrast, man-made threats tend to be rapid, directional and more intensely destructive, and often occur at a local scale. Burning, collection and misuse are all causes of orchid extirpation or extinction. Thus, *Mexipedium xerophyticum* was considered extinct for some years because of a massive fire at its only known locality. Fortunately, it survived the ordeal. The forest habitat of *Phragmipedium humboldtii* var. *exstaminodium* in southern Mexico was cleared by Guatemalan refugees fleeing oppression and it disappeared as a result.

The discovery on 2001 of *Phragmipedium kovachii*, with stunning royal purple flowers, would alone have merited press attention. But, in 2002, it featured in the press, on radio and television and in almost every other media-outlet for a different reason – orchid smuggling!

The tale of the discovery and importation of this amazing orchid has now been well-documented (Pittman, 2012). Indeed, numerous pages of titles on the world-wide-web testify to the sensation caused by its discovery, subsequent journey to Florida and hurried description. It was discovered by local orchid collectors in north-west Peru in 2001. In 2002, Michael Kovach saw a flowering plant on a roadside stall while he was collecting orchids in northern Peru and purchased some plants from a stall-holder (Figs. 17–19). On arrival in Florida, he took a plant to the Marie Selby Botanical Gardens in Sarasota where it was immediately recognised as new to science. The plant was illustrated and a description rushed into print. A few days later it was redescribed as *P. peruvianum* by Eric Christenson in *Orchids*, the Bulletin of the American Orchid Society. Of course, by the rules of priority of the International Code of Botanical Nomenclature, the earliest name given to a plant is the correct name, later names being reduced to synonymy. Selby publicised its exciting novelty to great acclaim but the climate rapidly changed when the origin of the type material was questioned by, amongst others, the Peruvian government. The case was heard by a Federal Grand Jury at which the Garden pleaded guilty and the press had another field day. The three localities, all close to each other, have been stripped of mature plants but the species survives there because seedlings were left to mature because of an enlightened attitude by the Peruvian government (see below and under *P. kovachii* account).

Conservation assessments
The data on distribution and abundance, together with assessments of the observed or inferred level of threat, have allowed assessment of the conservation status of all the orchids treated in this book. The methods employed are outlined below (after Barretto, Cribb & Gale, 2009).

IUCN Red Data Listing
The IUCN criteria (IUCN, 2012) are applicable at the regional level to assess the status of part of a global population defined by a political border (IUCN, 2003).

Using geo-referenced distribution and abundance data for all species, the slipper orchids were therefore assessed according to IUCN Criteria B (geographic range in the form of Extent of Occurrence (EOO), or Area of Occupancy (AOO), and D (very small or restricted population).

Criterion B: geographic range
EOO and AOO are both measures of distribution. EOO describes how widespread subpopulations are, and is defined as the area contained within the shortest continuous imaginary boundary that encompasses all known sites of the species being assessed (IUCN, 2001); a minimum of three sites are required to generate this estimate. AOO provides a measure of only the area known to host the species being assessed by counting the number of occupied imaginary contiguous grid cells laid across the entire geographic area being examined; the dimensions of the grid cells are clearly critical in generating this estimate. A grid cell size of 1 × 1 km was adopted for calculation of AOO as this is the minimum size recommended for use by the IUCN (2012). It also satisfies our assumption that orchids are generally habitat specific and that the availability of suitable habitat type is therefore one of the key factors limiting orchid distribution and abundance were measured using distribution data (Moat, 2007). The numeric thresholds for the categories under Criteria B and D are shown in Table 2.

Criterion D: very small or restricted population
Estimates of total population size (*i.e.* abundance) were made from the most recent data available based on current field knowledge. Where this was lacking, data recorded in field notes or on herbarium specimen labels were used. In all cases, however, these figures should be treated as strictly approximate because accurate assessments of the number of individuals are not possible without actual counts. Category VU (vulnerable) requires that assigned taxa also meet the Subcriterion of restricted AOO or number of localities. Data on these parameters were available from the assessment using Criterion B. Where a taxon was assigned to VU based on estimated population size but did not meet the threshold for this Subcriterion, the next rank down (near threatened, NT) was assigned.

Table 2. Categories and thresholds for IUCN Criteria B and D, and Subcriteria contained therein (IUCN, 2001).

		IUCN Criteria					
		Criteria B Ð Geographic range				Criteria D Ð Population size	
		1. Extent of occurrence (EOO)	2. Area of occupancy (AOO)	a. No. of locations	b. Continuing decline in any of the following	1. Population size (no. mature individuals)	2. Restricted AOO or no. of locations
IUCN Category	Vulnerable (VU)	< 20,000 km^2	< 2,000 km^2	≤ 10 locations	*i.* Extent of occurrence *ii.* Area of occupancy *iii.* Area, extent and/or quality of habitat *iv.* No. of locations or subpopulations *v.* No. of mature individuals	< 1,000 individuals	AOO < ca. 20 km^2 and/or ≤ 5 locations
	Endangered (EN)	< 5,000 km^2	< 500 km^2	≤ 5 locations		< 250 individuals	-
	Critically endangered (CR)	< 100 km^2	< 10 km^2	1		< 50 individuals	-

The Future

The prognosis for slipper orchids is more promising than many would accept. CITES has encouraged the propagation of *Phragmipedium* species from seed and by vegetative means and they have proved to be easy to propagate. Furthermore, *Phragmipedium* species and *Mexipedium xerophyticum* are easy to cultivate and grow rapidly in pot culture in suitable conditions. This has taken the pressure off desirable species, such as *P. besseae* and *P. kovachii*. Newly discovered species are vulnerable until horticulture is able to offer nursery-raised seedlings. The Peruvian government realised this and released mother plants of *P. kovachii* to two Peruvian nurseries and licensed them to sell seedlings. This has undoubtedly saved the remaining populations of the species in the wild that are now showing signs of recovery. The same is true for *P. besseae*, which is recovering in nature because nursery-raised plants are so relatively cheap to buy, good quality and easy to grow. Sites in the Andes that were thought to have been stripped have, today, healthy populations of flowering-sized plants because there is no money to be made by collecting and selling wild plants. The locals can now make more money from eco-tourists who want to see the plants in their native setting.

It would be helpful if more governments were as perspicacious as that of Peru so that legal seedlings can be sold, thereby protecting the wild sites and allowing these fabulous orchids to be seen in cultivation and used for breeding while, at the same time, protecting them in the wild.

THE SPECIES CONCEPT

The popularity of *Phragmipedium* in cultivation has resulted in many publications on the genus ranging from monographs to books on how to grow and hybridise them. As a consequence, naming plants can be difficult with more names applied to plants than there are species.

The main problem that underpins much of the misunderstanding that exists between horticulturists and botanists is the concept of the species. The horticulturist requires names for his plants and to be able to give different names to plants that look different. The botanist is more concerned that his classification reflects the relationships of species and their variability. In other words, the horticulturist is concerned with differences while the botanist is concerned with patterns and similarities.

The problem is discussed below with reference to some of the species in the genus.

DISTINCT SPECIES. Many of the species present no problems for the taxonomist and are easy to identify. A good example of this is *P. kovachii* which has a flower that is so distinct that it is unlikely to be confused with any other in the genus. True, variation exists within *P. kovachii* but its variants cause no problems for identification.

SPECIES PAIRS AND COMPLEXES. Within the genus *Phragmipedium*, there are a number of what might be called sister species, species pairs or species complexes. These are morphologically similar species differing in several characters and occupying distinct distributional ranges yet obviously closely related to each other. Thus *P. schlimii* and *P. fischeri* have similarly sized and coloured flowers but are reliably distinct in details of the flower. Similarly, *P. caudatum*, *P. warszewiczianum* and *P. humboldtii* are morphologically similar yet distinct enough to be readily separated on details of their floral morphology and colour.

The existence of these can be explained by reference to the geological history of the tropical Americas. We can postulate that species pairs have evolved from a common ancestor but that geographical isolation caused by the changes in climate, topography or catastrophic events has led to enough isolation and subsequent evolution to create distinct species. Sister species have been treated by some botanists as subspecies but, where several morphological characters and distinct distributions are involved, we see little justification for this.

VARIABLE SPECIES. The variable species are those that have most troubled botanists and horticulturists. In a variable species, individual plants and flowers can look different from each other even within a population but, more commonly, between populations. A good example is *P. longifolium* which is a wide-ranging species and varies over its range such that an experienced eye can sometimes determine the origin of a plant from its morphology. An orchid grower with two plants that look different will justifiably want different names for them. The botanist must bear this in mind when tackling such species.

Let us first consider the causes of variation. In *Phragmipedium*, the distribution of any species is seldom continuous but broken up or disjunct because suitable habitat is almost never contiguous. The populations can be thought of as growing on 'islands' surrounded by a 'sea' of unsuitable habitat. Where adjacent populations are nearby, cross-pollination can still occur but, where they are distant, cross-pollination becomes unlikely or impossible.

In these cases changes can occur within a population quite randomly through genetic drift that will make that population look a little different from the next. Mutations of genes or changes in chromosome number and size are known to produce such morphological changes. These phenomena appear to be common in the tropical American slipper orchids.

However, to treat each population that is a little distinct as a different species, as some would advocate, would quickly produce an unworkable classification and dozens of species. It seems better, therefore, to use the infraspecific categories of subspecies, variety and form allowed by the International Code of Botanical Nomenclature.

Hybridisation

Variability and consequently taxonomic confusion can also result from hybridisation in the wild. Again, new species might result from such hybridisation but, usually, the process is incomplete. *Phragmipedium* species will hybridise in nature. A recent example of a natural primary hybrid is that of *P. wallisii* and *P. longifolium* in the wild in Ecuador (Fessel *et al.*, 2000; Röth, 2013).

Teratologies

Teratologies can occur as individuals, e.g. flowers bearing a double lip or a split synsepal, but they can also be widespread and form populations of similar plants. This is the case with the best-known teratology in slipper orchids which is *Phragmipedium lindenii* which has a petaloid lip rather than a slipper-shaped one. As a result, it was initially described in the genus *Uropedium*, a complication because *Uropedium* predates *Phragmipedium* so that the latter name has had to be conserved over the former. Three taxa, namely *Phragmipedium exstaminodium*, *P. monstruosum* and *P. triandrum*, were initially described as a distinct species from Central America but are now considered to be variations of *Phragmipedium humboldtii*, the loss of the staminode possibly being related to self-pollination in the cases of the first and last. The recently described *Selenipedium chironianum* is another case where the lip is petaloid rather than slipper-shaped. The genus *Apedium* was unnecessarily erected for it.

Man-made problems

Far more perplexing for the botanist are the problems produced not by nature but by man. Many of these have accumulated over decades and a few are listed below:

1. ***Misidentifications***. Orchids are often collected from the wild without flowers and enter the trade with a guessed-at name attached to them. This can obviously lead to erroneous names. However, attaching a name to a plant without reference to the type specimen can also lead to errors.

2. ***Synonymy***. Some species have several Latin names. This is often the result of minor variants being described as species, or re-description from different parts of a species' range, or by a botanist who did not realise that the plant had already been described. Only one of the names is the correct one and that is usually the first given, the other names being termed synonyms.
3. ***Invalid names***. Over the years, several names coined by growers or authors have been widely used but never been validly published.

The botanist must treat orchids, including *Phragmipedium*, in a manner consistent with other plants. Our knowledge is increasing rapidly and inevitably new information will change previously held views. This might involve the sinking of some favourite names. A properly constructed classification of the genus including all the recent morphological, anatomical, cytological and DNA information will, we are certain, stand the test of time and will eventually gain broad acceptance.

PHYLOGENY

The subfamily Cypripedioideae is a well-defined group of five genera (Atwood, 1988; Albert & Chase, 1992; Albert, 1994) – *Selenipedium, Phragmipedium* and *Mexipedium* are tropical American; *Cypripedium* is widespread in north temperate zones extending into Central America (Cribb, 1997; Frosch & Cribb, 2012); and *Paphiopedilum*, which is Asiatic, extends from southern India into the islands of the south-west Pacific (Cribb, 1998).

By analogy with supposedly 'primitive' members of other subfamilies such as Apostasoideae, Orchidoideae and Epidendroideae (Dressler 1981) and Vanilloideae (Cameron & Chase, 2000), it is possible to identify certain expressions of some characters as ancestral, others as derived. Thus, in subfamily Cypripedioideae, features such as elongate stems, spirally-arranged plicate leaves, multi-flowered inflorescences, a trilocular ovary and a persistent perianth are considered ancestral traits. Using these and other characters, Atwood (1984) produced a phylogenetic tree using the Wagner Groundplan-divergence method which suggested that *Selenipedium* has the most ancestral features in the subfamily, approaching most closely an ancestral type, and that *Cypripedium* was linked through *C. irapeanum* to *Selenipedium*. He also suggested the splitting off of *C. arietinum* into a separate genus *Criosanthes*. The coriaceous conduplicate-leaved genera *Mexipedium*, *Phragmipedium* and *Paphiopedilum* have evolved from a common ancestor and Atwood has suggested a scenario whereby *Paphiopedilum* or its ancestors migrated from the Americas to Asia in warmer times, becoming finally separated when climatic changes pushed the tropical and subtropical forest belts south again.

Cladistic analyses by Atwood (1988) using morphological and anatomical characters and by Albert & Chase (1992) and Albert (1994) utilising both morphological and DNA data have confirmed the monophyly of *Cypripedium*, *Phragmipedium* and *Paphiopedilum*. *Selenipedium* and *Mexipedium* were represented by a single species in their analyses. For Albert & Chase (1992) and Albert (1994), *Criosanthes* nested well within *Cypripedium* in both analyses and also in the morphological analysis by Cribb (1997b). It is consequently considered conspecific with *Cypripedium* by most modern authors.

Albert & Pettersson (1994) proposed the amalgamation of *Phragmipedium, Mexipedium* and *Paphiopedilum* into an enlarged *Paphiopedilum* because *Mexipedium* is in many ways intermediate morphologically between the two better known genera. No one has followed this challenging view: *Paphiopedilum* and *Phragmipedium* have different basic chromosome numbers, different amounts of DNA and are difficult to hybridise. The few hybrids that have been raised are usually weak plants with little vigour.

Pfitzer (1903) and Garay (1979) produced sectional treatments of *Phragmipedium* based upon morphology. The infrageneric classification by Garay (1979) recognising five sections in the genus has been widely used: namely *Micropetalum*, *Platypetalum*, *Phragmipedium*, *Himantopetalum* and *Lorifolia*.

McCook (1989 a) provided a cladistic analysis of the genus based upon 27 morphological characters in her monograph of the genus. As a result she recognised only four sections, including section *Himantopetalum* in the typical section. She suggested that the presence of multicellular pubescence along the sides of the lip opening that borders the ribbed area on the claw face defined this section.

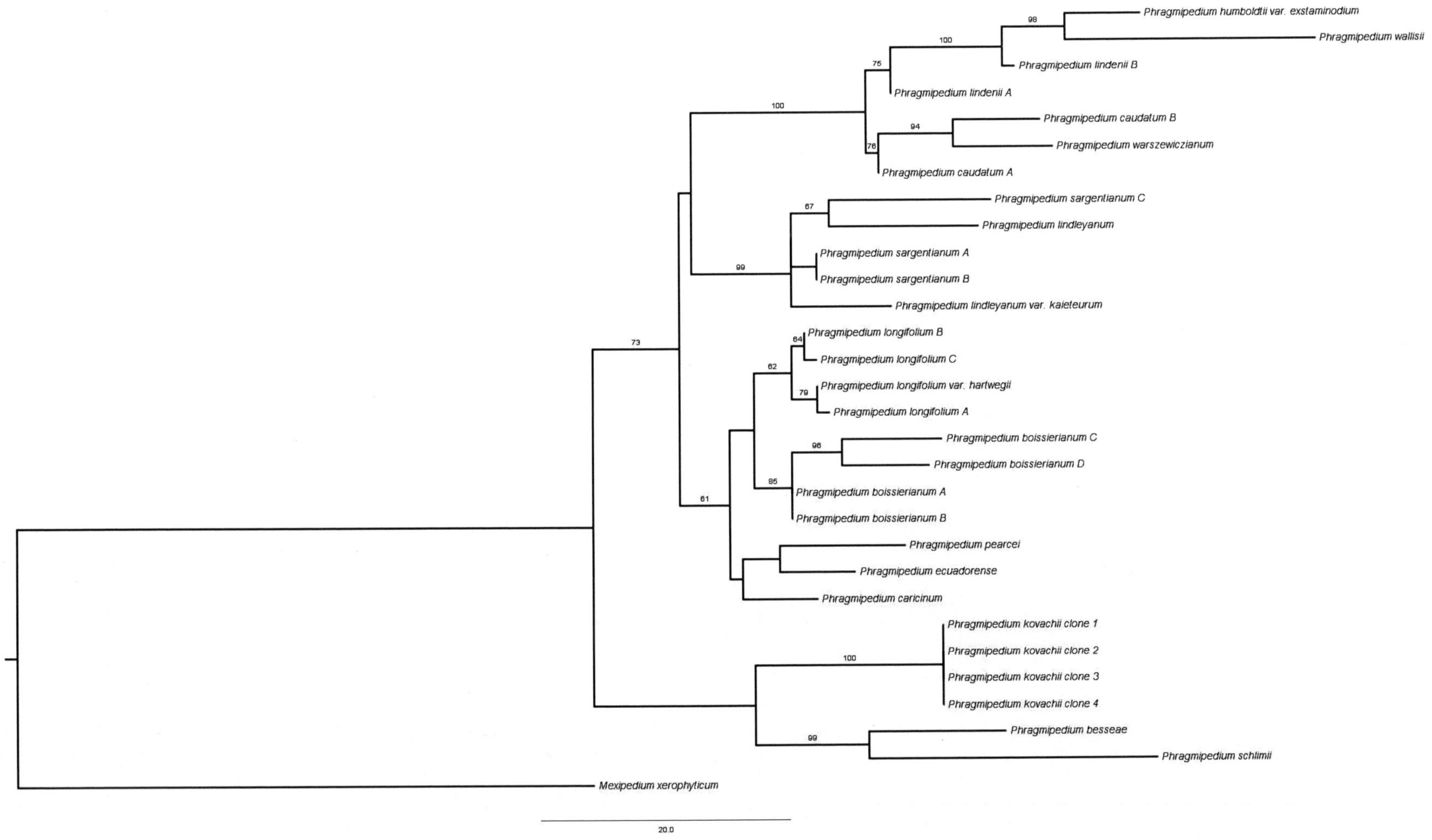

Fig. 20. Bayesian tree based on ITS sequences using of *Mexipedium* and *Phragmipedium* species from Genebank. Courtesy of Brian Yap

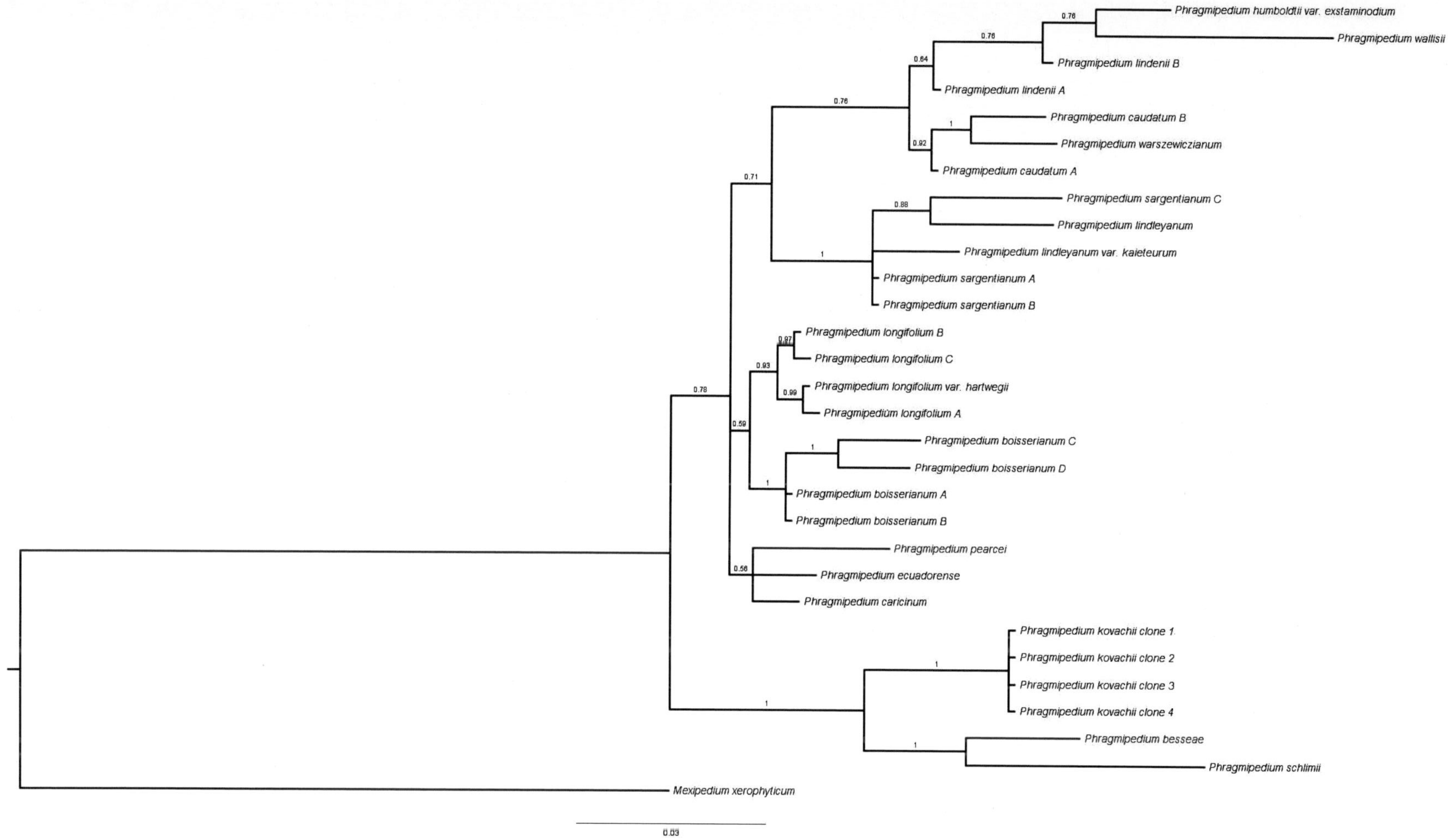

Fig. 21. Maximum parsimony tree using based on ITS sequences of *Mexipedium* and *Phragmipedium* species from Genebank. Courtesy of Brian Yap.

Cox *et al.* (1997) included 15 species of *Phragmipedium* and *Mexipedium* in their nuclear rDNA ITS sequence analysis of Cypripedioideae. *Phragmipedium* formed a clade sister to *Mexipedium* whereas the larger clade, including both genera, was sister to the tropical Asiatic *Paphiopedilum*. Within *Phragmipedium* the brightly coloured small-flowered *P. schlimii* and *P. besseae* formed a well-supported clade basal to the rest of the genus. Three further distinct clades could be identified comprising respectively the following groups of species: 1. *P. longifolium*, *P. pearcei*, *P. boissierianum* and *P. czerwiakowianum*; 2. *P. lindleyanum*, *P. sargentianum* and *P. kaieteuranum*; and 3. *P. caudatum*, *P. warszewiczianum*, *P. lindenii*, *P. exstaminodium*, *P. wallisii* and *P. caricinum*. Their treatment did not entirely confirm that of McCook: they found that *P. caricinum* was sister to species of the typical section whereas *P. pearcei* formed part of the same clade as *P. longifolium*. Since *P. pearcei* and *P. caricinum* are morphologically similar, possibly sister, species, misidentification of either of the specimens that they used cannot be ruled out.

Braem (2001) raised the sections of *Phragmipedium* to subgeneric rank and also established section *Viridiflorae* for *P. boissierianum* and subsections. These changes are unhelpful and an example of taxonomic inflation in such a small genus and are not accepted here. Finally, Braem (2004) established a new section, *Schluckebieria*, for the distinctive large purple-flowered Peruvian species *P. kovachii* that was discovered in 2001. Damian *et al.* (2005), using ITS sequence data suggested that section *Schluckebieria*, established on floral morphology and flower size criteria, could not be maintained as separate from section *Micropetalum*.

Rodriguez Salas *et al.* (2007) used Amplified Fragment Length Polymorphisms (AFLP) to analyse eight taxa of *Phragmipedium*, representing five of the six sections of the genus in Braem's treatment. Unfortunately, no species of sect. *Platypetalum* was included in their analysis. They identified four clades, one of which comprised *P. boissierianum*, *P. richteri* and *P. pearcei*, thereby supporting the conclusions of Cox *et al.* with regard to the last species. Section *Phragmipedium* appeared as basal to the other clades in the genus and section *Schluckebieria* formed a clade basal to one including in sections *Himantopetalum*, *Lorifolia* and *Micropetalum*.

Yap (pers. comm.) has re-run *Phragmipedium* sequences in Genebank (the DNA bank) where the DNA of 28 accessions of 14 taxa and one of *Mexipedium xerophyticum* are lodged (see Table 3 below). The resulting cladograms using the ITS gene sequences have been analysed using Bayesian (Fig. 20), Maximum Parsimony (Fig. 21) and Neighbour-joining algorhithms. The Bayesian analysis identifies five clades in *Phragmipedium* which correspond to sections *Phragmipedium*, *Platypetalum*, *Lorifolia*, *Himatophyllum* and *Micropetalum* (with section *Schluckebieria* sister to the other species in the section). The Maximum Parsimony analysis reveals four clades which are the same as those above but with section *Himantopetalum* included in section *Lorifolia*. The Neighbour-joining analysis distinguishes section *Micropetalum* (including *Schluckebieria*) as sister to a large clade including all the other sections.

In summary, we have retained much of the infrageneric classification suggested by Garay (1979) and McCook (1989 a & b, 1990, 1998) while dividing section *Himantopetalum'*, which is likely to be unnatural, between sections *Lorifolia* (*P. caricinum*, *P. pearcei*, *P. vittatum*) and *Phragmipedium* (*P. klotzschianum*). Section *Schluckebieria* has been included in section *Micropetalum*, its retention being, in our view, unnecessary. A more detailed DNA-based analysis of *Phragmipedium*, including all the known species, is required before a definitive phylogeny-based infrageneric classification can be proposed.

Some attention has been focused on the parallels found in the flowers of species in the different slipper orchid genera. Thus, similarities can be found in the small round pink flowers of *Paphiopedilum delenatii* and those of *Phragmipedium schlimii* from South America; species with more or less concolorous yellow flowers are found in *Selenipedium* (*S. aequinoctiale*), *Cypripedium* (*C. irapeanum*, *C. molle* and *C. dickinsonianum*) and *Paphiopedilum* (*P. armeniacum*); species with multi-flowered inflorescences of green, brown or purple flowers with spatulate petals are found in both *Paphiopedilum* (*P. lowii* and *P. haynaldianum*) and *Phragmipedium* (*P. lindleyanum* and *P. sargentianum*); and species with pendent elongated petals are found in *Phragmipedium* (*P. caudatum* and its allies) and *Paphiopedilum* (*P. sanderianum*, *P. adductum*, *P. philippinense*). Cribb (1987) suggested that such floral similarities are the result of adaptations to similar pollination syndromes, in the first two cases to solitary bee pollination and in the last two to hover-fly pollination. Molecular analyses (e.g. Cox et al., 1997) have confirmed that the concolorous species of *Cypripedium*, *Paphiopedilum* and *Phragmipedium* are basal in their genera and the hoverfly-pollinated species are derived.

Table 3. Accessions from DNA bank used to construct cladograms shown here. Note: Identifications are as per accession. The identity of closely related taxa, e.g. *P. caudatum*, *P. humboldtii* and *P. warszewiczianum*, cannot be confirmed.

Genus	Species	Variety	Accession number (DNA bank)
Phragmipedium	*kovachii*		AY918820
Phragmipedium	*kovachii*		AY918818
Phragmipedium	*kovachii*		AY918819
Phragmipedium	*kovachii*		AY918821
Phragmipedium	*besseae*		Z78513
Phragmipedium	*schlimii*		Z78514
Phragmipedium	*caricinum*		AY918822
Phragmipedium	*pearcei* (as *ecuadorense*)		AY918824
Phragmipedium	*pearcei*		Z78509
Phragmipedium	*lindleyanum*		Z78506
Phragmipedium	*lindleyanum*	*kaieteurum*	Z78504
Phragmipedium	*sargentianum*		Z78505
Phragmipedium	*sargentianum*		Z78506

Genus	**Species**	**Variety**	**Accession number (DNA bank)**
Phragmipedium	*sargentianum*		AY643428
Phragmipedium	*boissierianum*		Z78502
Phragmipedium	*boissierianum*		Z78503
Phragmipedium	*boissierianum*		AY643424
Phragmipedium	*boissierianum*		AY643425
Phragmipedium	*longifolium*		AY918825
Phragmipedium	*longifolium*		AY643426
Phragmipedium	*longifolium*		Z78510
Phragmipedium	*longifolium*	*hartwegii*	AY918825
Phragmipedium	*humboldtii*	*exstaminodium*	Z78511
Phragmipedium	*caudatum*		AY643429
Phragmipedium	*caudatum*		Z78501
Phragmipedium	*lindenii*		AY643430
Phragmipedium	*lindenii*		Z78507
Phragmipedium	*warszewiczianum*		Z78500
Phragmipedium	*warszewiczianum* (as *wallisii*)		Z78512
Mexipedium	*xerophyticum*		JQ91883

CLASSIFICATION

SUBFAMILY CYPRIPEDIOIDEAE

Subfamily Cypripedioideae Garay in Bot. Mus. Leafl. Harvard Univ. 19 (3): 86 (1960). Type: *Cypripedium calceolus* L.

Cypripedia Spreng., Anleit II, 1: 298 (1817). Type: *Cypripedium calceolus* L.

Cypripedieae Lindl., Orch. Sceletos: 1, 18 (1826). Type: *Cypripedium calceolus* L.

Cypripedilinae Pfitzer, Morph. Stud. Orch. Blüte: 108 (1886); Nat. Ord. Orchid.: 12, 95 (1888); In Engler & Prantl, Pflanzenf. II, 6: 765, 82 (1889); in Bot. Jahrb. Syst. 19: 31 (1894) & 25: 518, 525 (1898); in Engler, Pflanzenr. IV, 50. Orch. Pleonandr.: 9 (1901). Type: *Cypripedium calceolus* L.

DESCRIPTION: Small to large terrestrial, lithophytic or epiphytic *herbs* with a short to long rhizome; roots elongate, fibrous; rhizome short to elongate. *Shoots* unbranched, very short to long, erect, leafy, clustered or less frequently well spaced, glabrous, the lower part enclosed by two to several sterile sheathes, three- to many-leaved above. *Leaves* one to many, relatively thin-textured, fleshy or coriaceous, plicate or conduplicate, spreading or suberect, ovate, lanceolate, elliptic, ligulate, or oblong, obtuse, acute or acuminate, often tridenticulate at apex, yellow-green, green, or bluish green, plain, chequered or tesselated with dark and lighter green on upper surface or spotted with blackish maroon on upper surface, lighter green below, sometimes finely spotted or flushed with purple at base or all over on lower surface, glabrous or variously pubescent, ciliate or not on the margins. *Inflorescence* terminal, one- to many-flowered; rachis terete, hairy, glandular or glabrous; bracts conduplicate, linear, elliptic, lanceolate, ovate or oblong, green, sometimes spotted or flushed or striped with purple, glabrous, pubescent or glandular, ciliate or not. *Flowers* usually showy, concolorous or bicoloured; pedicel obscure to short; ovary unilocular or trilocular, three-ribbed, glabrous or hairy, placentation parietal or axial. *Dorsal sepal* erect to hooded over lip, ovate, lanceolate, obovate or elliptic, obtuse, acute or acuminate, glabrous or pubescent on the outer surface, sometimes pubescent within at base, ciliate or not, unmarked, spotted or streaked on inner surface. *Lateral sepals* usually fused to form a concave synsepal that is more or less similar to the dorsal sepal, rarely free to base, apex entire or bifid, sometimes two-keeled on outer surface. *Petals* free, flat, inflexed or reflexed, spreading or pendent, spiralling or not, elliptic, ovate, lanceolate, linear-lanceolate, linear or oblanceolate, rounded, obtuse, acute or acuminate at apex, unmarked, glabrous or rarely pubescent on outer surface, often pubescent in basal half on inner surface, usually ciliate on margins, sometimes warty on margins, spotted or streaked on inner surface, sometimes veins picked out in a different colour. *Lip* deeply pouched and inflated, slipper-shaped or urn-shaped, with or more rarely without more or less pronounced incurved side lobes which are sometimes reduced to auricles, hairy within especially on lower surface, sometimes spotted within, glabrous or hairy on outer surface; front margin incurved or not, sometimes with short marginal horns; side lobes sometimes warty. *Column* short, stalked, porrect; anthers two, bilocular, borne on short obtuse to acute filaments; pollen powdery or viscid; staminode terminal on column, sessile or shortly stalked, often shield-like, rarely reduced or absent, transversely reniform, oblong, ovate, obcordate or linear, flat, convex or longitudinally conduplicate, entire, bifid or tripartite at apex, glabrous to papillose or finely pubescent, ciliate or not; stigma stalked or subsessile, dependent, tripartite, more or less papillose. *Capsule* erect to pendent, three-ribbed, cylindrical to almost ellipsoidal, sometimes beaked at apex. *Seeds* spherical to ellipsoidal, with a thin testa except in *Selenipedium* where the testa is hard and dark coloured.

DISTRIBUTION. A subfamily comprising five genera. Widespread in the temperate regions of Eurasia across to the Philippines, New Guinea and the Solomon Islands, and in North, Central and tropical America south to Brazil and Bolivia. Four genera are found in the Americas, namely *Selenipedium*, *Cypripedium*, *Mexipedium* and *Phragmipedium*.

DIAGNOSTIC FEATURES. Garay (1960) characterised this subfamily by its zygomorphic flowers with dissimilar perianth parts; slipper-shaped lip; two anthers of the inner whorl fused to the style, a third from the outer staminal whorl being a more or less shield-like staminode at the apex of the column; viscid granulose pollen; monosulcate microspores aggregated into pollinia; and trilobed stigma.

KEY TO THE GENERA OF SLIPPER ORCHID IN THE AMERICAS

1. Stems usually elongate and leafy; leaves plicate with convolute vernation; flowers not deciduous, persistent on fruit 2
   Stems short with a fan of leaves; leaves conduplicate with duplicate vernation; flowers deciduous, not persistent on fruit 3

2. Ovary trilocular; seeds subglobose with a crustose coat .... **Selenipedium**
   Ovary unilocular; seeds fusiform with a thin seed coat .... **Cypripedium**

3. Plants with short rhizomes; ovary trilocular; petals elliptic, subcircular or tapering, spreading to pendent .... **Phragmipedium**
   Plants with long rhizomes; ovary unilocular; petals oblong, incurved .... **Mexipedium**

SELENIPEDIUM

Selenipedium Rchb. f. in Bonplandia, 2: 116 (1854) & in Xenia Orch. 1: 3 (1854); *emend.* Pfitz. in Engl. & Prantl, Naturl. Pflanzenfam. 2, 6: 84 (1888); Cribb, Curtis's Bot. Mag. n.s. 26: 5–20; Szlachetko & Kolanowska, Syst. Bot. 41, 1: 142–159 (2016). Type species: *S. chica* Rchb.f., selected by Pfitzer (1888).

Solenipedium Beer, Prakt. Stud. Orch. 310 (1854), orth. var.

Selenipedilum Pfitz. in Engler & Prantl, Nat. Pflanzenfam. 2, 6: 84 (1888) & in Engl., Pflanzenr. 4, 50, 12: 27 (1903), orth. var.

Apedium Chiron, Sambin & Braem in Richardiana 15: 253 (2015) **syn. nov.** Type species: *Apedium chironianum* (Sambin & Braem) Chiron, Sambin & Braem (based upon *Selenipedium chironianum* Sambin & Braem).

DESCRIPTION. Tall, slender, terrestrial *herbs*. *Roots* elongate, fibrous, branched, arising from a short to elongate rhizome; *rhizomes* horizontal, short, creeping, rather woody. *Stems* cane-like, erect, 2–5 m tall, clustered or well spaced, simple or branching in upper half, rather woody, leafy, terete, enclosed by hairy, sheathing sterile bracts below and leaf bases above, several-leaved; *pseudobulbs* absent. *Leaves* plicate, distichous or subspiralled, spreading or suberect, several-many, plicate, many-veined, elliptic-lanceolate to linear-lanceolate, acute or acuminate, green, glabrous, sparsely hairy or glandular, ciliate or not on the margins, not articulate, contracted at the base to form a cylindrical sheath. *Inflorescence* terminal on main stem or branches, racemose or one-branched from base, many-flowered, erect, arching or pendent; *rachis* terete, densely hairy or glandular; bracts usually leaf-like but much smaller than the uppermost leaf, lanceolate to ovate, hairy and/or glandular. *Flowers* with a persistent perianth, short-lived, small to medium-sized, resupinate, concolorous or bicoloured; *pedicel* obscure to short, hairy; *ovary* trilocular, with axile placentation, three-ribbed, more or less triangular in cross-section, hairy and/ or glandular, crowned with a marcescent perianth. *Sepals* valvate. *Dorsal sepal* erect to hooded over lip, ovate or elliptic, obtuse or acute, pubescent and/or glandular on the outer surface, ciliate or not; *lateral sepals* usually fused to form a concave synsepal that is similar to the dorsal sepal, but free at the tips, pubescent on outer surface, ciliate/glandular or not. *Petals* free, spreading or incurved, lanceolate or linear-lanceolate, blunt, acute or acuminate, often pubescent in basal half within, rarely ciliate. *Lip* deeply pouched and inflated, slipper-shaped, glabrous on outer surface; *side lobes* obscure to well developed, incurved, entire; *mid-lobe* deeply saccate, calceiform or urceolate, incurved on front margin, hairy and often spotted within especially on lower surface. *Column* porrect, short, stalked; *anthers* two, bilocular, borne on short obtuse filaments; *pollen* compact-powdery or viscid when mature; *staminode* terminal on column, more or less sessile or stalked and more or less perpendicular to the column, simple, ovate or trullate, somewhat conduplicate, glabrous; *stigma* very shortly stalked, dependent, triangular, trisulcate, papillose or not, wider than the staminode. *Capsule* spreading to dependent, three-ribbed, cylindrical to fusiform, smelling of vanilla; seeds subglobose, crustose.

DISTRIBUTION: A small genus of perhaps nine closely related species in the tropical Americas from Panama and Trinidad, south to Venezuela, the Guyanas, Brazil, Colombia, Ecuador and possibly Peru.
HABITAT: In lowland and lower montane rain forest and head-high shrubberies with grasses on sandy loam; sea level–1800 m.
DERIVATION OF NAME. The name derives from the Greek *selene*, the moon or a crescent, and *pedilon*, a slipper or shoe, in allusion to the flowers with a crescent-shaped slipper-like lip.

H.G. Reichenbach established the genus *Selenipedium* in 1854 to accommodate the tropical American slipper orchids with trilocular ovaries with parietal placentation. He recognised two sections: one with tall leafy stems and pleated leaves, the other apparently stem-less with a fan of conduplicate leaves. In 1882, Pfitzer removed the latter and united it with the Old World hard-leaved species in his genus *Paphiopedilum*, because of their conduplicate leaves and deciduous perianth. Rolfe (1896) reviewed the evidence and redefined *Selenipedium* for those species with tall stems, pleated leaves, a persistent perianth, a trilocular ovary with parietal placentation, valvate sepals and subglobose seeds. He removed all of the conduplicate-leaved tropical American species to his new genus *Phragmipedium*.

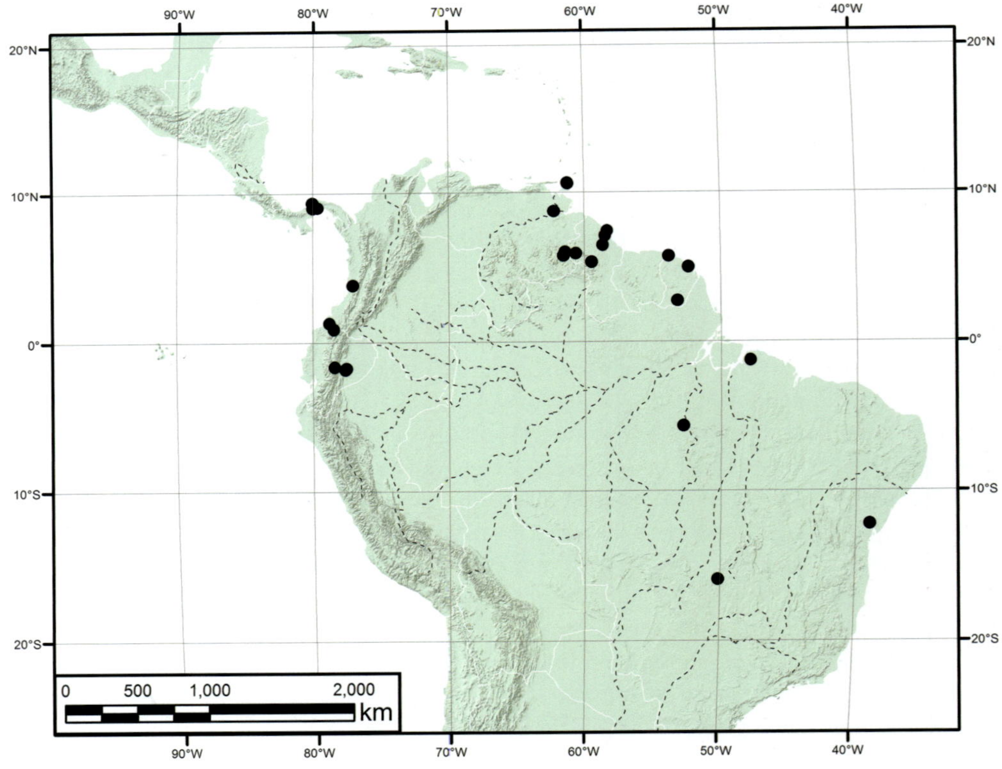

Map 2. Distribution of the genus *Selenipedium* in Central and South America based upon herbarium records.

Rolfe recognised three species in *Selenipedium*: *S. chica*, *S. palmifolium* and *S. isabelianum*. Since then further species have been described by Barbosa Rodrigues (1937), Foldats (1961), Garay (1978), Sambin & Braem (2015), Cribb & Schuiteman (2015) and Szlachetko and Kolanowska (2016). The genus ranges from Panama and Trinidad across to eastern Brazil and south to north-western Ecuador and possibly northern Peru. The currently accepted species are morphologically similar, but distinguishable by plant size, leaf shape and indumentum, flower size and colour, ovary indumentum and petal venation. However, more work needs to be undertaken to define the species. Photographs of plants from across the range of the genus can be viewed on the internet but the identifications should be taken with some scepticism. Part of the problem is that they are seldom collected, probably because of their resemblance in habit to large grasses or *Sobralia*, a common orchid growing in similar conditions throughout its range. The flowers do not preserve very well when pressed and their thin texture makes it difficult to reconstruct their three-dimensional shapes. We recommend that flowers should be preserved in spirit to assist future studies.

Atwood (1984) postulated that *Selenipedium*, based upon morphological evidence, was the most basal lineage in the Cypripedioideae. Albert (1994) published the first cladistic treatment of the subfamily Cypripedioideae to include DNA data. He included data from a single collection of *S. chica* in both his morphological and DNA matrices. It consistently came out sister to a clade comprising all of the other slipper orchid genera in each of the morphological, DNA and combined cladograms. Based upon nucleotide sequence divergence and divergence time estimates he calculated that *Selenipedium* had diverged from an apostasioid-like ancestor in the mid- to late-Miocene (49.3–69 million years ago), or possibly earlier if divergence data from the tropical American hard-leaved slipper orchids are taken into account.

Chiron *et al.* (2015) established the new genus *Apedium* to accommodate *S. chironianum*, supporting their decision with a morphological analysis. They consider *Apedium* the ancestral clade to slipper orchids. However, an alternative and more likely suggestion is that the strange flowers of *S. chironianum* with three sepals, three petals and three anthers are based upon a mutation in a gene or genes in flower development, much as is seen in *Phragmipedium lindenii*.

Szlachetko and Kolanowska (2016) have recently revised the genus, recognising nine species, including two new species, *S. garayanum* and *S. olgae* and a new subspecies of *S. chica*. Their classification emphasises staminode structure and leaf shape and pubescence as the most reliable characters. However, deducing these from herbarium material alone is fraught with problems, especially as only about 35% of the specimens that they examined bore flowers.

The recognised species, which are very similar morphologically, except for the tetratological *S. chironianum*, are undoubtedly closely related. Their present distribution and the disjunct nature of the populations of the more widespread species suggest strongly that extant populations represent relicts of a formerly more widespread distribution of a single species. The current species have probably evolved in isolation through genetic drift.

Selenipedium fruits have much the same fragrance as vanilla and are used regionally for flavouring. Local names reflect this: 'Vanillia chica' in Panama; 'Baunilha' in E Brazil.

Pollination by small bees has been recorded for *S. aequinoctiale* (Dodson & Dodson, 2004).

CULTIVATION. *Selenipedium* species have seldom been successfully grown in cultivation. They have proved to be difficult to transplant from the wild because their roots dry out rapidly once a plant is lifted. Some success has been obtained by removing a plant with the soil around its roots intact. However, this is an illegal practice in most countries where the species are found and is to be strongly discouraged. No success has been obtained by growing plants from seed; indeed, I know of no attempt ever having been made. Given the ease with which other slipper orchids can be raised from seed in vitro, it would certainly be worth a try.

Selenipediums grow in nutrient-poor well-drained soils in moist tropical regions where rainfall is regular. A number of composts should be tried if seedlings can be obtained. Other slipper orchids, particularly the plicate leaved *Cypripedium* species, have proved to be easily grown provided that the natural humus-poor, well-drained soils in which they grow are mimicked.

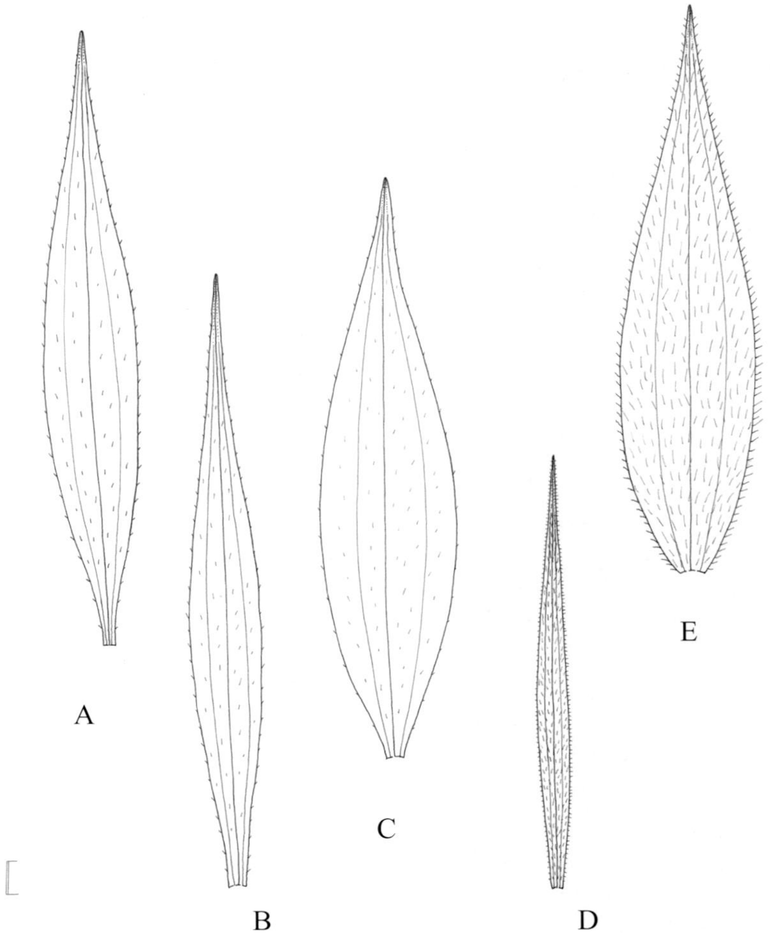

Fig. 22. *Selenipedium* leaves. **A**, *S. aequinoctiale*; **B**, *S. chica*: **C**, *S. isabelianum*; **D**, *S. steyermarkii*; **E**, *S. palmifolium*. All drawn by Judi Stone from Kew herbarium specimens. Double line = 10 mm.

KEY TO THE SPECIES OF SELENIPEDIUM

1. Flowers yellow-green; lateral sepals free, not united into a synsepal; lip petal-like, linear .... 9. ***S. chironianum***
   Flowers not pure yellow-green; lateral sepals united into a synsepal; lip saccate, not linear 2

2. Lip lilac; petals ciliate .... 8. ***S. vanillocarpum***
   Lip yellow, brown or orange, often marked on rim of lip with dark purple; petals not ciliate 3

3. Largest leaves 6–10 times as long as broad 4
   Largest leaves less than 6 times as long as broad 6

4. Leaves coriaceous, 8–10 times as long as broad; bracts ovate; staminode lanceolate, acute, stalk half length of staminode; lip orifice angular across base and broadest in basal half .... 7. ***S. steyermarkii***
   Leaves thin-textured, 6–7.5 times as long as broad; bracts lanceolate; staminode ovate or cordate, shortly apiculate; lip orifice elliptic-subcircular 5

5. Sepals and petals flushed with brown; lip side lobes touching for c. 10 mm under column; staminode broadly ovate; filament tips obtuse; bracts lanceolate, acuminate .... 3. ***S. chica***
   Sepals and petals yellow, faintly red-spotted near base; lip side lobes touching for 2–3 mm below column; staminode cordate; filament tips globose; bracts ovate-lanceolate .... 2. ***S. buenaventurae***

6. Lip incurved side lobes overlapping beneath the staminode 6
   Lip incurved side lobes not overlapping beneath the staminode 8

7. Staminode rhombic, with a stalk longer than the staminode, a quarter the width of the lip base, yellow .... 1. ***S. aequinoctiale***
   Staminode trullate with an upcurved tip and with a very short stalk, a third the width of the lip base, green .... 4. ***S. dodsonii***

8. Leaves densely villose below, glabrescent above, ciliate; sheaths pubescent; bracts elongate glandular villose; ovary shortly pubescent, glandular at apex; petals 5-nerved; lip 1.25–1.52 times as long as broad, yellow or brownish yellow maroon-marked around the apical two-thirds of the orifice; staminode triangular-trullate, as broad as long .... 7. ***S. palmifolium***
   Leaves sparsely pubescent beneath, glabrous above and on margins; basally ciliate; sheaths glabrous; petals 1- to 3-nerved; bracts not glandular; ovary pubescent on ribs, not glandular at apex; lip 1.6 times or more longer than broad, orange with two maroon marks near the apex of the orifice; staminode deltoid, longer than broad .... 6. ***S. isabelianum***

1. SELENIPEDIUM AEQUINOCTIALE

Leslie Garay described *S. aequinoctiale* in 1978 based on examination of two herbarium collections from near Lita in northern Ecuador, distinguishing it on its leaf and ovary morphology. Another of the distinguishing features given by Garay was the height of the plant which he stated as being only 1 metre tall. This is certainly erroneous, living material reaches 3 m or more in the wild.

Szlachetko and Kolanowska (2016) described *S. olgae* based on a specimen from close to the Ecuadorian border in south-west Colombia. They compared it with *S. aequinoctiale*, which it resembles, but distinguished it by its villose leaf bases, hirsute ovary, lip lateral lobes, linear petals, elliptic-obovate dorsal sepal with a cuneate base and linear-lanceolate staminode. Examination of their dissections does not support their thesis. Furthermore, we believe that the staminode shape is an artefact of the pressing of the flower. Photographs of the side view of the staminode of *S. aequinoctiale* shows a very similar shape that could easily be achieved by compression in the pressing process. We, therefore, reduce *S. olgae* to the synonymy of *S. aequinoctiale* here.

Selenipedium aequinoctiale was known for many years only from the type locality in north-west Ecuador where a colony of perhaps 50–100 plants has survived for many years. It grows there on a steep, west-facing laterite bank amongst ferns and *Miscanthus* grasses at about 690–700 m elevation. It is apparently absent from many similar sites in the vicinity.

Tom Dodson, Cal Dodson's son, photographed a small bright metallic blue-gold anthrophorine bee, seen to be pollinating flowers of *S. aequinoctiale* at Lita (Dodson & Dodson, 2004).

DESCRIPTION. A terrestrial *herb* with erect, terete, leafy stems up to 3 m tall, often sparsely branching above; stems dark green, covered by coarsely and densely glandular-villose leaf-bases. *Leaves* lanceolate to elliptic-lanceolate, acuminate, cuneate at base of lamina, 11–16 × 2–3.5 cm, dark green, glabrous above, sparsely pilose beneath, about

Fig. 23. *Selenipedium aequinoctiale.* Growing on a hillside near Lita, N Ecuador. (SD)

Plate 8. *Selenipedium aequinoctiale*. Drawn by Carol Woodin for *Curtis's Botanical Magazine* n.s.t. 636 (2009), from a plant cultivated by Harry and Rosemarie Zelenko.

Fig. 24 (left). *Selenipedium aequinoctiale* from near Lita in N Ecuador (PC). **Fig. 25** (right). *S. aequinoctiale* from northern Ecuador (SD).

Fig. 26. Alex Hirtz holding a plant of *Selenipedium aequinoctiale*, near Lita, N Ecuador. (PC).

8-nerved, sheathing at base; sheaths minutely puberulent. *Inflorescence* terminal or from axils of upper branches, up to 20 cm long, densely many-flowered; bracts chartaceous, ovate, acute or subacuminate, 10–20 × 6 mm, glandular ciliate. *Flowers* small, produced in succession; sepals and petals light straw yellow, lip yellow, finely spotted with purple within and with two large maroon marks on the basal margin of the rim; pedicel and ovary 50–70 mm long, pubescent. *Dorsal sepal* elliptic, acute, 20–30 × 18 mm. *Synsepal* elliptic-

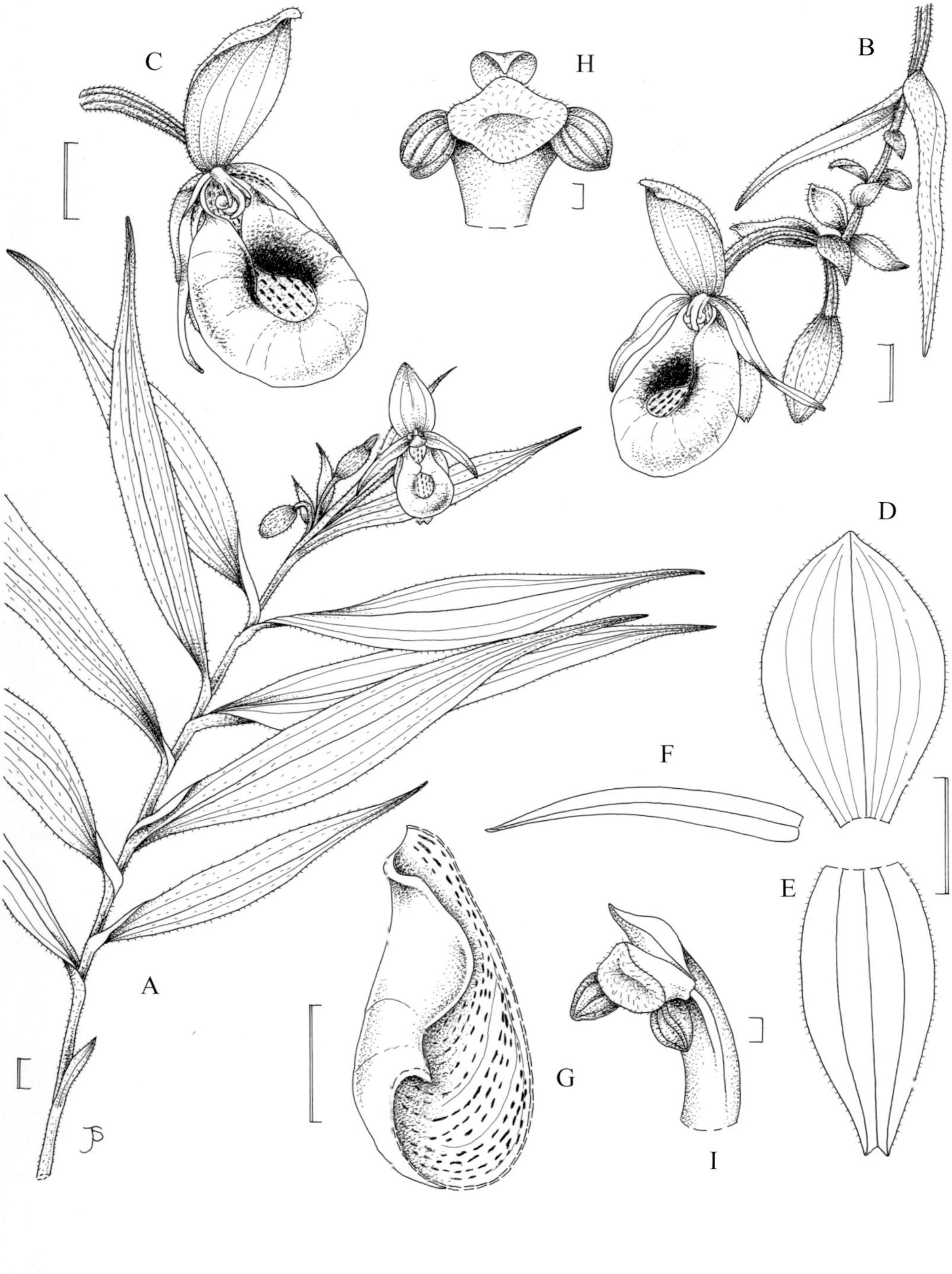

Fig. 27. ***Selenipedium aequinoctiale.*** **A**, habit; **B**, inflorescence; **C**, flower; **D**, dorsal sepal; **E**, synsepal; **F**, petal; **G**, lip, longitudinal section; **H**, column; **I**, column, oblique view. All drawn by Judi Stone from *Harling* 4304 and photographs. Single bar = 1 mm; double bar = 10 mm.

ovate, bidentate at apex, 20–25 × 16 mm. *Petals* obliquely linear-lanceolate, acute, up to 25 × 4 mm. *Lip* calceolate, broadly elliptic in outline, up to 30 × 25 mm; inflexed basal margins rounded. *Column* 5 mm long; staminode stalked, rhombic, acute, up to 5 mm long, longer than the triangular, papillate stigma. *Fruits* triangular-cylindrical, slightly arcuate, up to 5 cm long, round in cross-section. **Plate 8. Figs. 23–27.**

S. aequinoctiale Garay in Op. Bot., B (Fl. Ecuador), 9(225: 1): 13 (1978). Type: Ecuador, Lita, *Harling* 4304 (holo. S).

S. olgae Szlachetko & Kolanowska, Syst. Bot. 41, 1: 157 (2016), **syn. nov.** Type: Colombia, Nariño, Tumaco Municipality, Corrigimiento de Llorente, Vereda El Carmen, 210 m, 11 June 1986, *de Benavides* 6684 (holo. PSO).

DISTRIBUTION. NW Ecuador and south-western Colombia (Nariño region). **Map 3.**

HABITAT. On steep banks on laterite with ferns, grasses and low shrubs surrounded by wet forest; up to 700 m elevation.

PHENOLOGY. Flowering in October–November.

CONSERVATION STATUS. Endangered (EN): B2ab(ii,iii,v); C1+2a(i,ii); D.

It is a very local, narrowly endemic species, restricted to very few localities. It appears to be a primary coloniser of bare soil, such as roadside banks, the subpopulations have small numbers of mature individuals (possibly less than 300) and little sign of seedling regeneration. The species is vulnerable to deforestation, human disturbance and habitat degradation. The estimated extent of occurrence and estimated area of occupancy of the species are less 5000 km^2 and 500 km^2, respectively with an inferred decline in the number of mature individuals and the quality of habitats as competition increases from other plants.

ETYMOLOGY. From the Latin *aequinoctialis*, relating to the equinox and, by inference, Ecuador which is on the equator.

Map 3. Distribution of *Selenipedium aequinoctiale* in the northern Andes.

2. SELENIPEDIUM BUENAVENTURAE

Günter Gerlach photographed a *Selenipedium* near Buenaventura in the coastal region of Colombia. The plants have flowers with a lip with a yellow base and more densely flushed all around the smaller orifice with deep purple that are closer to those of typical *S. aequinoctiale*. The staminode is also similar to *S. aequinoctiale* but slightly larger with a shorter stalk and distinctive globular tips to the filaments, while the latter has less colourful flowers with less extensive purple markings. This had been mis-identified as a Colombian form of *S. chica*.

Szlachetko and Kolanowska (2016) have recently described it as subsp. *buenaventurae* of *S. chica*. However, it differs significantly from typical *S. chica*, which is endemic to Panama. In particular, the sepals of the Colombian taxon are slightly obovate-elliptic and yellow (rather than ovate-elliptic and brown), the lip side lobes are rounded and touch at the base only (rather than being oblong and touching for 10 mm or more), the petals are falcate, ovate-lanceolate and yellow (rather than linear-tapering and brown), the stamens have globose free tips (rather than blunt) and the staminode is stalked and cordate (rather than ovate). Given the morphological similarity of the currently recognised species of *Selenipedium*, we have no hesitation in raising it to species rank as *S. buenaventurae*.

Fig. 28 & 29. *Selenipedium buenaventurae*. Flowers, Buenaventura, W Colombia. (GG)

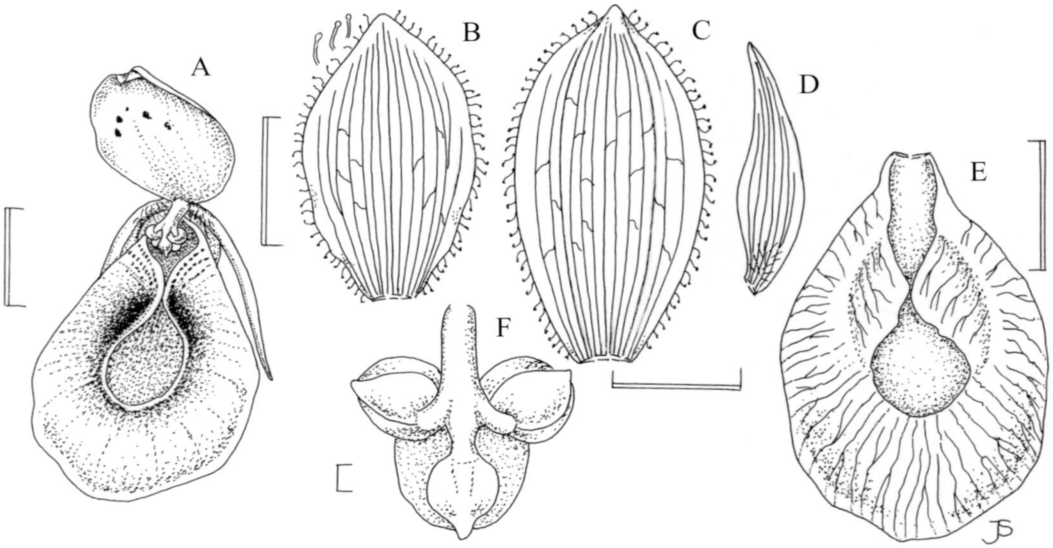

Fig. 30. ***Selenipedium buenaventurae.*** **A**, flower; **B**, dorsal sepal; **C**, synsepal; D, petal; **E**, lip, from above; **F**, column from above. Drawn by Judi Stone after Szlachetko and Kolanowska (2016) and Guenter Gerlach's photographs. Double bar = 10 mm; single bar = 1 mm.

DESCRIPTION. A terrestrial *herb* with erect, terete, leafy stems up to 3 m tall, often sparsely branching above; stems dark green, covered by coarsely and densely glandular-villose leaf-bases. *Leaves* lanceolate to elliptic-lanceolate, acuminate, cuneate at base of lamina, 16–30 × 2–4 cm, dark green, glabrous above, pilose beneath; leaf sheaths pubescent. *Inflorescence* terminal, up to 23 cm long, many-flowered; bracts chartaceous, ovate-lanceolate, subacuminate, up to 15 × 6 mm long pubescent. *Flowers* small, produced in succession; sepals and petals light yellow, speckled with red in basal half, lip yellow, with two large maroon marks on the basal margin of the rim and flushed red around the rim; pedicel and ovary 50–70 mm long, glandular-pubescent. *Dorsal sepal* elliptic-ovate, obtuse, 20–25 × 13–15 mm, densely glandular-pubescent on outer surface. *Synsepal* elliptic, subacute, 25–29 × 12–16 mm. *Petals* falcate ovate-lanceolate, acuminate, up to 20 × 4 mm. *Lip* calceolate, broadly elliptic in outline, up to 32 × 19–25 mm; inflexed basal margins rounded and just touching. *Column* 8.5 mm long; anther filaments globose at tips; staminode stalked, cordate, acute, up to 3 mm long, as long as the papillate stigma. *Fruits* triangular-cylindrical, slightly arcuate, up to 5 cm long, round in cross-section. **Figs. 28–31.**

S. buenaventurae (Szlachetko & Kolanowska) P.J. Cribb, **comb. et stat. nov.**
S. chica subsp. *buenaventurae* Szlachetko & Kolanowska, Syst. Bot. 41, 1: 146 (2016), **syn. nov.** Type: Colombia, Valle del Cauca, Buenaventura, 3 May 1926, *Cork s.n.* (holo. US; iso. AMES).

DISTRIBUTION. Western Colombia (Choco region). **Map 4**.
HABITAT. In scrub surrounded by wet forest; up to 300 m elevation.
PHENOLOGY. Flowering in October–November.

CONSERVATION STATUS. Data Deficient but possibly Endangered (EN).

It is a very local, narrowly endemic species, restricted to very few localities. The species is vulnerable to deforestation, human disturbance and habitat degradation. It has an estimated extent of occurrence and estimated area of occupancy of less than 5000 km^2 and 500 km^2.

ETYMOLOGY. From Buenaventura, a coastal city in western Colombia.

Fig. 31. *Selenipedium buenaventurae*. Buenaventura, W Colombia. (GG)

Map 4. Distribution of *Selenipedium buenaventurae*.

3. SELENIPEDIUM CHICA

H.G. Reichenbach (1854) described *Selenipedium chica* based upon a collection by Duchaissaing from Panama. He distinguished it from *S. palmifolium*, the only other species known at the time, by its habit, up to 5 m tall, narrow leaves 6–7 times as long as broad, 3-nerved petals, an eglandular ovary and relatively longer deltoid-trullate staminode.

It is endemic to a small area on the Pacific slopes of the main mountain chain in Panama. Records from Colombia are unconfirmed (see *S. aequinoctiale*).

It is closest to *S. aequinoctiale* from Ecuador, both having flowers with a similarly shaped lip with a more or less elliptic orifice and a deltoid staminode. In *S. chica* the purple marking on the lip covers runs from the apex of the incurved side lobes around the apical part of the orifice. The dorsal sepal is brown and both it and the petals are also purple-flecked. In *S. chica* the anther filaments are shorter than the stamens, but as long as those in *S. aequinoctiale*.

Dressler (1993b) provided an excellent photograph of *S. chica* in his account of Panamanian and Costa Rican orchids. Escobar (1996) published Dressler's photograph of a flower of a Panamanian plant in his *Native Colombian Orchids*.

DESCRIPTION. A terrestrial *herb* with leafy stems up to 5 m tall covered by sheathing leaf bases. *Leaves* somewhat membranaceous, distichous, lanceolate to elliptic-lanceolate, acuminate, 15–30 × 1.5–5 cm, sparsely pubescent dorsally especially along the veins, with a cylindrical sheathing base, 2–4 cm long. *Inflorescence* terminal, many-flowered, densely pubescent; bracts lanceolate, acute to acuminate, 10–15 mm long, shortly pubescent. *Flowers* small to medium-sized, opening in succession, soon deciduous if not fertilised, pubescent on outer surface of sepals; sepals and petals greenish brown, lip yellow flushed with brown and stained with red-purple on incurved margin of lip orifice; pedicel and ovary 15–23 mm long, pubescent. *Dorsal sepal* erect, elliptic, acute, c. 20 × 10 mm. *Synsepal* elliptic, slightly bifid at apex, c. 20 × 10 mm. *Petals* deflexed, obliquely linear-lanceolate, acute, up to 20 × 2 mm, glabrous. *Lip* calceolate, with an incurved margin, 20–30 × 10–18 mm. *Column* 5–7 mm long; stigma twice as long as the stigma, papillate; staminode spatulate, acute, much narrower than column apex, triangular. *Fruit* cylindrical, somewhat arcuate, pubescent. **Figs. 32–36.**

Fig. 32. *Selenipedium chica.* Flower. Panama (KD).

Fig. 33. ***Selenipedium chica*****.** **A**, habit; **B**, flower; **C**, dorsal sepal; **D**, synsepal; **E**, petal; **F**, lip, longitudinal section; **G**, column, side view; **H**, column, dorsal view; **I**, column, ventral view; **J**, ovary, transverse section. All drawn by Judi Stone from *C.W. Powell 258* and Kew Spirit Collection no. 48733. Single bar = 1 mm; double bar = 10 mm.

Fig. 34 & 35 (above & below left). *Selenipedium chica.* Inflorescence and flower, Panama. (KD). **Fig. 36** (below right). *Selenipedium chica* being examined by Robert Dressler, Panama. (KD)

S. chica Rchb.f., Xenia Orch. 1: 3, t. 2 (1854); Pfitzer in Engl., Pflanzenr. IV, 50: 27 (1903); Ames, Sched. Orch. 2: 1, t. 1 (1923). Type: Panama, *Duchassaing s.n.* (holo. W!; iso. P!).
Selenipedilum chica (Rchb.f.) Pfitzer in Engler & Prantl, Nat. Pflanzenfam. 2, 6: 84 (1888), *orth. var.*

DISTRIBUTION. Panama, endemic to Pacific slope and upper Chagres basin. **Map 5**.
HABITAT. Terrestrial at sea level–100 m.
PHENOLOGY. Flowering from October to December and possibly longer.
CONSERVATION STATUS. Endangered (EN): B1ab(ii,iv,v)+2ab(ii,iii,v). It is a local species with a large but very fragmented distribution. Its subpopulations in Panama are not protected, the subpopulations are well separated but not threatened by collection for horticulture. Threats include agriculture intensification, fire and deforestation. The estimated area of occupancy of the species is 790 km^2 and the estimated area of occupancy is 12 km^2 with an estimated continuing decline in the quality of habitats and the number of mature individuals in all the locations. Dressler (1993b) stated that this was one of Panama's rarest and most endangered species.
ETYMOLOGY. From the Spanish word *chica*, a girl, according to Berthold Seeman the local name used for it in Panama (Lindley, 1853).

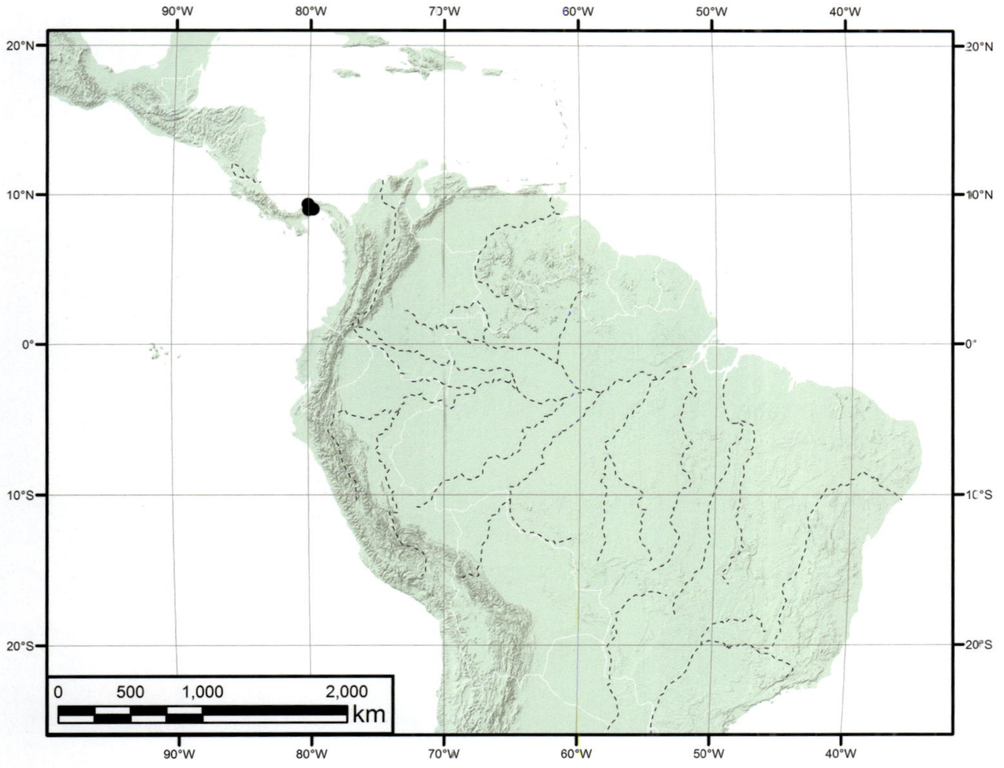

Map 5. Distribution of *Selenipedium chica* in Panama.

4. SELENIPEDIUM DODSONII

Calaway Dodson (1999) recognised two species of *Selenipedium* in his checklist of the Orchidaceae for Jørgensen & Leon-Yánez's *Catalogue of the Vascular Plants of Ecuador*. One of these, *Selenipedium aequinoctiale*, was described from northern Ecuador, whereas the other, based on a specimen from Succumbios Province in north-eastern Ecuador (cited as *Davis* 1048 at AMES), he identified as *S. chica*, a species previously considered endemic to Panama. Later, Dodson (pers. comm.) sent me an updated manuscript in which he listed a further two collections of this taxon from Pastaza Province, to the south of Succumbios. Perez Garcia and Mo Mo (2015a) recently published a photograph of this orchid, also under the name *S. chica*. It is certainly not *S. chica* which has much longer and narrower leaves with short hairs only on the veins of the underside, pubescent but not glandular bracts, and distinct flowers in which the ovary is villose but not glandular, the petals are tri-nerved and the staminode is spatulate not trullate. It differs from *S. aequinoctiale* in having a flower with a lip with a narrower margin around the orifice and longer side lobes and a much larger trullate staminode that is about a third the width of the lip base (rather than less than a quarter the width). The only other candidate for this species is the widespread *S. palmifolium*, but that differs in having a dorsal sepal that is not ciliate, a distinctive lip with scarcely defined incurved side lobes and a large elliptic orifice marked with purple around the rim and a small triangular staminode on a distinct stalk.

Szlachetko and Kolanowska (2016) drew a flower from another of the specimens cited by Dodson and described it as a new species *S. garayanum*. They distinguished it from *S. palmifolium* in having obliquely triangular side lobes to the lip and from other species in having almost glabrous leaves, leaf sheaths and short floral bracts and a spatulate-obovate staminode. They did not examine the type material of *S. dodsonii*, but distinguished their taxon from it by the shape of the staminode. We are sure that leaf pubescence and the shape of the staminode can be deceptive, especially in dried, pressed material. Also, leaves that are glabrescent often appear glabrous in older material. Both *S. dodsonii* and *S. garayanum* have a broad, shortly stalked staminode but in other characters and, having seen photographs of living plants, we do not consider the variation exhibited to be significant.

DESCRIPTION. A terrestrial *herb* with erect, terete, leafy stems up to 3 m tall, often sparsely branching above; stems dark green, covered by coarsely and densely villose leaf-bases. *Leaves* narrowly elliptic-lanceolate, acuminate, cuneate at base of lamina, 17–20 × 2.8–3.8 cm, dark green, glabrous above, sparsely pilose at base, about 8-nerved, ciliate, sheathing at base; sheaths

Fig. 37. *Selenipedium dodsonii*. Flower. Northern Ecuador (FT).

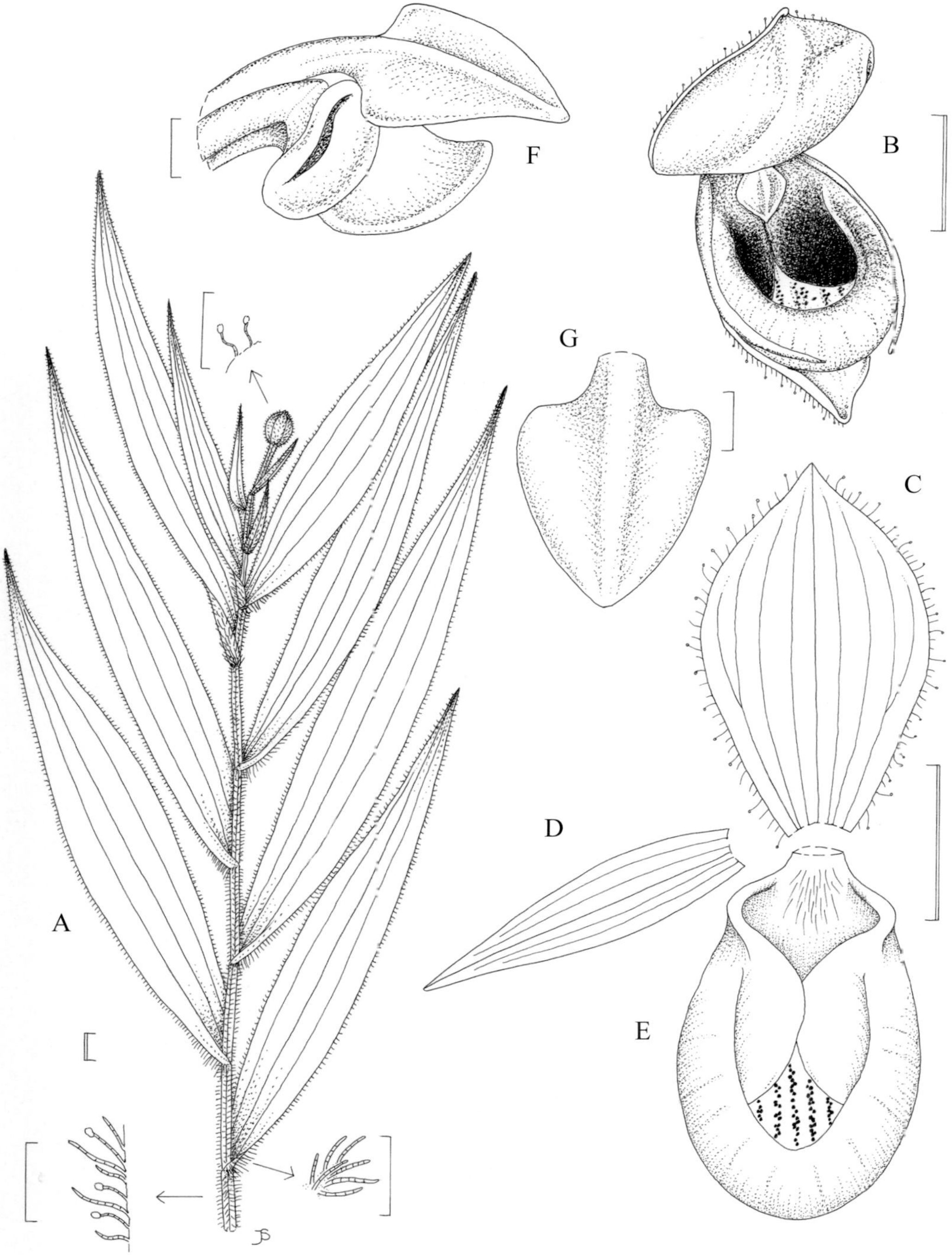

Fig. 38. ***Selenipedium dodsonii.*** **A**, habit; apical part of stem, leaves and inflorescence; **B**, flower; **C**, dorsal sepal; **D**, petal; **E**, lip; **F**, column; **G**, staminode. All drawn by Judi Stone. Single bar = 1 mm; double bar = 10 mm.

minutely puberulent. *Inflorescence* terminal or from axils of upper branches, up to 15 cm long, densely few-flowered; bracts chartaceous, ovate, acute or subacuminate, 10–30 × 3–4 mm, pubescent, ciliate. *Flowers* small, produced in succession; sepals and petals light dull yellow, lip yellow, finely spotted with purple within and with two large maroon marks on the incurved side lobes; pedicel and ovary 30–50 mm long, densely glandular-pubescent. *Dorsal sepal* elliptic, acute, 22–25 × 13–14 mm. *Synsepal* elliptic-ovate, shortly bidentate at apex, 20–24 × 16 mm. *Petals* slightly obliquely narrowly lanceolate, acute, up to 24–25 × 4 mm, 5-veined. *Lip* calceolate, broadly elliptic in outline, 17–20 × 12–13 mm; inflexed basal margins rounded; side lobes broadly semi-elliptic, overlapping on incurved margins, longer than broad. *Column* 5 mm long; staminode stalked, trullate, acute, 4 mm long, 2 mm broad, longer than the obtriangular, papillate stigma; anther filaments truncate at tips; anthers large, 2 mm long. *Fruits* not seen. **Figs. 37–42.**

Fig. 39 (left). *Selenipedium dodsonii.* Apex of inflorescence and flower, N Ecuador (FT). **Figs. 40-42** (right). *Selenipedium dodsonii.* Column from below, obliquely below and from side. (FT)

Selenipedium dodsonii P.J. Cribb in Lankesteriana 15, 3: 181 (2015). Type: Ecuador, Pastaza, *Holguer Lugo S.* 5538 (holo. AMES).

S. garayanum Szlachetko & Kolanowska in Syst. Bot. 41, 1: 151 (2016), **syn. nov.** Type: Ecuador, Napo, confluence of Quiwado and Tiwaeno rivers, 28 April 1981, *Davis, Yost, Keno & Tomo* 1048 (holo. AMES!; iso. QCA).

DISTRIBUTION. N Ecuador (on eastern side of Andes). **Map 6.**

HABITAT. In damp scrub near rivers in rain forest; 200–300 m elevation.

PHENOLOGY. Flowering April-November.

CONSERVATION STATUS. Endangered (EN): B1ab(ii,iii,v)+2ab(ii,iii,v). It is a rare and very local species with a restricted distribution in northern Ecuador. It is protected by its remote habitat but the development of the forests on the eastern side of the Andes in Ecuador is increasing with anthropogenic changes to the forest and rivers where it occurs. The estimated extent of occurrence and estimated area of occupancy of the species are less than 1000 km^2 and 20 km^2, respectively, with an estimated continuing decline in the quality in its habitats.

ETYMOLOGY. Named in honour of Dr Calaway Dodson (1928–), the author of numerous accounts of Andean orchids. He was Executive Director of the Marie Selby Botanical Gardens in Sarasota, Florida from 1973 and is now Honorary Curator of Orchidaceae there. He also established the Rio Palenque reserve in western Ecuador. The reserve comprises one of the few surviving patches of tropical forest on the western side of the Ecuadorian Andes.

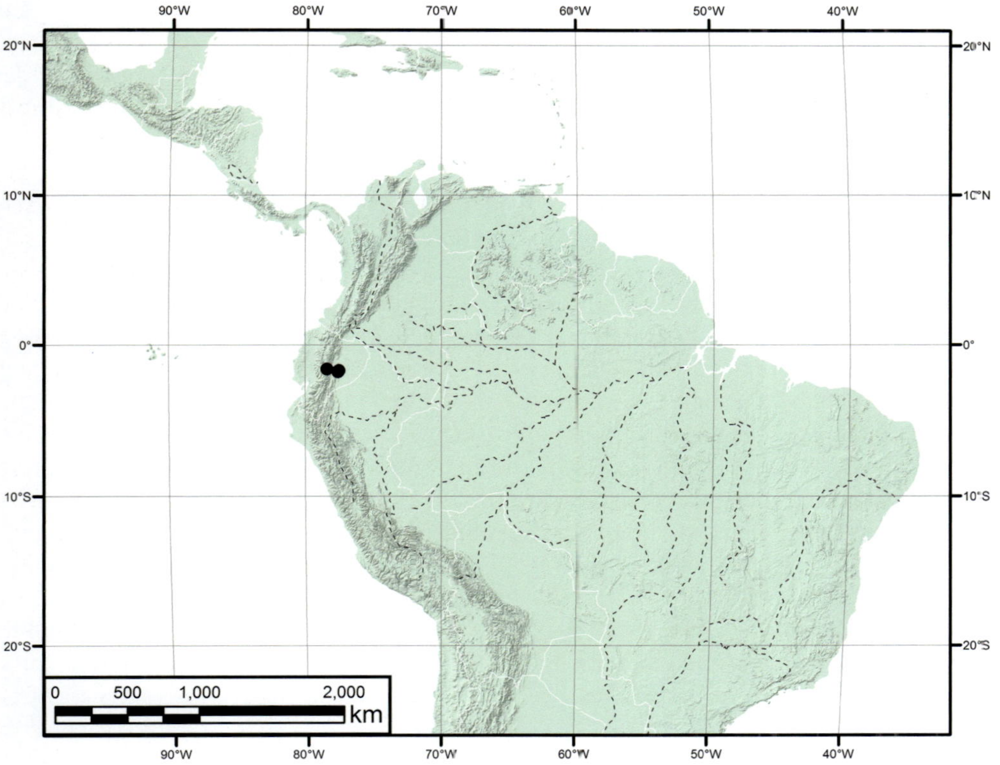

Map 6. Distribution of *Selenipedium dodsonii* in northern Ecuador.

5. SELENIPEDIUM ISABELIANUM

When he described *S. isabelianum*, Barbosa Rodrigues (1877) failed to compare it with any other species in the genus, but noted the vanilla-fragrance of the seed-pods which were used in Para by the local women to scent their hair. His description and drawing (Fig.), reproduced in Sprunger *et al.* (1996), suggest that *S. isabelianum* is very similar to *S. palmifolium* but a much smaller plant with an orange-yellow lip. Like the latter, it has triangular ovaries that are longer than the bracts and a blunt, somewhat spatulate staminode, whereas in most other features it also agrees well with preserved material I have examined of *S. palmifolium*. It differs most in its narrower obovate, shallow orange-yellow lip with maroon markings confined to two well-defined marks near the apex of the rim of the orifice, and in its ovary that is hairy on the ribs rather than all over. The illustration labelled as *S. palmifolium* by Pabst & Dungs (1975) is more likely to be this species.

Barbosa based his description of *Selenipedium isabelianum* on his own collection from near Belem in Para state. A small colony of 57 plants has recently been re-discovered by João Batista (pers. comm.) but the flowers have a somewhat ovate dorsal sepal and obovoid, somewhat tear-shaped lip. Another population, found in the state of Bahia, has flowers that agree well with Barbosa's illustration (Sprunger *at al.*, 1996), having an elliptic dorsal sepal and somewhat elliptic lip (van den Berg, pers. comm.)

A photograph on the world-wide web, labelled *S. palmifolium* from Piste Lassouraille in French Guyana, shows a plant with flowers with a bright orange lip and a staminode atypical of *S. palmifolium* (lachauseesetterrouge.orchidees.over-blog.com). It is close to the Brazilian *S. isabelianum* and should be compared with it if a specimen can be located. As far as we can determine, no herbarium specimen of it exists.

Some photographs of *S. isabelianum* on the world-wide web show flowers with a lilac base to the lip. Comparison of these with *S. vanillocarpum* would be useful to confirm if the latter is really distinct.

DESCRIPTION. A terrestrial *herb* with 1–4 erect leafy stems 50–100 cm tall, often sparsely branching above; internodes 3–6 cm long; stems dark green, covered by coarsely and densely glandular-villose leaf-bases. *Leaves* oblanceolate to elliptic-lanceolate, acuminate, cuneate at base of lamina, 10–25 × 2–7 cm, dark green, glabrous above, hairy beneath, ciliate, about 8-nerved, sheathing at base; sheaths glabrous. *Inflorescence* terminal, 5–27 cm long, one-branched, densely many-flowered, with 1–2 open at a time; bracts erect-spreading, chartaceous, ovate-lanceolate to ovate, acute, 4–18 mm long, coarsely shortly hairy, ciliate. *Flowers* small; sepals and petals light yellow-green to dull pale fawn, lip orange-yellow, marked with two maroon spots near apex of rim of orifice and spotted within with purple; pedicel and ovary triangular in cross section, 15–40 mm long, glandular pubescent on ridges. *Dorsal sepal* narrowly elliptic-ovate, obtuse, 16–20 × 8–10 mm. *Synsepal* elliptic-ovate, bidentate at apex, 15–20 × 8–12 mm, pilose on outer surface. *Petals* obliquely linear-lanceolate, acute, 16–20 × 1.5–2 mm. *Lip* calceolate, broadly elliptic in outline, 19–23 × 10–15 mm. *Column* 5–6 mm long; staminode narrowly ovate-deltoid, truncate, slightly longer than the stigma. *Fruit* cylindrical, slightly arcuate, 5–9 cm long, triangular in cross-section, pubescent on the ridges. **Figs. 43–45.**

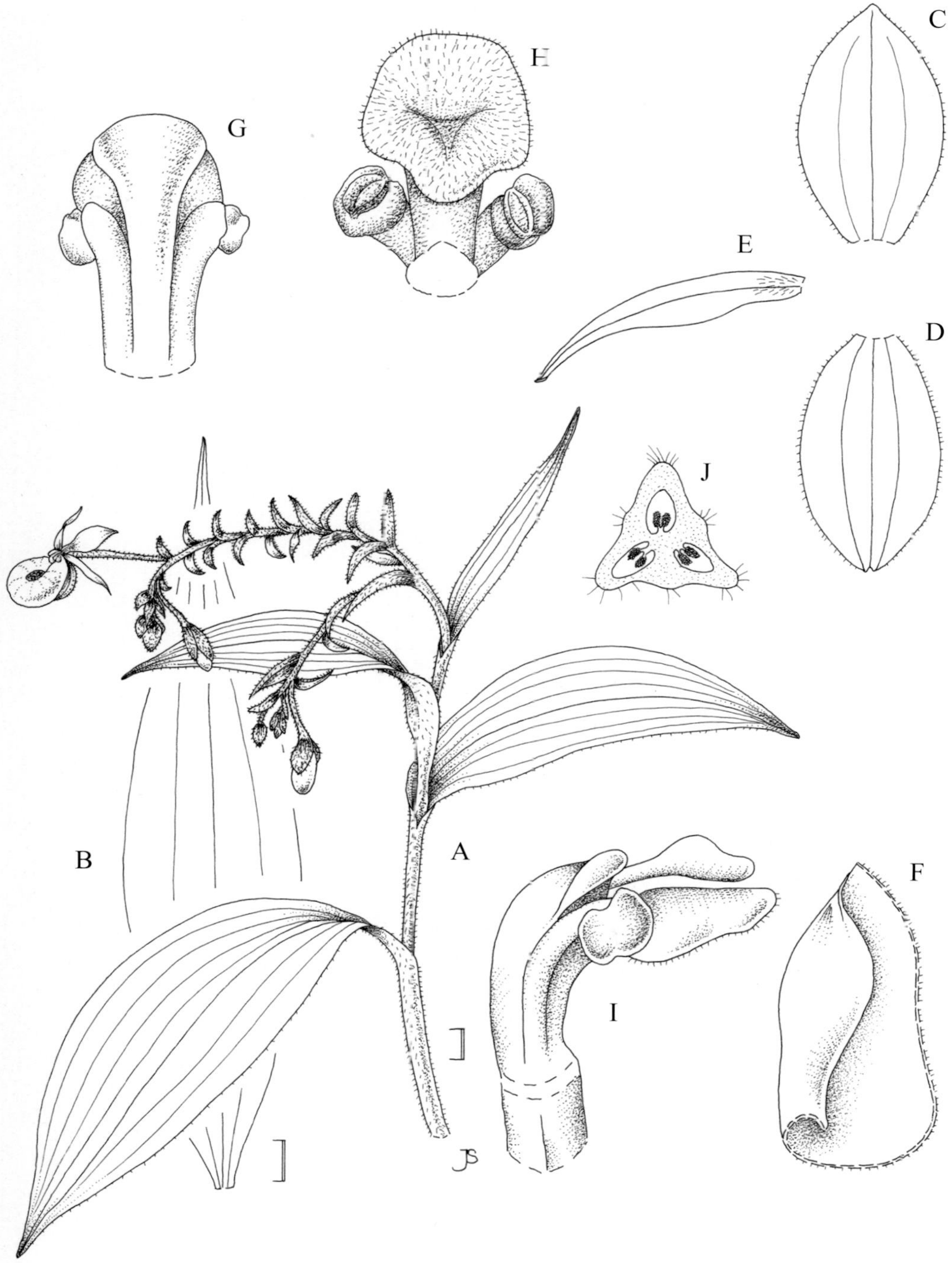

Fig. 43. ***Selenipedium isabelianum.*** **A**, habit; **B**, leaf; **C**, dorsal sepal; **D**, synsepal; **E**, petal; **F**, lip, longitudinal section; **G**, column, dorsal view; **H**, column, ventral view, **I**, column, side view; **J**, ovary, transverse section. All drawn by Judi Stone after *Sprunger et al.* eds, *J. Barbosa Rodrigues, Iconographie des Orchidées du Brésil*. Double bar = 10 mm.

Fig. 44 & 45. *Selenipedium isabelianum.* Inflorescence and flower, Bahia, Brazil. (CB).

Selenipedium isabelianum Barb. Rodr. in Diario do Grão Pará, 285 (8 Dec. 1874) & Gen. Sp. Orchid. Nov. 1: 201 (1877). Type: Brazil, Pará, *Barbosa Rodrigues s.n.* (lecto. illustration in Barb. Rodr. Icon. 1: t. 2).

DISTRIBUTION. E Brazil (Bahia and Pará States). **Map 7**.

HABITAT. In shady, slightly disturbed forest.

PHENOLOGY. Flowering in May–November.

CONSERVATION STATUS. Endangered (EN): B2ab(ii,iii,v). It is a rare and very local species with a distribution restricted to three small and widely separated areas in north-eastern Brazil. It is threatened by factors such as anthropogenic habitat degradation, deforestation and climate change. The estimated extent of occurrence and estimated area of occupancy of the species are 6500 km^2 and 10 km^2, respectively, with a continuing decline in the quality of habitats in all the locations.

ETYMOLOGY. Named after Dona Isabel (1846–1921) Princess Imperial and former Regent of Brazil, she was the elder daughter of Dom Pedro II; her husband, Gaston of Orléans, Count d'Eu, was considered, after the death of Dom Pedro II, the *de jure* "emperor-consort".

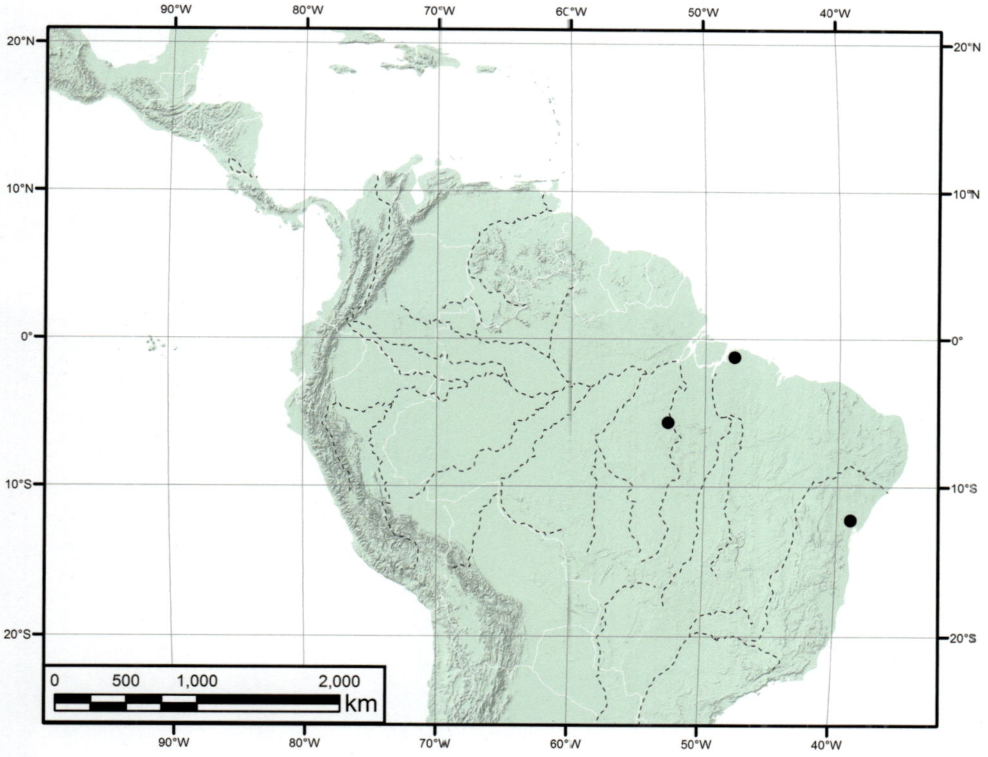

Map 7. Distribution of *Selenipedium isabelianum* in Brazil.

6. SELENIPEDIUM PALMIFOLIUM

Robert Schomburgk (later Sir Robert) discovered this strange bamboo-like slipper orchid in British Guiana (now Guyana) in the late 1830s and John Lindley described it in 1840 as *Cypripedium lindleyanum*. Schomburgk explored British Guiana and the course of the Orinoco between 1835 and 1839, and from 1840 until 1844 was a member of the Committee that considered the delimitation of the border between British Guiana and Venezuela (a border still part-contested today). Schomburgk was the inspiration from Arthur Conan Doyle's Professor Challenger and his adventures in *The Lost World* (Hodder & Stoughton, 1912). H.G. Reichenbach transferred it to his genus *Selenipedium* in 1854 in the first volume of his *Xenia Orchidacea*, distinguishing the genus from *Cypripedium* by its trilocular ovaries and *Vanilla*-like seeds. It is a lowland plant, often found in marshy areas, and the most widespread of the five species of *Selenipedium*.

This species always has a shallow lip, broadest around the rim and shaped much like a bed-pan, in contrast to the deeper lips of *S. aequinoctiale*, *S. chica* and *S. steyermarkii* and an almost triangular staminode as long as broad. It is variable in flower colour as demonstrated by examining the images posted on line (under *Selenipedium* images). The Trinidad form has a yellow lip whereas those from the Venezuelan shield have brownish ones. The purple markings are on the sides of the lip orifice but often extend around the front of the rim. A photograph on the world-wide web, labelled *S. palmifolium* from Piste Lassouraille in French Guyana with flowers with a bright orange lip, is close to the Brazilian *S. isabelianum* and it needs further study (see under *S. isabelianum*).

The Ecuadorian *S. dodsonii*, the Brazilian species *S. isabelianum* and the enigmatic *Selenipedilum vanillocarpum* are closest to *S. palmifolium* but the first is found in the Ecuadorian Andes, and the others do not occur north of the Amazon. Morphologically, they are similar to *S. palmifolium*, although they differ in details of their lip shape and colour and staminode structure.

Fig. 46. *Selenipedium palmifolium*. Inflorescence. Trinidad (PH).

Plate 9. *Selenipedium palmifolium*. Drawn by Everard Im Thurn on Mt Roraima, Guyana/Venezuela border, in December 1884. (Kew collection)

Fig. 47. *Selenipedium palmifolium*. Flower. Trinidad (PH).

DESCRIPTION. A terrestrial *herb* with erect leafy stems 0.5–2.2 m tall, often sparsely branching above; stems dark green, covered by coarsely and densely glandular-villose leaf-bases. *Leaves* oblanceolate to elliptic-lanceolate, acuminate, cuneate at base of lamina, 10–25 × 2–7 cm, about 6 times as long as broad, dark green, glabrous above, sparsely pilose on veins and ciliate, about 8-nerved, sheathing at base; sheaths hairy. *Inflorescence* terminal, 10–27 cm long, densely many-flowered; peduncle densely pubescent; bracts chartaceous, ovate-lanceolate to ovate, acute, 6–20 mm long, with long, coarse, glandular hairs, ciliate. *Flowers* small; sepals and petals translucent yellow to purple-brown to dull pale fawn, lip lemon-yellow or dark yellow, often flushed orange or brown, marked with maroon around rim of orifice and spotted within with purple; anthers green; stigma light green; bracts 10–15 mm long, puberulent; pedicel and ovary 15–40 mm long, twice as long as the bracts, glandular pubescent. *Dorsal sepal* narrowly elliptic-ovate, obtuse, 18–26 × 8–15 mm. *Synsepal* elliptic-ovate, bidentate at apex, 15–29 × 10–16 mm, pilose on outer surface. Petals obliquely linear-lanceolate, acute, 20–32 × 1.5–3 mm. *Lip* calceolate, broadly elliptic in outline, 23–35 × 15–18 mm. *Column* 5–6 mm long; staminode ovate, truncate, slightly longer than the stigma. *Fruit* cylindrical, slightly arcuate, 5–9 cm long, triangular in cross-section, glandular pubescent. **Plate 9, Figs. 46 & 47.**

Selenipedium palmifolium (Lindl.) Rchb.f., Xenia Orch. 1, 5, t. 2, f. 6–9 (1854); Pfitzer in Engler, Pflanzenr. Orch. Pleon.: 27 (1903). Type: Guyana, *Schomburgk s.n.* (holo. K!).
Cypripedium palmifolium Lindl., Gen. Spec. Orch. Pl. 527 (1840).

DISTRIBUTION. Venezuela, Guyana, Surinam, French Guyana, Trinidad, Amazonian Brazil. **Map 8**.

HABITAT. Wet low evergreen forest bordering savannas, in bushes on sandy loam, often in marshy areas; 200–400 m.

PHENOLOGY. Flowers from July until November.

CONSERVATION STATUS. Near Threatened (NT): B2b(ii,iii,v). It is a local species with a broad distribution area in northern South America. The trend of the population is stable. Populations are often widely separated and the estimated extent of occurrence is more than 20000 km^2 but the area of occupancy is less than 2000 km^2. The species is of no horticultural interest but populations can be threatened by competition, anthropogenic habitat degradation, deforestation, destruction of watershed areas and erosion.

ETYMOLOGY. Named from the resemblance of its leaves to those of some palms.

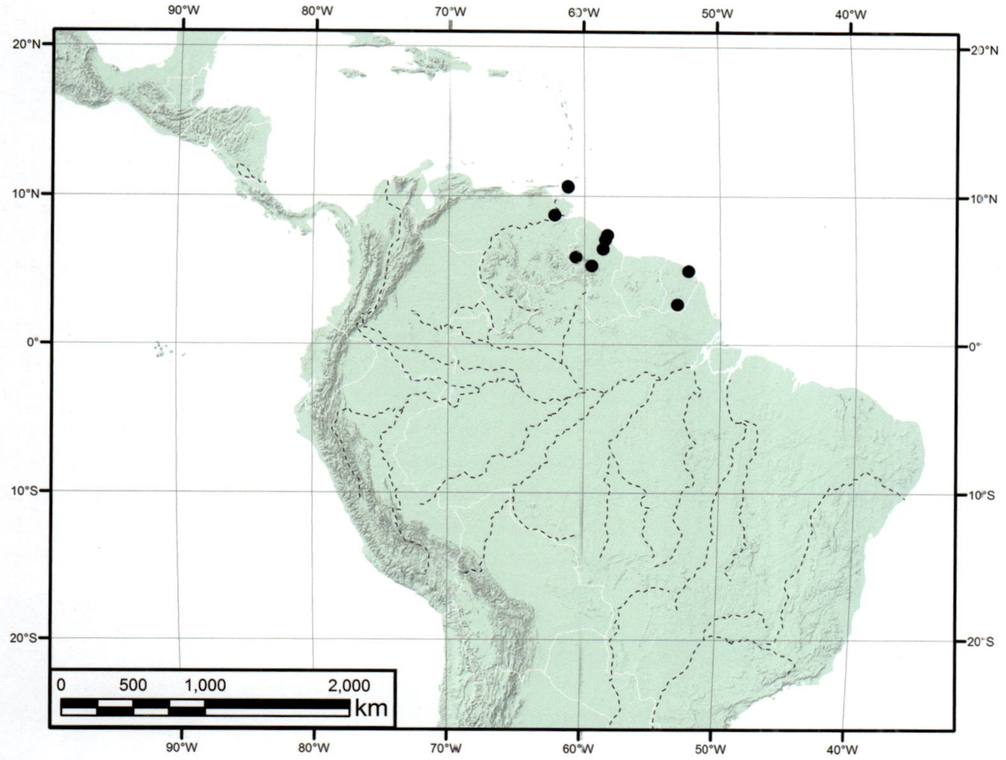

Map 8. Distribution of *Selenipedium palmifolium* in northern South America.

7. SELENIPEDIUM STEYERMARKII

Selenipedium steyermarkii was described by the Venezuelan botanist Ernesto Foldats based upon a collection made in 1960 by Steyermark and Nilsson at 1600 m elevation at Fila de la Danta, between Luepa and Cerro Venamo, 132 km from El Dorado on way to Santa Elena de Uairén, Estado Bolivar, Venezuela. Julian Steyermark, after whom it was named, was a prolific botanist and plant collector with over 110,000 collection numbers to his name. He collected extensively throughout Venezuela from 1943 onwards.

Dunsterville & Dunsterville (1971) provided a detailed account of its habitat in south-eastern Venezuelan Guyana between the flat-topped mountains Auyantepui and Cerro Venamo. It grows in extremely sandy soil mixed with leaf mould typical of the region. It is found in vegetation of tall grasses and scrub so that it is difficult to see even when in flower. It is distinguished from *S. palmifolium* which is also found in Venezuela by its narrower coriaceous leaves, more than eight times as long as broad, its sepals and petals which are larger than the lip, and its bracts which are usually longer than the pedicel and ovary.

DESCRIPTION. A terrestrial *herb* with slender, erect, leafy, glandular-pubescent stems 2–3.5 m tall, branching in the apical part. *Leaves* subcoriaceous, linear-lanceolate or lanceolate, acuminate, 10–20 × 0.8–2.3 cm, about 8 times longer than broad, the upper leaves progressively narrower, very finely pubescent on upper surface, glandular on lower surface and margins. *Inflorescence* terminal, flowering successively on a nutant raceme; rachis densely pubescent; bracts lanceolate, acute or acuminate, 15–32 mm long, glandular-pubescent. *Flowers* variable in size, pubescent on outer surface of sepals, glabrous within, subnutant, sepals maroon-bronze, the lateral sepals somewhat lighter, the petals olive-green, the lip yellow-brown with a paler base or shiny chestnut-brown with a yellow-green base, marked with a dark maroon spot on either side of the orifice; pedicel and ovary 13–20 mm long, one and a half times as long as the bracts, densely pubescent. *Dorsal sepal* erect to hooded over lip and column, ovate-elliptic, obtuse, 25–42 × 13–25 mm, slightly pubescent on outer surface. *Synsepal* elliptic or ovate-elliptic, usually minutely bifid at obtuse apex, 25–35 × 13–23 mm, pubescent on outer surface. *Petals* spreading-deflexed and somewhat incurved, linear-lanceolate, subacute, 24–32 × 4–5 mm. *Lip* calceiform, somewhat dorsi-ventrally flattened, subnutant, 20–36 × 14–22 mm, sometimes with an inflated rim, margins incurved forming an obovate orifice, somewhat

Fig. 48. *Selenipedium steyermarkii.* Flower, Venezuela. (GD)

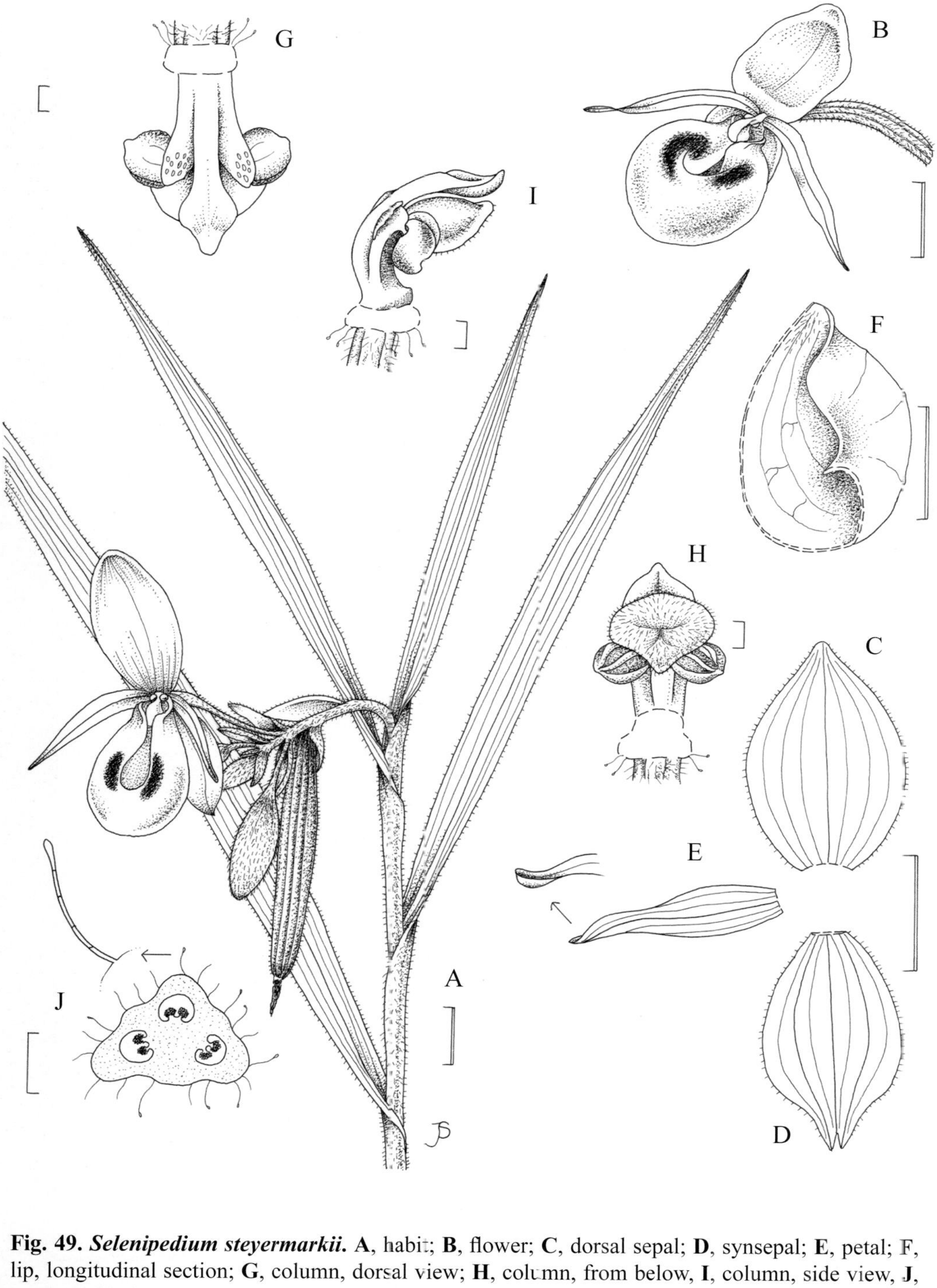

Fig. 49. ***Selenipedium steyermarkii.*** **A**, habit; **B**, flower; **C**, dorsal sepal; **D**, synsepal; **E**, petal; **F**, lip, longitudinal section; **G**, column, dorsal view; **H**, column, from below, **I**, column, side view, **J**, ovary and hair from ovary. All drawn by Judi Stone from *Dunsterville s.n.* (Kew Spirit Collection no. 41809). Single bar = 1 mm; double bar = 10 mm.

angular near the base, with incurved rounded side lobes. *Column* 5–7 mm long, yellow-green; stigma, papillate; staminode ovate, acute, 2 mm long, stalked. **Figs. 48 & 49.**

Selenipedium steyermarkii Foldats in Bol. Soc. Venez. 22: 253 (1961); Garay & Dunsterville, Venez. Orch. Ill. 3: 284–285 (1965). TYPE: Venezuela, Estado Bolivar, Fila de la Danta, between Luepa and Cerro Venamo, 132 km from El Dorado on way to Santa Elena de Uairén, *Steyermark & Nilsson* 273 (holo. VEN, iso. NY!).

DISTRIBUTION. SE Venezuela and adjacent Guyana only. **Map 9**.

HABITAT. In cloud forest, often in marshy areas of fairly thick grass and head-high shrubs, in very sandy loam; 800–1650 m.

PHENOLOGY. Flowering late February to May and possibly at other times.

CONSERVATION STATUS. Vulnerable (VU): B2ab(ii,iii,v). It is local over a small area of eastern Venezuela. The species is not collected and is not of interest in horticulture because it is a large straggling plant with small flowers. Its small subpopulations are vulnerable to habitat change and destruction from competition, deforestation, habitat degradation and human disturbance. The estimated area of occupancy of the species is less than 200 km^2.

ETYMOLOGY. Named for the American botanist Julian Steyermark (1909–1988) who made over 130,000 collections on numerous expeditions in the American tropics.

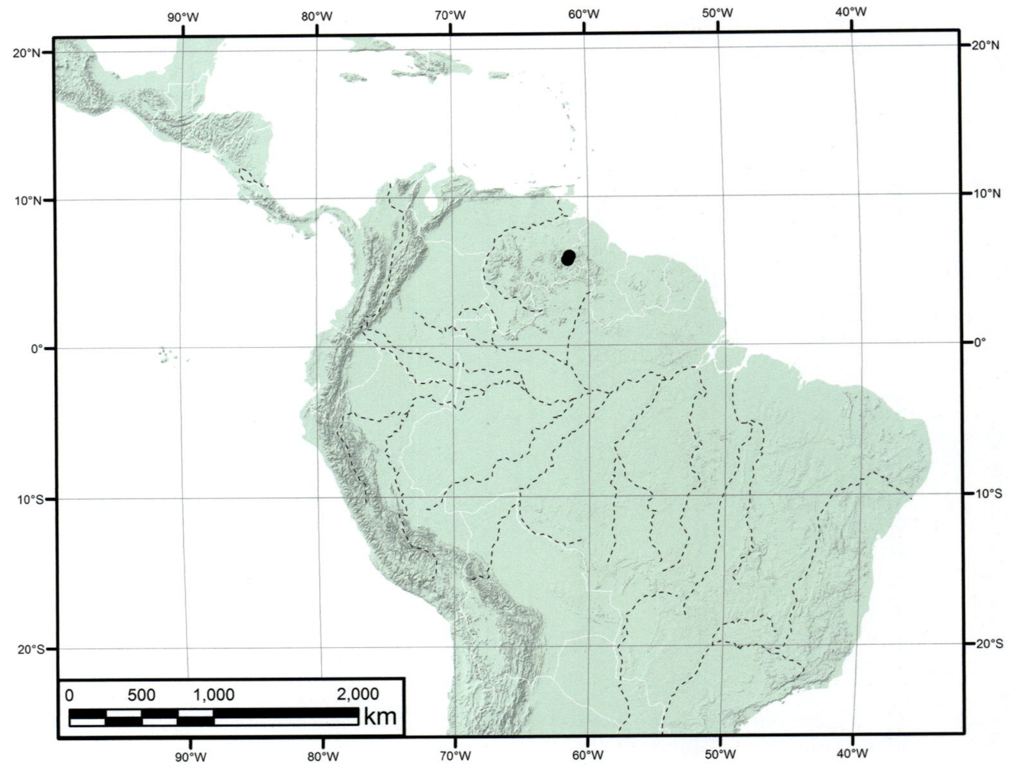

Map 9. Distribution of *Selenipedium steyermarkii* in Venezuela.

8. SELENIPEDIUM VANILLOCARPUM

Selenipedium vanillocarpum is an enigmatic species. Barbosa Rodrigues (1937) described it, as *Selinopedilum vanillocarpum*, based upon a specimen collected by Henrique Silva near Antas in Goias State. This locality is inland from the Bahian and Para localities of *S. isabelianum* and it may well represent a distinct taxon. The name *Selenipedium vanillocarpum* has also been used for this species (see IPNI database) but it does not appear in Barbosa's original text.

Unfortunately, his type specimen was destroyed in the 1940s. His description of it is rudimentary as is the accompanying illustration, selected here as the lectotype, showing the upper stem, leaves and inflorescence with flower buds and a floral dissection. Barbosa distinguished it by its lilac-coloured lip and ciliate petals. In its glandular pubescent bracts and ovary it agrees with *S. palmifolium* and *S. isabelianum* but it has bracts more reminiscent of *S. chica* and *S. aequinoctiale* in shape. It should be critically compared with other species when authentic material becomes available.

DESCRIPTION. A terrestrial *herb* with 1–4 erect leafy stems; internodes 3–6 cm long; stems dark green, covered by coarsely and densely glandular-villose leaf-bases, 3–6 cm long. *Leaves* erect-spreading, elliptic-lanceolate, acuminate, cuneate at base of lamina, 20–23 × 5–6 cm, green, glabrous above, hairy beneath, ciliate, about 8-nerved, sheathing at base; sheaths glabrous. *Inflorescence* terminal, simple, 5–7 cm long, densely several-flowered, with 1–2 open at a time; rachis densely pubescent; bracts erect-spreading, chartaceous, lanceolate, acuminate, 23–50 mm long, densely coarsely shortly hairy, ciliate. *Flowers* small; sepals and petals light yellow-green to dull pale fawn, lip lilac; pedicel and ovary triangular in cross section, 15–40 mm long, densely glandular pubescent. *Dorsal sepal* narrowly elliptic-ovate, obtuse, 26 × 13 mm, densely pilose on outer surface. *Synsepal* elliptic-ovate, bidentate at apex, 29 × 16 mm, densely pilose on outer surface. *Petals* obliquely linear-lanceolate, acute, 20–32 × 1.5–3 mm, ciliate. *Lip* calceolate, broadly elliptic in outline, 30 × 15 mm, the incurved side lobes narrow and not meeting beneath the column. *Column* 7–8 mm long; staminode not stalked, ovate, truncate, 3 mm long, slightly longer than the stigma. *Fruits* cylindrical, slightly arcuate, 5–9 cm long, triangular in cross-section, glandular pubescent. **Fig. 50.**

Selenipedium vanillocarpum Barb. Rodr. in Rodriguesia 2, 8: 39, t. 36 A (1937), as '*Selenipedilum vanillocarpum*'. Type: Brazil, Goias, near Antas, *Silva s.n.* (holo. lost); lectotype selected here, illustration (t. 36A) cited above.

DISTRIBUTION. Brazil (Goias State). **Map 10.**

HABITAT. In woods.

PHENOLOGY. Flowers in May.

CONSERVATION STATUS. Data Deficient (DD). This species has not been seen since its description.

ETYMOLOGY. The fruit has a vanilla fragrance and produces vanillin, hence the scientific name and vernacular name of 'Baunilha' (Barbosa Rodrigues, 1937).

Tab XXXVI

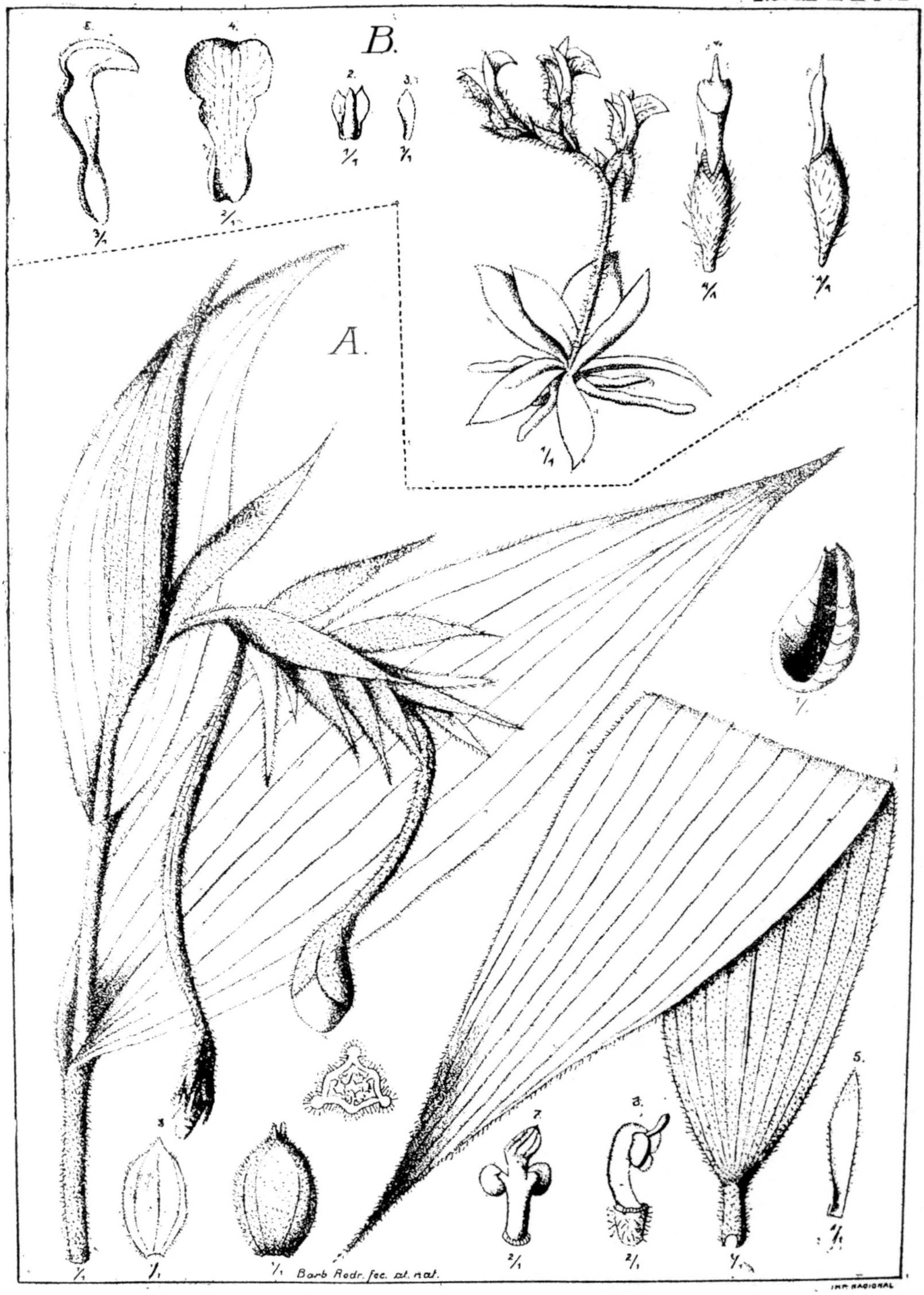

A-SELENIPEDILUM VANILLOCARPUM BARB. RODR.
B-STENORHYNCHUS CORNUTUS BARB. RODR

Fig. 50. *Selenipedium vanillocarpum.* Drawing of the type material by J. Barbosa Rodrigues from *Rodriguesia* 2: t. 34 (1937).

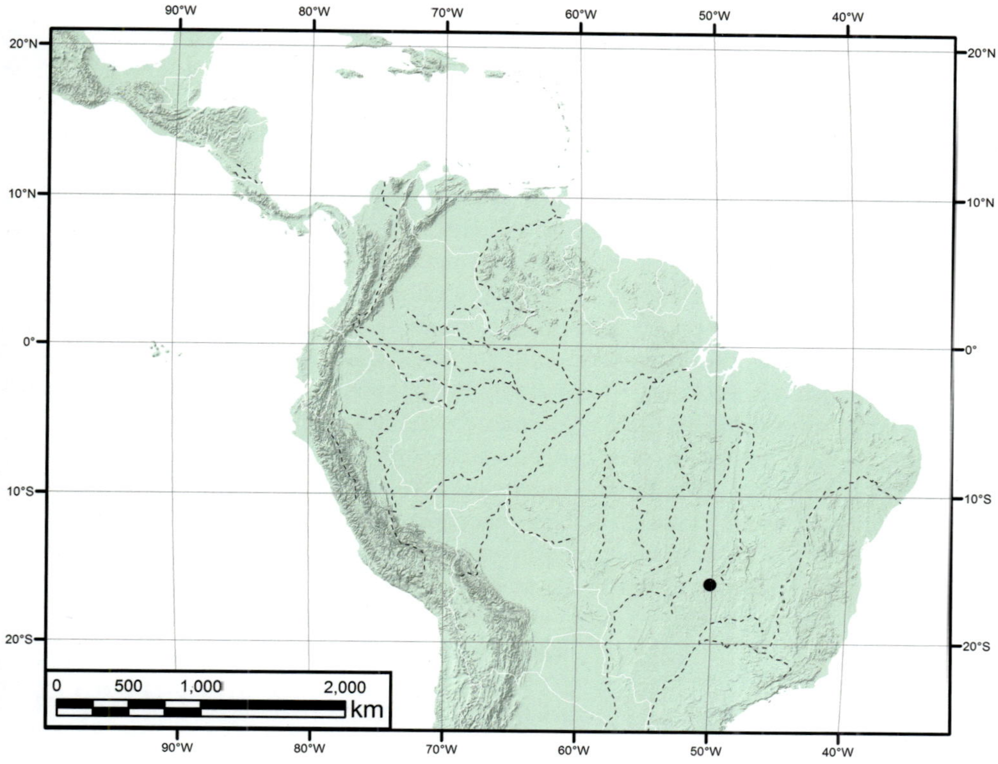

Map 10. Distribution of *Selenipedium vanillocarpum* in Brazil.

9. SELENIPEDIUM CHIRONIANUM

Sambin and Braem (2015a) described this unusual and abnormal plant based upon a collection made by the first author in 2014 that flowered in February the following year. Occasionally, slipper orchids, particularly *Paphiopedilum* and *Phragmipedium* species, are known to produce teratological flowers in which the lip is petaloid or the petals resemble a lip. Rarely, the flowers may have a split synsepal and/or a column bearing three stamens or even three stigma lobes. These plants usually do not persist as more than a single or few individuals. However, the tropical American *Phragmipedium lindenii*, a teratological derivative of *P. warszewiczianum*, with a flower with a petaloid lip, has been known for almost a century and a half. It is almost certainly self-pollinating and is a common plant in the northern Andes, forming extensive colonies.

Selenipedium chironianum has an extremely limited distribution, being known from an area of 1–1.5 hectares in French Guyana. It seems likely that it is a semi-peloric form of the widespread *S. palmifolium*. Thus, we recognise it at specific rank because it is very distinctive and readily identified.

DESCRIPTION. A terrestrial *herb* up to 110 cm tall; rhizome short, subterranean. *Stems* cylindrical, 2.5–8 mm in diam., strongly pubescent. *Leaves* plicate, oblong-lanceolate, acuminate, 5.5–20.5 × 1.5–4.9 cm, 7-nerved, with ciliate margins. *Inflorescence* terminal, branching, arcuate, reaching 7 cm long, with up to 11 flowers, opening successively; bracts similar to the leaves but only 30 × 6 mm. *Flowers* 5.5 cm in diameter, yellow-green to whitish, externally pubescent; pedicel and ovary 42 × 2 mm, slightly arcuate, ovary triangular, strongly pubescent, green. *Sepals* subsimilar, 7-nerved, pubescent; dorsal sepal elliptic to obovate, acute, 32–38 × 18–20 mm; lateral sepals free, not forming a synsepal, elliptic, shortly concave at the tip, 31 × 10 mm. *Petals* strongly arcuate, narrowly linear, 30 × 3.2 mm, the margins slightly incurved, pubescent at the base, with a fleshy mid-vein. *Lip* petaloid, arcuate, linear, 30–3.8 mm, apical part abruptly incurved and somewhat fleshy, subrectangular, obscurely bilobed on the upper face. *Column* 7 × 2 mm, yellow-green; staminode broadly elliptic, 4 × 4.9 mm, echinate; anthers 3, bilocular. *Fruit* cylindrical, 63 × 4 mm, triangular, sulcate, green, more or less arcuate or sigmoid. **Fig. 51.**

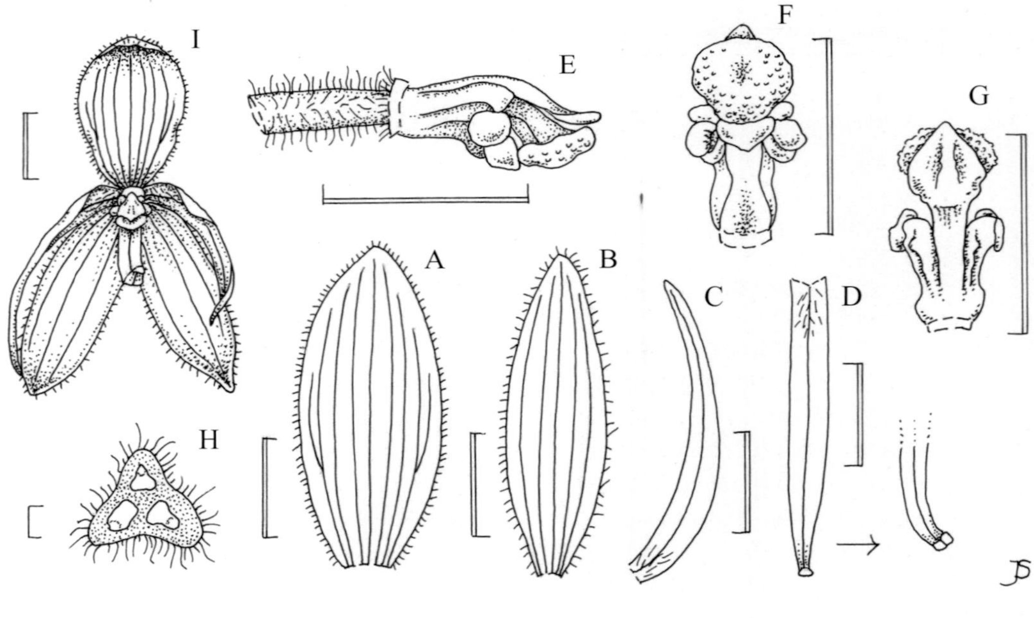

Fig. 51. ***Selenipedium chironianum.*** **A**, dorsal sepal; **B**, lateral sepal; **C**, petal; **D**, lip, from front and close-up of tip; **E**, column, side view; **F**, column from below; **G**, column from above; **H**, ovary, transverse section, **I**, flower from front. All drawn by Judi Stone after drawings of type by A. Sambin. Single bar = 1 mm; double bar = 10 mm.

Selenipedium chironianum Sambin & Braem in Richardiana 15: 185 (2015). Type: French Guyana, Mana, near Angoulême, 2014, fl. in cult. 16 Feb. 2015, *Aurélien Sambin* in BL AG S 001 AS 01 (holo. CAY).

Apedium chironianum (Sambin & Braem) Chiron, Sambin & Braem in Richardiana 15: 243–254.

DISTRIBUTION. North-west French Guyana. **Map 11**.

HABITAT. Terrestrial in primary forest on white sand; 10–30 m elevation.

PHENOLOGY. Flowers in cultivation in February.

CONSERVATION STATUS. Critically Endangered (CR): B1ab(ii,iv,v)+2ab(ii,iii,v); C1+2a(i,ii); D. It is a very local, narrowly endemic species, known only from the type locality, an area of 1–1.5 hectares in North-west French Guyana. The population is low (possibly less than 100). Emmanuel Ravet found eight flowering specimens at the type locality between 2001 and 2015 (Sambin & Braem, 2015). It is not recorded if they were the same or different plants and how many sterile plants were growing with them. The abundance of the species is low while deforestation, human disturbance and habitat degradation are a threat. The estimated extent of occurrence and estimated area of occupancy of the species are minute, less than 100 km^2 and 10 km^2

ETYMOLOGY. Named in honour of Guy Chiron (born 1944), editor of the journal *Richardiana*.

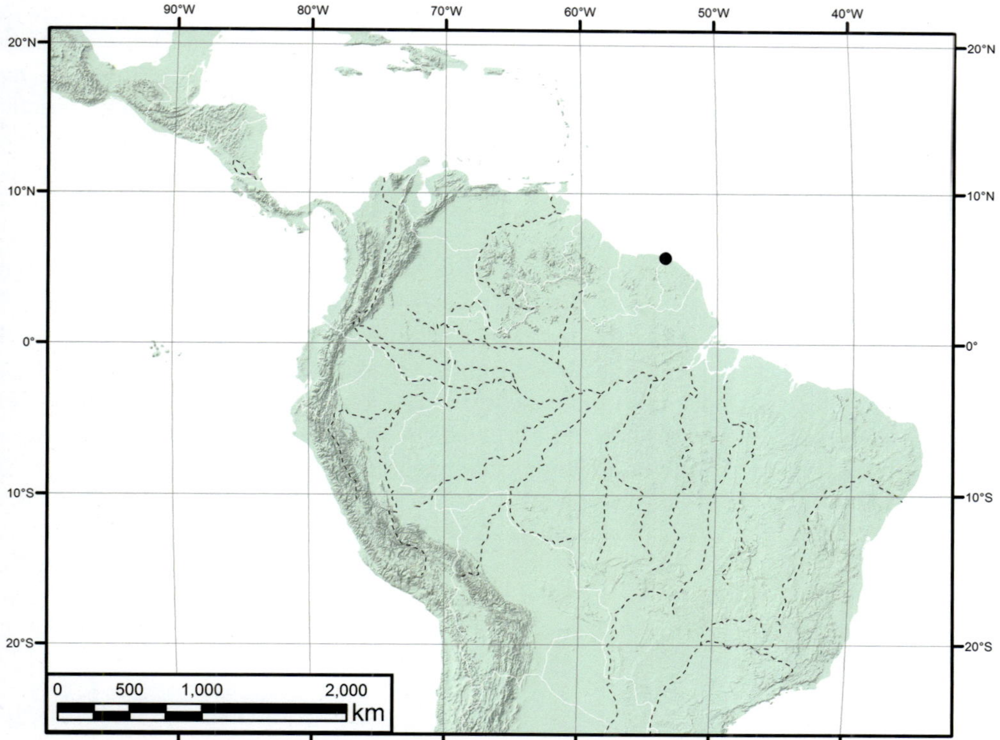

Map 11. Distribution of *Selenipedium chironianum* in French Guyana.

MEXIPEDIUM

Mexipedium V.A. Albert & M.W. Chase in Lindleyana 7, 3: 173 (1992). Type: *M. xerophyticum* (Soto Arenas, Salazar & Hagsater) V.A. Albert & M.W. Chase (basionym: *Phragmipedium xerophyticum* Soto Arenas, Salazar & Hagsater).

A small rupicolous *herb*, with an elongate creeping stout rhizome with well-spaced growths and unbranched roots. *Stems* very short bearing a fan of leaves. *Leaves* up to 8, coriaceous, conduplicate, ligulate, obtuse and mucronate at apex, green. *Inflorescence* terminal, erect, 1- to 2-branched, several-flowered but producing flowers in succession; peduncle cylindrical; floral bracts imbricate, conduplicate, hairy and ciliate. *Flowers* deciduous, very small in comparison with other slipper orchids; pedicel very short; ovary elongate, densely pubescent with multi-cellular reddish hairs. *Dorsal sepal* elliptic, slightly concave; *synsepal* suborbicular, obtuse or slightly bilobed at tip. *Petals* linear-ligulate, acute, incurved. *Lip* saccate, subglobose; side lobes very small. *Column* short; stigma curved, fleshy, minutely papillose; staminode somewhat convex, transversely oblong but somewhat bluntly mucronate on apical margin. *Capsule* slenderly cylindrical.

DISTRIBUTION. Mexico (endemic to Oaxaca State).

Mexipedium is a monotypic genus endemic to a small region of the state of Oaxaca in Mexico. Albert and Chase (1992) established the genus to accommodate *Phragmipedium xerophyticum*, a dwarf Mexican slipper orchid that has morphological features intermediate between those of the tropical American *Phragmipedium* and the tropical Asiatic *Paphiopedilum*. It was originally included by Soto-Arenas *et al.* (1990) in *Phragmipedium* because of its possession of sepals with valvate aestivation, the absence of sinuous epidermal cells in the perianth, the fusion of the side lobes of its lip and the ventral synsepal being larger than the dorsal sepal. Albert and Chase added a further character, its branched inflorescence which is found in *Phragmipedium* but is absent from *Paphiopedilum* but noted that only the first criterion listed by Soto-Arenas *et al.* was consistent in distinguishing *Phragmipedium* (including *P. xerophyticum*) from *Paphiopedilum*. Because of its unusual combination of characters, Albert and Chase chose to erect a new genus for this strange little orchid. Two years later, with the benefit of data from DNA analysis, Albert (1994) found that *Mexipedium* was basal in the *Phragmipedium* clade. It has been maintained as distinct because, apart from its morphological singularities, it does not easily breed with *Phragmipedium*.

Because *Phragmipedium*, *Mexipedium* and *Paphiopedilum* form a clade that is monophyletic, Albert and Pettersson (1994) sunk the first two into *Paphiopedilum*, the earliest name. This treatment has not been taken up by other botanists or horticulturists, mainly because of the distinctiveness of the chromosome number and breeding behaviour of these genera.

1. MEXIPEDIUM XEROPHYTICUM

Although the spectacular *Phragmipedium besseae* and *P. kovachii* grabbed the headlines when they were discovered and first burst upon the horticultural scene, the discovery in southern Mexico of a tiny slipper orchid is, if anything, more remarkable in that it proved to be an evolutionary link between the tropical American and Asian conduplicate-leaved slipper orchids.

Miguel Soto Arenas, Gerardo Salazar Chávez and Eric Hagsater (1990) announced its discovery in the journal *Orquidea* (Mexico) based upon a collection from the eastern lowlands of Oaxaca State. Soto and Salazar and two companions discovered it when exploring a little known area. The climate in this region is distinctive with a seasonal character so that the plant xerophytic character means that they can survive lengthy droughts. Its size, dwarf habit, inflorescence and tiny flowers are distinctive amongst the slipper orchids. Only the some species of *Cypripedium*, such as *C. debile*, *C. fasciculatum*, *C. palangshanense* and *C. wardii*, have flowers approaching it in size, but none has a branching inflorescence nor conduplicate leaves.

Its elongate rhizome and rupicolous habit were compared by the authors to those of the Indian *Paphiopedilum druryi* that, otherwise, bears it little resemblance.

When discovered in 1988, seven plants were found and growths from two of them were brought into cultivation where they thrived. Two further plants were taken after its discovery by an amateur grower but their survival is uncertain. On a return trip, Salazar collected two further cuttings which also survived in cultivation. Four clones were propagated in Mexico City and distributed to growers in Mexico and abroad. As a result, plants of *M. xerophyticum* are now widely grown around the world. It is, in truth, a novelty,

Figs. 52 & 53. *Mexipedium xerophyticum*. Inflorescence and flower of cultivated plant. (PC)

Figs. 54–56. *Mexipedium xerophyticum*. Growing on boulders in xerophytic forest, Oaxaca State, Mexico. (ES)

and not a showy species. Nevertheless, these plants are of considerable significance as the type locality was reported destroyed by a wild fire in 1998. It was with some relief that Perez-Garcia (2011) and Perez Garcia & Mo Mo (2015b) reported the rediscovery of it at the type locality in the Los Chimalapas region in the Isthmus of Tehuantepec of Oaxaca state which was revisited in 2009. Nearby, at least six groups of plants were discovered growing on the walls of a ravine.

DESCRIPTION. A small rupicolous *herb*, up to 20 cm tall, forming large clumps up to a metre in diam., with an elongate creeping stout rhizome 3–8 cm long between growths and 10–12 mm in diam., covered in sheaths; roots unbranched, up to 11 cm long, 8 mm in diam., glabrous or slightly hairy. *Stems* very short, bearing a fan of leaves. *Leaves* up to 8, coriaceous, conduplicate, ligulate, obtuse and mucronate at apex, 3.5–12 × 1.2–1.8 cm, green. *Inflorescence* terminal, erect, 1- to 2-branched, several-flowered but producing flowers in succession; peduncle cylindrical, 6.5–13.5 cm long, bearing many multi-cellular brownish red hairs; floral bracts imbricate, conduplicate, 4–5 mm long, 5 mm wide, hairy and ciliate. *Flowers* deciduous, white touched with pink on the staminode and with reddish brown backs to the sepals; pedicel and ovary 3–3.3 cm long, densely pubescent with multi-cellular reddish hairs. *Dorsal sepal* elliptic, slightly concave, subacute, up to 14 × 5–6.5 mm; synsepal suborbicular, obtuse or slightly bilobed at tip, 8–9.5 × 8.5–12 mm. *Petals* linear-ligulate, acute, incurved, 11–15 × 2.5–3 mm. *Lip* saccate, subglobose, 10–14 mm long, 7–9 mm wide and 6–8 mm deep; side lobes very small. *Column* short, 1–2 mm long; stigma curved, fleshy, minutely papillose; staminode somewhat convex, transversely oblong but somewhat bluntly mucronate on apical margin, 3 mm long, 4–5 mm wide. *Capsule* slenderly cylindrical, 4–5 cm long. 2n = 26. **Plate 10. Figs. 52–57.**

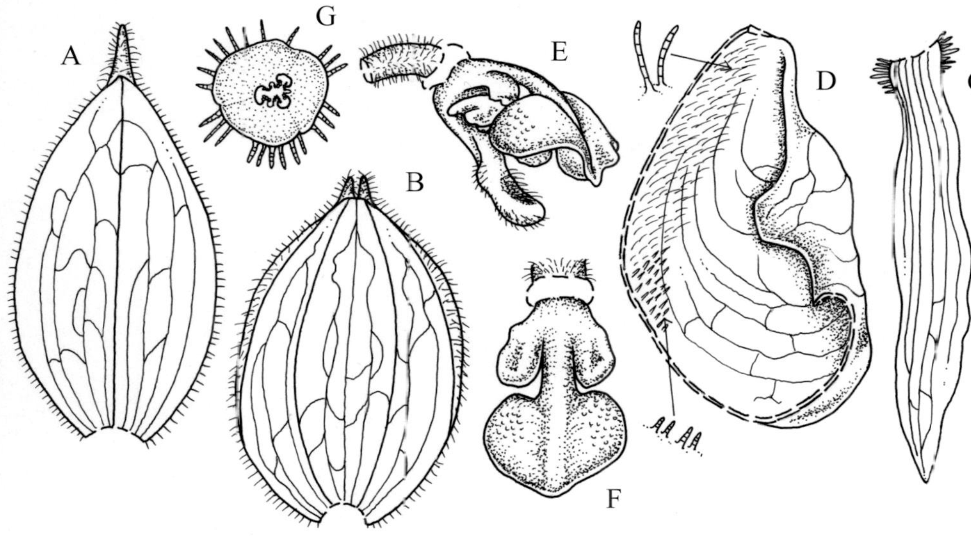

Fig. 57. ***Mexipedium xerophyticum.*** **A**, dorsal sepal; **B**, synsepal; **C**, petal; **D**, lip, longitudinal section; **E**, column, side view; **F**, staminode; **G**, ovary, transverse section. All drawn by Judi Stone from cult. Levy (Kew Spirit Collection no. 60896). A–D, × 4.5; E, F, × 6; G, ×15.

Plate 10. *Mexipedium xerophyticum*. Drawn by Carol Woodin for *Curtis's Botanical Magazine* n.s.t. 297 (1996), from a cultivated plant.

Mexipedium xerophyticum (Soto Arenas, Salazar & Hagsater) V.A. Albert & M.W. Chase in Lindleyana 7, 3: 173 (1992); Perez in Lankesteriana 9, 3: 557–563 (2010); Perez & Mo in Die Orchidee 66, 2: 128–133 (2015). Type: Mexico, Oaxaca, *Salazar, Soto, Yanez & Hernandez* 3740 (holo. AMO!; iso. K!).

Phragmipedium xerophyticum Soto Arenas, Salazar & Hagsater in Orquidea (Mex.) 12, 1: 2 (1990).

Paphiopedilum xerophyticum (Soto Arenas, Salazar & Hagsater) V.A. Albert & Borge Borge Pett. in Lindleyana 9, 2: 138 (1994).

DISTRIBUTION. Mexico (endemic to Oaxaca State). **Map 12.**

HABITAT. On north-facing limestone rocks and cliffs in karst in lowland areas in seasonally dry scrub in a tropical rainforest area where the annual rainfall is 2500 mm.

PHENOLOGY. Flowers in August and September, until November in cultivation in Mexico (Perez-Garcia, 2011).

CONSERVATION STATUS. Critically Endangered (CR): B1ab(i,ii,iii,iv,v)+2b(i,ii,ii i,iv,v); C1+2a(i,ii); D. It is a very local, narrowly endemic species, known only from the type locality where two adjacent subpopulations of a few plants have been found. The population is low (possibly less than 100 mature plants). A serious fire was thought to have exterminated it from the type locality not long after its discovery but a few surviving plants were discovered during a subsequent visit as reported above Perez Garcia (2011) and Perez Garcia & Mo Mo (2015b). The abundance of the species is very low while deforestation, human disturbance, fire and habitat degradation are a continuing threat. The estimated extent of occurrence and estimated area of occupancy of the species are very small, less than 100 km^2 and 10 km^2 respectively.

ETYMOLOGY. Named for its xerophytic habit.

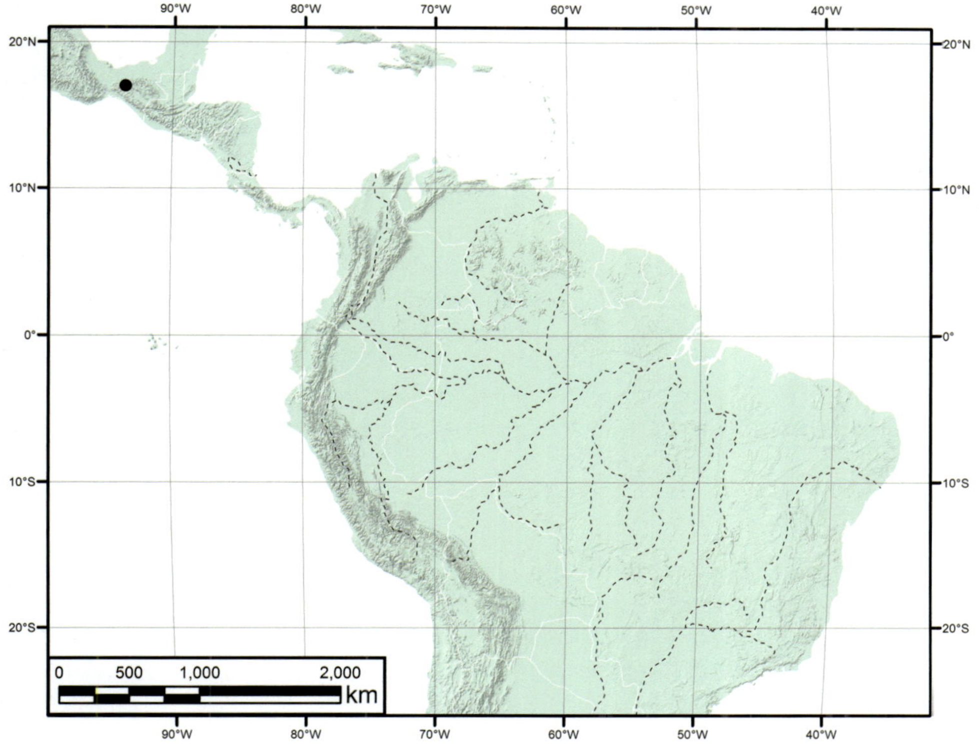

Map 12. Distribution of *Mexipedium xerophyticum* in southern Mexico.

PHRAGMIPEDIUM

Phragmipedium *Rolfe* in Orchid Rev. 4: 331 (1896), *nom. conserv.* Type: *Cypripedium caudatum* Lindl. [= *Phragmipedium caudatum* (Lindl.) Rolfe].
Uropedium Lindl., Orch. Linden. 28 (1846). Type. *Uropedium lindenii* Lindl. [= *Phragmipedium lindenii* (Lindl.) L.O.Williams].
Uropedilum Lindl. *emend.* Pfitzer, Entwurf Anordn. Orch. 95 (1887), *orth. var.*
Phragmopedilum Rolfe *emend.* Pfitzer in Bot. Jahrb. Syst. 25: 527 (1898), *orth. var.*
Phramipedilum Rolfe in Orchid Rev. 9: 175 (1901), *orth. var.*

DESCRIPTION. Small to large terrestrial, lithophytic or epiphytic *herbs* with elongate, fibrous roots arising from a short to elongate rhizome. Erect *shoots* leafy, clustered or less frequently well spaced, glabrous, the base enclosed by two to four sheathing sterile bracts, three- to several-leaved above. *Leaves* several, coriaceous, conduplicate, spreading or suberect, ligulate, elliptic, or oblong, obtuse to acute, often tridenticulate at apex, mid to dark green on upper surface, lighter green below, glabrous, ciliate or not on the margins. *Inflorescence* terminal, sometimes branching, few- to many-flowered; rachis terete, hairy, glandular or glabrous; bracts conduplicate, elliptic, lanceolate, ovate or oblong, green, sometimes spotted or flushed or striped with purple, ciliate or not. *Flowers* deciduous, usually showy, concolorous or not, vernation imbricate; pedicel obscure to short; ovary trilocular, three-ribbed, glabrous or hairy. *Dorsal sepal* erect to hooded over lip, ovate, lanceolate, obovate or elliptic, obtuse, acute or acuminate, glabrous or pubescent on the outer surface, sometimes pubescent within at base, ciliate or not. *Lateral sepals* usually fused to form a concave synsepal that is more or less similar to the dorsal sepal, sometimes keeled on outer surface. *Petals* free, spreading or pendent, flat, reflexed or spiralling, elliptic, ovate, lanceolate, linear-lanceolate, linear or oblanceolate, rounded, obtuse, acute or acuminate at apex, often pubescent in basal half within, usually ciliate. *Lip* deeply pouched and inflated, slipper-shaped or urn-shaped, with more or less pronounced incurved side lobes, rarely petaloid, hairy within especially on lower surface, glabrous or hairy on outer surface; front margin incurved or not. *Column* short, stalked, porrect; anthers two, bilocular, borne on short obtuse to acute filaments; pollenia 4 but somewhat obscure, powdery not viscid; staminode terminal on column, sessile or shortly stalked, transversely reniform, oblong, ovate, obcordate or linear, flat, convex or longitudinally conduplicate, glabrous to papillose or finely pubescent, ciliate or not; stigma stalked, dependent, tripartite, more or less papillose. *Capsule* erect to pendent, three-ribbed, cylindrical to almost ellipsoidal.

DISTRIBUTION. Southern Mexico and Guatemala through Central America to South America, south to Bolivia and Brazil. A genus of about 21 species and several varieties. **Map 13**.

INFRAGENERIC TREATMENT

Hallier (1896) produced the first sectional names, *Micropetalum* and *Himantopetalum* of *Cypripedium* subgenus *Phragmopedilum*, relevant to *Phragmipedium*. Kraenzlin (1897) placed all *Phragmipedium* species in *Paphiopedilum* and established the sections *Lorifolia* and *Caudata* in the latter genus. Pfitzer (1903) revised *Phragmipedium* (as

Phragmopedilum) and recognised five sections: *Micropetalum*, *Himantopetalum*, *Ceratopedilum*, *Desmopedilum* and *Platyphyllum*, establishing the last for the species with spatulate petals.

Phragmipedium was revised by Garay (1979) who recognised five sections in the genus: *Micropetalum*, *Platypetalum*, *Phragmipedium* (incl. *Caudata* and *Desmopedilum*), *Himantopetalum* and *Lorifolia* (incl. *Ceratopedilum*).

McCook (1989 a & b, 1990) published the first detailed monograph of the genus and provided a cladistic analysis of the genus based upon 27 morphological characters. As a result she recognised only four sections, including section *Himantopetalum* in the typical section. She suggested that the presence of multicellular pubescence along the sides of the lip opening that borders the ribbed area on the claw face defined this section. This treatment was not entirely confirmed by Cox *et al.* (1997) who analysed 14 taxa, including *P. caricinum*, *P. pearcei* and *P. caudatum*, in their DNA-analysis of the Cypripedioideae. They found that *P. caricinum* was sister to species of the typical section whereas *P. pearcei* formed part of the same clade as *P. longifolium*. Since *P. pearcei* and *P. caricinum* are morphologically similar, possibly sister species, misidentification of either of the specimens that they used cannot be ruled out. Rodriguez Salas *et al.* (2007), using Amplified Fragment Length Polymorphisms (AFLP) to analyse eight taxa of *Phragmipedium*, found four clades, one of which comprised *P. boissierianum*, *P. richteri* and P. *pearcei*, thereby supporting the conclusions of Cox *et al.* with regard to the last species. Unfortunately, no species of sect. *Platypetalum* was included in their analysis.

Braem (2001) raised the sections of *Phragmipedium* to subgeneric rank and also established section *Viridiflorae* for *P. boissierianum* and several subsections. This seems to be unnecessary and taxonomic inflation in such a small genus and is not accepted here. Finally, Braem (2004) established a new section, *Schluckebieria*, for the large purple-flowered Peruvian species *P. kovachii*.

The treatment followed here is largely based on those of Garay (1979) and McCook (1989a) but places *P. klotzschianum* in the typical section and the other species of section *Himantopetalum* into sect. *Lorifolia*:

1. Sect. **Micropetalum** (Hallier) Garay in Orch. Dig. 43 (4): 133 (1979). Type species: *Selenipedium schlimii* Linden & Rchb.f. [= *Phragmipedium schlimii* (Linden & Rchb.f.) Rolfe].
Paphiopedilum sect. *Micropetalum* Hallier in Ann. Jard. Bot. Buit. 14: 46 (1897).
Plants relatively small except for *P. kovachii*. Leaves relatively thin-textured and broad. Inflorescence with flowers blooming successively. Floral bracts short, fused on basal margins. Flowers small to large and brightly coloured, concolorous, buds covered with multi-cellular hairs. Petals short and broad, more or less oblong to elliptic or oval. Lip lacking side lobes and gibbous hollows at the base. Species 1–7.

2. Sect. **Platypetalum** (Pfitzer) Garay in Orch. Dig. 43, 4: 133 (1979). Lectotype: *Cypripedium lindleyanum* Lindl. [= *Phragmipedilum lindleyanum* (Lindl.) Rolfe].
Phragmopedilum sect. *Platypetalum* Pfitzer in Engler, Pflanzenr. Heft 12: 43 (1903).
Inflorescences several-flowered. Flowers with spatulate petals and a staminode with a hairy surface. Species 8–9.

3. Sect. **Lorifolia** (Kraenzl.) Garay in Orch. Dig. 43, 4: 141 (1979). Lectotype: *Cypripedium longifolium* Rchb.f. & Warsc. [= *Phragmipedium longifolium* (Rchb.f. & Warsc.) Rolfe]. *Cypripedium* sect. *Lorifolia* Kraenzl., Orch. Gen. Spec. 1: 13 (1897).
Plants small to large. Inflorescence with flowers blooming successively. Floral bracts equalling or longer than the ovary. Flowers large. Petals elongate and tapering. Lip with tubercles on the infolded margins of the claw; lip opening spurred or not on the sides. Column with a staminode that is more or less transversely elliptic and usually has a basal margin of dense short hairs that extend onto the lateral margins. We include here all the species of section *Himantopetalum*, except for *P. klotzschianum*, but appreciate that further work is needed to confirm this realignment. Species 10–15.

4. Sect. **Phragmipedium**.
Plants large, usually epiphytic (except *P. lindenii* and *P. klotzschianum*). Inflorescence with flowers blooming simultaneously. Bracts with glands on adaxial surface, marginally fused at base. Flowers large. Petals elongate, much longer than the lip, tapering and pendent. Lip rim pubescent, claw face ribbed. *Phragmipedium klotzschianum* is included here: it has a staminode and lip apex resembling that of *P. caudatum* (McCook, 1989a) and morphologically belongs in the typical section. Species 16–21.

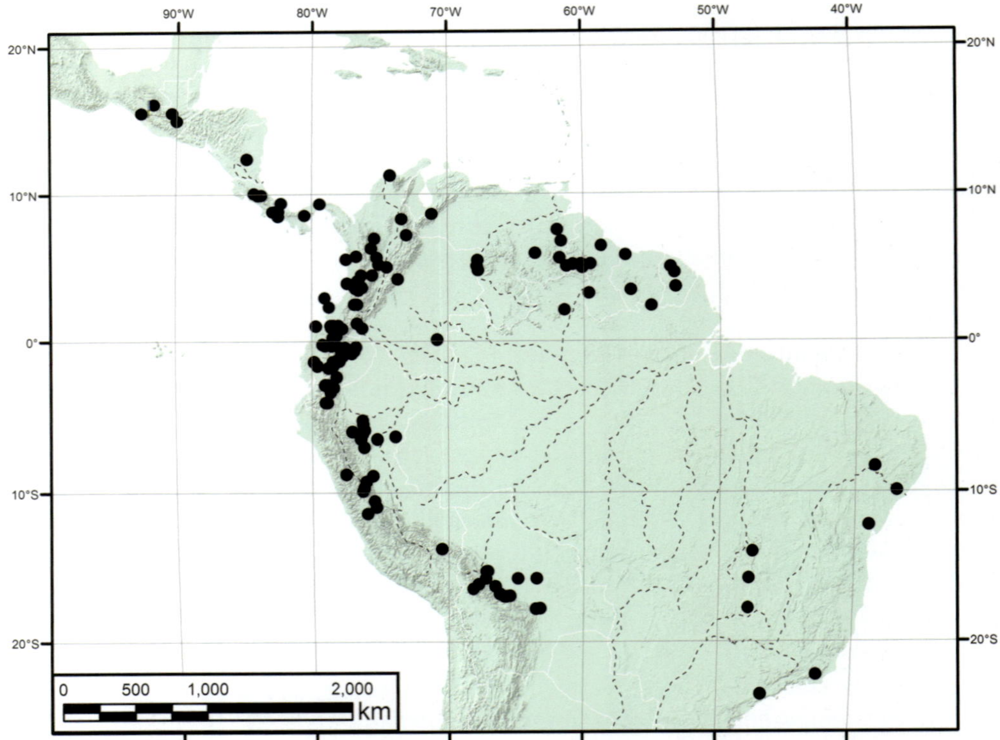

Map 13. Distribution of the genus *Phragmipedium* in Central and South America.

KEY TO THE SECTIONS OF PHRAGMIPEDIUM

1. Petals and sepals concolorous, white, pink or reddish purple, royal purple, rarely pure yellow, broad, elliptic to suborbicular, neither tapering to the tip nor spatulate .... section ***Micropetalum***
   Petals and sepals not concolorous; petals slender, tapering to the tip or spatulate 2

2. Petals spatulate .... section ***Platypetalum***
   Petals tapering to the tip 3

3. Staminode transversely rhombic with slightly twisted sides and bearing sparse long hairs on its basal margin; scape lacking sheaths; flowers produced simultaneously; petals more or less pendent, usually many times longer than the sepals, elongating to maturity for several days after the bud has opened .... section ***Phragmipedium***
   Scape bearing sheaths; flowers opening in succession; petals two to three times longer than the sepals, spreading not pendent, fully developed at anthesis .... section ***Lorifolia***

KEY TO PHRAGMIPEDIUM

1. Petals elliptic, less than three times as long as broad 2
   Petals linear-tapering or spatulate, more than four times longer than broad 9

2. Flowers 15 cm or more across, rich royal purple .... 1. ***P. kovachii***
   Flowers 7 cm or less across, white, pink, rose-pink, red or yellow 3

3. Flowers red, orange-red or rarely yellow 4
   Flowers white, pink, rose-pink, or rarely pink with a yellowish dorsal sepal 5

4. Inflorescence branched; petals decurved, less than 12 mm broad .... 3. ***P. dalessandroi***
   Inflorescence simple; petals spreading more or less horizontally .... 2. ***P. besseae***

5. Dorsal sepal yellow or yellowish green ....7a. ***P. schlimii*** var. ***manzurii***
   Dorsal sepal white to rose-pink 6

6. Flower pale pink, not marked around the rim; staminode longer than wide, white, emarginate ....4. ***P. andreettae***
   Flowers white to rose-pink, usually with darker purple markings around the rim of the lip; staminode wider than long, not white, variously marked with yellow and purple 7

7. Staminode broadest in apical half, narrower across the base, transversely obcordate-ovate, lacking a central ridge, yellow with two purple marks at the tip; lip unevenly streaked and flushed with pink or rose-pink around rim ..... 7. ***P. schlimii*** var. ***schlimii***
   Staminode broadest in middle and with a central ridge to apex 8

8. Staminode broadly cordate-oblong to subquadrate, with undulate basal margins, white with a yellow base and purple V-shaped mark at the tip; lip evenly flushed rose-purple .... 6. ***P. fischeri***
Staminode rhombic, obtuse, pubescent and ciliate, lacking undulate margins, white with yellow central mark .... 5. ***P. anguloi***

9. Petals spatulate; inflorescence branching 10
Petals linear-tapering; inflorescence simple 11

10. Staminode broadest in apical part, much broader than long; lip notably inflated below rim; leaves with a yellow margin .... 8. ***P. lindleyanum***
Staminode almost subquadrate, only slightly broader across the apex than the base, as long as broad; lip not notably inflated below rim; leaves green ...... 9. ***P. sargentianum***

11. Petals pendent, five times or more longer than the lip; sepals twice as long as the lip 12
Petals spreading to arcuate-subpendent, usually twice as long as the lip or less; staminode broadly trullate-triangular, often with moustache-like hairy basal margin of short dark purple hairs 16

12. Staminode absent or knob-like .... 18a. ***P. humboldtii*** var. ***exstaminodium***
Staminode present, broadly rhombic-elliptic 13

13. Lip not urn-shaped, petal-like, linear-tapering like the petals .... 19. ***P. lindenii***
Lip urn-shaped or ovoid 14

14. Lip margins incurved around mouth, dark chocolate-brown flushed .... 18. ***P. humboldtii*** var. ***humboldtii***
Lip margins not incurved, slightly flared, not chocolate-brown flushed 15

15. Lip pink, margin yellow-flushed .... 20. ***P. warszewiczianum***
Lip yellow or green, rim green-flushed 16

16. Dorsal sepal 90–170 mm long; petals 440–880 mm long, pendent; lip 50–70 mm long .... 17. ***P. caudatum***
Dorsal sepal 45 mm long; petals 72–103 mm long, spreading at 45 degrees below the horizontal, not pendent; lip 45 mm long .... 21. ***P. guianense***

17. Opening of lip with an entire margin 18
Opening of lip with a pair of horn-like protuberances on the sides 20

18. Scape and ovary pubescent .... 16. ***P. klotzschianum***
Scape and ovary glabrous 19

19. Staminode broadly triangular-rhombic or broadly cordate, with dense hairs along the basal margins .... 13. ***P. caricinum***

Staminode transversely elliptic, with dense hairs on the basal and lateral margin .... 14. ***P. pearcei***

20. Petals with heavily undulate-crisped margins 21
Petals with flat to slightly undulate margins, never undulate-crisped 22

21 Staminode with a basal fringe of dense short black-purple hairs .... 12. ***P. longifolium*** var. ***longifolium***
Staminode with a basal fringe of sparse white hairs .. 12a. ***P. longifolium*** var. ***hartwegii***

22. Leaves with distinct yellow margins; sepals with smooth or flat margins .... 15. ***P. vittatum***
Leaves with green margins; sepals with undulate-crisped margins; synsepal larger than the lip and cupping it 23

23. Lip ivory-coloured to pinkish; synsepal more or less as long as the lip ...... 11. ***P. hirtzii***
Lip green; synsepal much longer than the lip and cupping it .... 10. ***P. boissierianum***

Fig. 58. Habitat of *P. kovachii*, Northern Peru. (BB)

PHRAGMIPEDIUM SECTION MICROPETALUM

Phragmipedium section **Micropetalum** (Hallier) Garay in Orchid Dig. 43, 4: 133 (1979). Type species: *Selenipedium schlimii* Linden & Rchb. f. [= *Phragmipedium schlimii* (Linden & Rchb. f.) Rolfe].

Paphiopedilum section *Micropetalum* Hallier in Ann. Jard. Bot. Buitenzorg 14: 46 (1897).

Phragmipedium subgenus *Micropetalum* (Hallier) Szlachetko in Polish J. Bot. 46, 1: 11 (2001), **syn. nov.**

Phragmipedium subgenus *Micropetalum* (Hallier) Braem in Richardiana 12, 1: 19 (2011), isonym.

Phragmipedium section *Schluckebieria* Braem in Richardiana 4, 3: 102 (2004), **syn. nov.** Type: *P. kovachii* Atwood, Dalström & Fernandez.

Phragmipedium subgenus *Schluckebieria* (Braem) Braem in Richardiana 12, 1: 20 (2011), nom. inval.; Braem in Richardiana 15, 1: 289 (2015), **syn. nov.**

An Andean section of six species, five of which have a dwarf stature and small flowers, characterised by their predominantly small red, pink, rose-coloured or white flowers with elliptic petals and a small, hairy globose, ovoid or ellipsoidal lip with incurved margins and a small, subquadrate to rhombic hairy staminode. *Phragmipedium kovachii* is a much larger plant with very large, royal purple flowers. However, it has largely concolorous flowers and floral segments of a similar shape to those of the dwarf species. DNA analysis shows clearly that *P. kovachii* lies in the same clade as, and sister to, *P. schlimii* and the other small-flowered species. The species range from northern Colombia to northern Peru.

1. PHRAGMIPEDIUM KOVACHII

Few plants can claim to have graced the front pages of leading national newspapers, and national news broadcasts not once, but twice. The sensational *Phragmipedium kovachii* featured in the *New York Times*, on NBC and in almost every other media-outlet in 2002. Then, in 2012 it was the subject of a book devoted to its discovery and the subsequent sensational events surrounding its importation into the USA and description (Pittman, 2012: *The Scent of Scandal*) and the trial that followed. Its first appearance followed its description by John Atwood, Stig Dalström and Ricardo Fernandez in a supplement to the journal *Selbyana*, based on a plant brought from northern Peru by Michael Kovach after whom it was named. Its amazing flower, over 20 cm in diameter and a stunning royal purple, is unique in the genus which specialises in green, yellow, pink, red and brown flowers. Of the other species, only the pink-flowered *P. schlimii* (Linden & Rchb.f.) Rolfe, the rosy pink *P. fischeri* Braem & Mohr and the orange-scarlet *P. besseae* Dodson & J.Kuhn and *P. dalessandroi* Dodson & Gruss are at all similar in flower shape, but not in size and colour. Its second appearance followed an indictment by the US Federal authorities and a court case in which the Marie Selby Botanical Gardens and the Director of its systematic

Plate 11. *Phragmipedium kovachii*. Drawn by Carol Woodin for *Curtis's Botanical Magazine* n.s. t. 511 (2005), from a plant cultivated by Manuel Arias.

botany unit pleaded guilty to not informing the US authorities of the illegal importation of the original material upon which the description was based before they returned it without the necessary permit to Peru. For this they lost their Scientific CITES permit, received a $5000 fine and had a six-month restriction order placed on their staff member, Wesley Higgins. The importer, Michael Kovach, was found guilty of its illegal importation and was fined.

The tale of the discovery and importation of this amazing orchid has now been well-documented (Pittman, 2012). Indeed, numerous pages of titles on the world-wide web testify to the sensation caused by its discovery, subsequent journey to Florida and hurried description. It was discovered by local orchid collectors in north-west Peru in 2001. In 2002, Michael Kovach saw a flowering plant on a roadside stall while he was collecting orchids in northern Peru and purchased some plants from a stall-holder. On arrival in Florida, he took a plant to the Marie Selby Botanic Gardens in Sarasota where it was immediately recognised as new to science. The plant was illustrated by Stig Dalstrom and a description rushed into print in a specially prepared supplement of their house journal *Selbyana*. A few days later it was redescribed as *P. peruvianum* by Eric Christenson in *Orchids*, the Bulletin of the American Orchid Society. Of course, by the rules of priority of the International Code of Botanical Nomenclature, the earliest name given to a plant is the correct name, later names being reduced to synonymy.

Figs. 59 & 60. *Phragmipedium kovachii*. Plant and flower, Moyobamba region, N Peru. (PC)

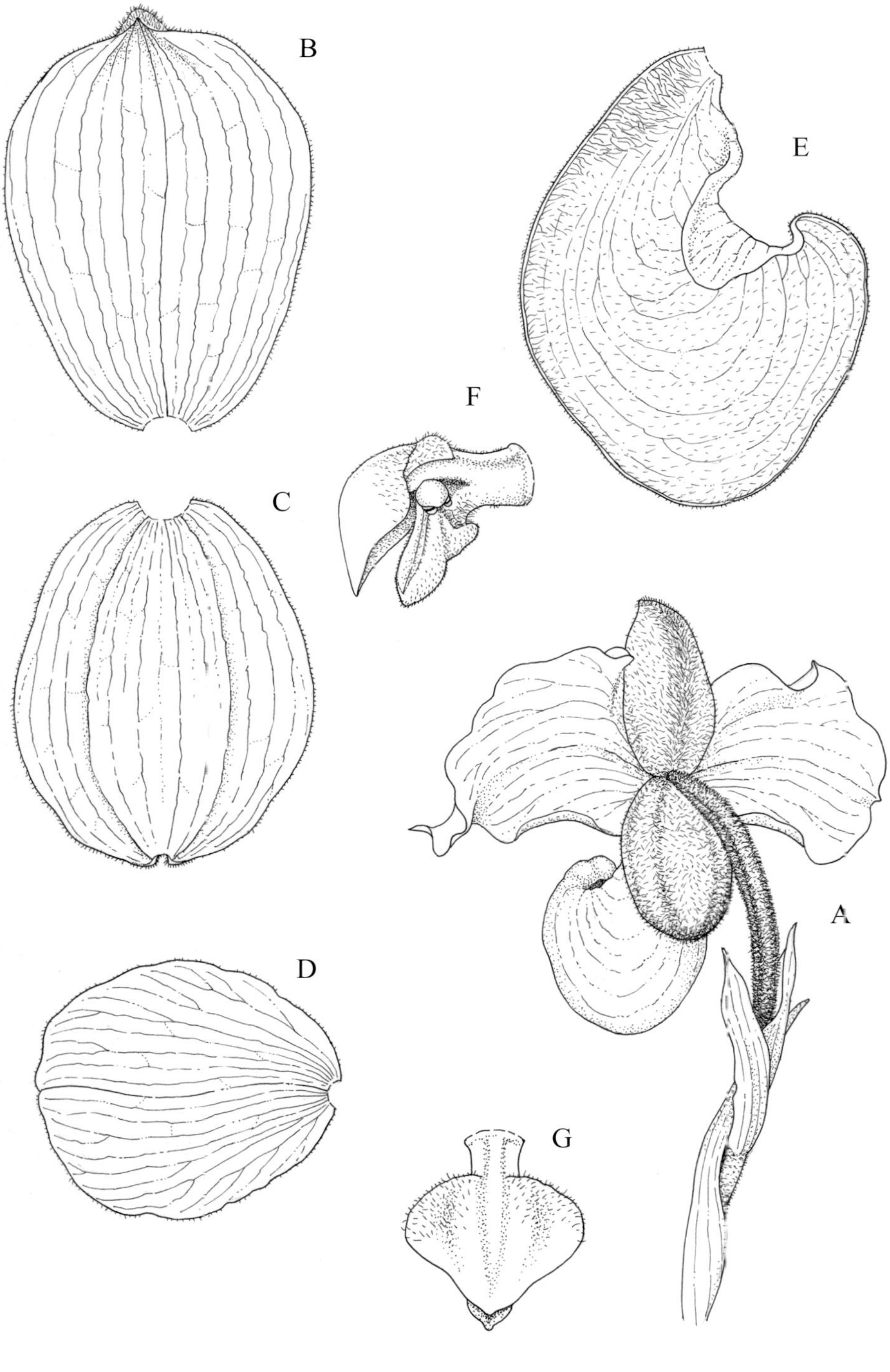

Fig. 61. ***Phragmipedium kovachii.*** **A**, flower, from reverse; **B**, dorsal sepal; **C**, synsepal; **D**, petal: **E**, lip, longitudinal section; **F**, column, side view; **G**, staminode. All drawn by Carol Woodin from cult. Arias. A, × 0.5; B, C, × 1; D, × 0.5; E, × 1; F, G, × 1.

Its distribution is limited to a small area of northern Peru where it has been found in about four localities. Koopowitz (2003) graphically described its habitat based upon a visit to the only surviving population in 2003. He found it growing on unshaded, southwest-facing, limestone cliffs, about 85 m high, in primary montane forest from about 1900 to 2000 m. The plants formed large clumps, often with 15 to 20 growths, and grew covered with brown moss in soil on ledges on the cliff. The thin, brown, gravelly soil was almost neutral with a pH in the range 6.8–7.1. No doubt the high rainfall leaches the soil which is derived from decaying plant material and the limestone of the cliff above the plants.

The excitement and clamour in Florida somewhat overshadowed events in Peru where the orchid's existence was severely threatened by the activities of local collectors and orchid growers. Its original habitat was rapidly cleared of plants, almost certainly before it was described (Oakeley, 2004). Three further localities in the same general region suffered the same fate in the months following its description.

On the positive front, the Peruvian government granted licences to two Peruvian nurseries in 2003 to take five plants each for propagation purposes. They have succeeded in growing seedlings in flasks from the mother plants. The Peruvian government has already licensed the sale of these artificially produced seedlings to take the pressure off the wild population that survives. Unfortunately, nearly every orchid grower in Peru and, reportedly, many abroad, already had plants. Outside Peru, plants have reputedly changed hands for between $1000 and $10000, depending upon which article is consulted.

One of us (PC) visited the type locality reported by Koopowitz and a second site in November 2005. The known localities lie in a deep valley on the eastern side of the Andes. The valley has been cleared of forest for a kilometre or more below the sites, but the forest above remains more or less intact apart from a little selective logging. At the first site, which is a cliff at 1830 m elevation by a mountain river, some twenty or more seedlings survived, the largest just large enough to bear a flowering spike in bud. If left undisturbed, it would undoubtedly recover at this locality. The second site that we visited was at a higher elevation, a steep slope overhung by a rock-face at just over 2000 m elevation in wet montane forest. The colony still contains many large plants but is known to local collectors who have made a trail to the base of the cliff.

Local collectors had all their plants confiscated by the government agency IRENA in 2004–5, but assured us that they had recollected plants which they now kept hidden in the forest. In fact, we saw large plants for sale in two roadside nurseries in the area. The natural populations remain under threat but we suspect that the local people would earn more money from eco-tourism, from orchid lovers keen to see one of the world's most spectacular orchids, than from stripping plants for sale.

Alfredo Manrique (pers. comm.) has provided detailed analyses of the habitat. The soils are crushed limestone with clay with a pH of 7.9 and a conductivity of 0.41dS/m. He measured the Calcium carbonate at 85.7%, Phosphorus at 1.8 ppm, Potassium at 60 ppm and water pH of 6–6.5. However, some plants were seen growing in a thick layer of wet humus whose pH was certainly more acidic. The temperature never drops below 50°F and peaks in April at 100°F. However, in the deep valley on the steep west-facing cliffs where it grows amongst grasses, it is shaded for much of the day.

Koopowitz (2003) commented on the large basal apertures of the lip that lie behind the staminode, suggesting that the pollinator was larger than in most species in the genus. He also suggested, based upon the vivid purple colour of its flowers, that it might be a mimic of

Figs. 62 & 63. Fine examples of *Phragmipedium kovachii*. Flowers. Cult. in Jersey and Lima, Peru respectively (PC, HO)

the woody *Meriana* (Melastomataceae) which has large flat flowers of a similar colour and bright yellow stamens: the large petals of the slipper orchid each mimicking a *Tibouchina* flower and the lip a bud (Figs. 4 & 5). However, the yellow longitudinal stripes on the lateral margins of the lip suggest that the whole flower might mimic a *melastomataceous* flower with the yellow lip margins resembling its yellow anthers, the pollinator, presumably a bee, being attracted more by colour than form. A similar mimicry of Melastomataceae flowers has been suggested for *Eulophiella roempleriana* in Madagascar.

Although originally placed in section *Micropetalum* (Hallier) Garay, a section characterised by its brightly coloured, pink, purple, orange or red flowers, oblong to elliptic petals, usually less than three times longer than broad, and by its small more or less obovoid to ellipsoid, pubescent lip with an incurved margin, *Phragmipedium kovachii* was placed in its own section *Schluckebieria* Braem (Braem, 2004). It differs from the species of section *Micropetalum* by its habit, the size and colour of its flowers, petal and staminode shape. The flowers are also notable for the indumentum of the flowers which is of long russet-brown hairs.

The horticultural world was, understandably, very keen to see legal plants of *P. kovachii* on sale. Its potential for breeding is difficult to underestimate. The impact of the orange- to scarlet-flowered *P. besseae* from Peru and Ecuador, which was described in 1982, has been immense, reviving interest in hybrid phragmipediums, a number of which have won the highest accolades in the orchid world, namely First Class Certificates from the American Orchid Society and Royal Horticultural Society and Gold Medals at World Orchid Conferences. A delegation from the former visited Peru in 2003 and, at a show in

Fig. 64. *Phragmipedium kovachii*, growing on ledges on forested cliff at the type locality in N Peru. (BB)

Lima, awarded a plant a Certificate of Horticultural Merit (94 pts), a score that would give it a First Class Certificate in the USA. Many further awards are to be anticipated when legal plants eventually appear elsewhere (Fighetti *et al.*, 2004).

The superb painting reproduced here was made by Carol Woodin who flew to Peru in September 2004 to illustrate it in the nursery of Alfredo Manrique, one of two collections in Peru with official permission to grow it. This plant has the largest flowers of any that have so far been seen in cultivation, the span being almost 20 cm from petal-tip to petal-tip. However, a flower measured in the field in 2005 had a span of 24.5 cm from petal-tip to petal-tip.

CULTURE. Alfredo Manrique (pers. comm.) recommends growing it in a well-drained compost of 60% gravel, 10% fir bark, 10% charcoal, and 20% perlite in wire baskets. It needs plenty of water, high humidity of 70% and good ventilation and should be grown in an intermediate glasshouse. He feeds the plants with a 15/5/15 fertiliser with added Calcium and Magnesium at 150–200 ppm every 15–20 days in the growing season. He grows the seeds in vitro on a standard medium suitable for phragmipediums. They germinate in the dark in 30–45 days.

DESCRIPTION. A terrestrial or lithophytic *herb*, often forming large compact clumps; rhizomes short, stout; roots thick, white, hairy. *Stems* short, bearing up to 9 leaves. *Leaves* coriaceous, linear-lanceolate, acute or acuminate, 35–55 ×1.5–5 cm, glossy dark green or yellow-green above, paler below, imbricate at the base. *Inflorescence* erect, bearing 2–4 flowers in succession, 20–55 cm tall; peduncle terete, russet-brown-pubescent, bearing a sheathing green acuminate sterile bract in middle; fertile bracts sheathing, lanceolate, acuminate, brown. Flower buds covered in dense red-brown hairs. *Flowers* showy, 16–24.5 cm in diameter, lasting up to 21 days, expanding in size after anthesis until maturity, lacking fragrance, with off-white to rose-pink sepals, bright orchid-purple or rose-purple petals and an orchid-purple to fuchsia-red lip, the base of the lip white, the rim of the lip white with a yellow band at the base on each side below the staminode, with lines of purple spots within; the staminode white with a pink margin; buds maroon-brown covered with russet-brown hairs. *Dorsal sepal* concave, elliptic, obtuse, 45–65 × 30–45 mm; synsepal similar but broader. *Petals* spreading or somewhat deflexed, oblong-elliptic to obovate, obtuse, 65–120 × 60–80 mm. *Lip* porrect, deeply cup-shaped, obovate to almost ovoid, 65–75 × 43–48 mm, white, spotted with red within, apical and lateral margins incurved, basal margins connate, the basal openings large and somewhat triangular. *Column* short, 20–25 mm long; staminode convex, transversely elliptic, 15–20 × 25–30 mm, glabrous; anthers elongate 2.5–3.5 mm long. *Capsule* 8–10.5 × 8–9 mm in diam.; seeds large, black. 2n = 32 (fide Fischer). **Plate 11. Figs. 58–65.**

Phragmipedium kovachii Atwood, Dalström & Fernandez in Selbyana 23, Suppl.: 1–2 (2002); Koopowitz in Orchid Dig. 67, 4: 248–255 (2003); Fighetti, Eloe & Bechtel in Orchids 73, 1: 40–42 (2004); Oakeley in Orchid Soc. Great Britain J. 53: 24–25 (2004). Type: Peru, Dept. San Martin, 2 km NE of Aguas Claras between Moyobamba and Nueva Cajamarca, 5 June 2002, *M. Kovach s.n.* (holo. USM (photo!)).

P. peruvianum Christenson in Orchids 71, 7: 620 (2002). Type: Peru, Dec. 2002, *Villena* in *Bennett* 7496–1 (holo. USM (photo!)).

DISTRIBUTION. N Peru, on the eastern side of the Andes. **Map 14**.

HABITAT. On rocks and ledges of cliffs amongst sedges and other herbs in montane forest; 1900–2100 m.

PHENOLOGY. Flowering March to November.

CONSERVATION STATUS. Critically endangered. Its original habitat was rapidly cleared of plants, almost certainly before it was described (Oakeley, 2004). Three further localities in the same general region suffered the same fate in the months following its description. On the basis of its limited range and the considerable collection for horticulture that has already eliminated four of its five known populations it has been given a conservation rating of CR (A1acd; B1ab(i–v); C 1 + 2a(i–ii); D; E) using the latest Red List criteria (IUCN, 2001).

ETYMOLOGY. Named for Michael Kovach, an American orchid enthusiast and collector who introduced it to the USA.

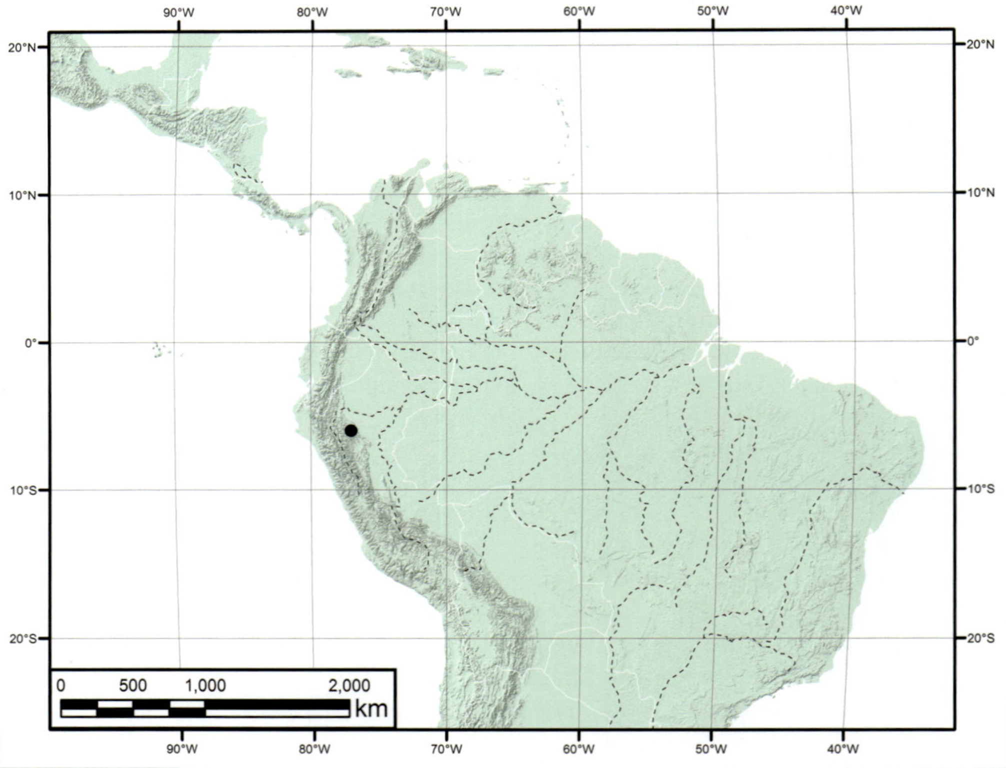

Map 14. Distribution of *Phragmipedium kovachii* in northern Peru.

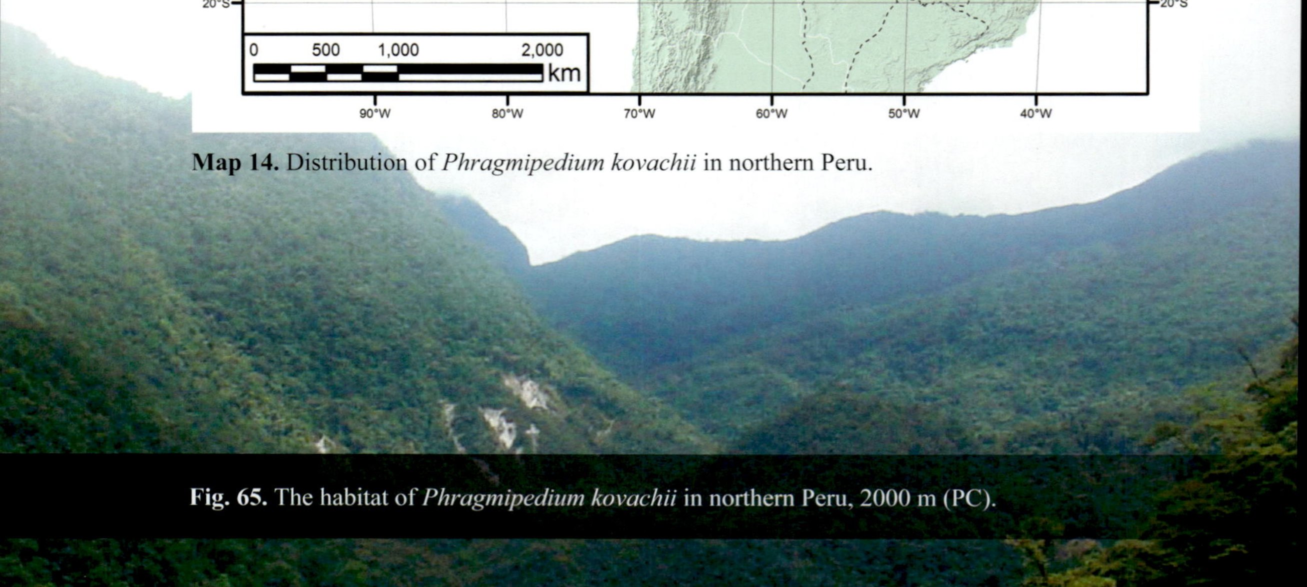

Fig. 65. The habitat of *Phragmipedium kovachii* in northern Peru, 2000 m (PC).

2. PHRAGMIPEDIUM BESSEAE

The brilliant orange-scarlet-flowered *Phragmipedium besseae* was discovered in northern Peru by J. Halton and Libby Besse in 1979. Calaway Dodson and Lee Kuhn named it in Besse's honour in 1981 in the American Orchid Society's Bulletin. Its discovery surprised the orchid world and, when plants first flowered in cultivation, they created a sensation. Several clones received First Class Certificates and Awards of Merit from the American Orchid Society, Royal Horticultural Society and others.

Dodson in Robbins (1989) claimed that Hoppe discovered this species in 1924 in Colombia because Schlechter, who examined the specimen, noted its 'richtig rosenrot' flower colour. Given its provenance, we suspect that this plant was *P. fischeri* rather than *P. besseae*.

Phragmipedium besseae has been collected in flower in the months of March and July, plants having an extended flowering period in the wild. As previously observed by Hegedus and Stermitz (1986), the labellar opening is very small, requiring a small pollinator. Cross-fertilisation is almost always successful whereas self-fertilised flowers fail to produce fruits. The red flowers might suggest bird pollination, but the structure of the flower suggest that it is a trap flower and bird-pollination is highly unlikely. *P. besseae* grows on cliffs where other red-flowered plants, such as begonias and fuchsias, are often found.

The discovery of *P. besseae* started a passion in the early 1980s for tropical American hard-leaved slipper orchid species and their hybrids that has continued unabated in orchid-growing circles. Its brilliant colour, relatively easy culture and propensity to hybridise freely with other *Phragmipedium* species ensured that it became the most significant species in the genus, particularly after Don Wimber, at the Eric Young Orchid Foundation in Jersey, used colchicine to chromosome double its hybrid progeny to produce some spectacular red-flowered hybrids, whose substantial flowers lasted for much longer than earlier diploid hybrids.

This species has not been refound in Peru where the type originated but several populations survive in Ecuador where, despite initial over-collection, they seem to have recovered well.

DESCRIPTION. Lithophytic *herb* with well-spaced and often scrambling growths; rhizomes elongate; roots striate. *Leaves* elliptic-oblong, minutely trifid and oblique at acute apex, 13–30 × 2.2–5 cm, dull green, glabrous, midvein strongly keeled and pointed, margins revolute. *Inflorescences* 1–6-flowered, flowering in succession, racemose, up to 40 cm tall; peduncle brown, pilose; sterile bracts 1–2, elliptic, apex minutely trifid, 4–5 cm long, 1.6 cm wide, lowest 5 cm from plant base; floral bracts widely ovate, apex minutely bifid or trifid, 2–3 cm long, 1.6–2 cm wide, shorter than ovary, glabrous, fused 0.7–1 cm from base. *Flowers* orange-red, the sepals and petals paler at the base, the lip yellow marked with red or orange stripes on the veins, the outer surfaces mostly velutinous, the inner surfaces mostly puberulous, the margins ciliate; ovary, including pedicel, 50–80 mm long, velutinous, green. *Dorsal sepal* ovate-elliptic, obtuse, 18–23 × 7–13 mm, 15-nerved, margins ciliate; synsepal elliptic, obtuse, 20–22 × 9–15 mm, yellow-green, margins ciliate. *Petals* borne at a 80–90° angle from lip, ovate, acute, 22–30 × 15–24 mm, nerves branching, inner surface puberulent, basal hairs arachnoid, margins ciliate. *Lip* ovate-elliptic, pointed apically, 20–22 ×10–14 mm at the widest point, with translucent areas between the veins

Plate 12. *Phragmipedium besseae* and forma *flavum*. Drawn by Carol Woodin from plants cultivated by Lynn and Bob Wellenstein, Candor, New York.

on sides and back of lip; lip opening ovoid, tomentose, the apical rim scalloped, shallowly incurved, the lateral rim glabrous to tomentose; claw interior pubescent. *Column* yellow, pubescent; staminode transversely oblong-subrhombic, apex acute, minutely emarginate, 6–8 × 6–7 mm, entire surface puberulent, orange or red with yellow margins and mid-vein; stigma conical. *Capsule* brown, velutinous, slightly curving apically. 2n = 24, 26. **Plate 12. Figs. 66–70 & 75.**

Phragmipedium besseae Dodson & Kuhn in Amer. Orchid. Soc. Bull. 50: 1308 (1981); Dodson & Dodson in Icon. Plant. Trop. Ser. 2, t. 0567 (1989). Type: Peru, San Martin, road from Tarapoto to Yurimaguas, *Halton & Besse* 62 (holo. SEL!).

Paphiopedilum besseae (Dodson & Kuhn) V.A.Albert & Borge Pett. in Lindleyana 9: 137 (1994).

DISTRIBUTION. S Ecuador (Azuay) and N Peru. **Map 15.**

HABITAT. On ledges on cliffs on the eastern flanks of the eastern Andes, amongst herbs and sedges; 800–1500 m.

PHENOLOGY. Flowering December to July.

CONSERVATION STATUS. Endangered (EN): B1ab(ii,iii,v)+2ab(ii,iii,v); C1+2a(i). It is a very local species with a restricted distribution. The population has decreased, the number of mature individuals dropping dramatically after discovery and the earliest known population has been completely destroyed (Bennett & Christenson, 1995). The

Figs. 66–68. Flowers of *Phragmipedium besseae* from cultivated plants. (PC, EYOF).

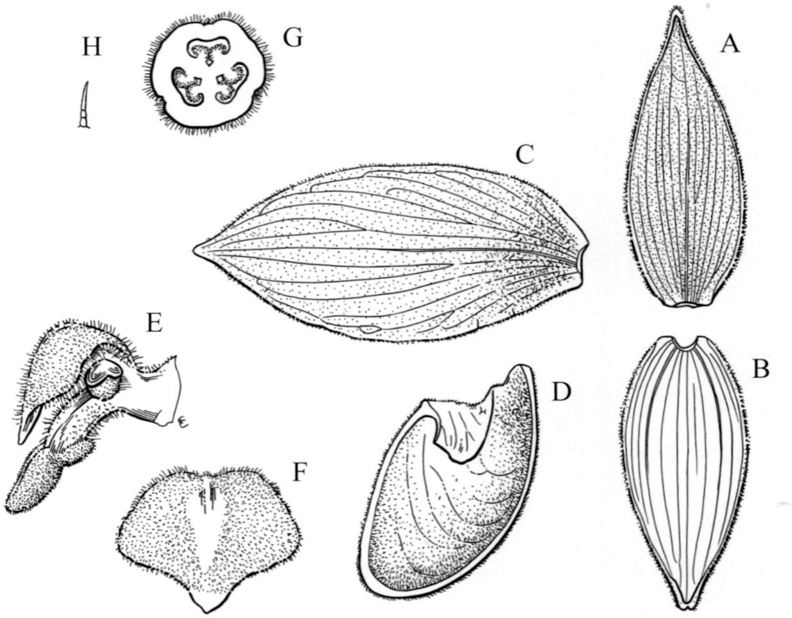

Fig. 69. ***Phragmipedium besseae.*** **A**, dorsal sepal; **B**, synsepal; **C**, petal; **D**, lip, longitudinal section; **E**, column, side view; **F**, staminode; **G**, ovary, transverse section; **H**, hair on ovary. All drawn by Paul Harwood from cult. Kuhn. A–D, × 2.25; E, F, × 4.5; G, × 12; H, × 60.

abundance of the species was significantly reduced in the years following its discovery due to many threats, particularly collection for regional and international horticultural trade, human disturbance and deforestation and other habitat degradation. The estimated extent of occurrence and estimated area of occupancy of the species are 6500 km^2 and 50 km^2, respectively.

Several localities in Ecuador that suffered from over-collection following their discovery have, however, recovered since seedlings that eluded the collectors subsequently flowered and produced seed.

ETYMOLOGY. Named for Libby Besse, an American orchid enthusiast who collected the type specimen.

P. besseae f. **flavum** (Braem) O.Gruss & Roeth in Caesiana, 12: 65 (1999). Type: ex Peru, cult. Braem 88/17b (holo. SCHLE).
P. besseae var. *flavum* Braem in Schlechteriana 1990, 2: 55 (1991), (as var. *flava*).

Differs from the typical form in having pure yellow flowers. Illustrations are given by Braem (Schlechteriana 1, 2: 54. 1990) and Gruss (Die Orchidee 47, 1: 217. 1996) and here.

This form is important in the breeding of yellow-flowered and pale-flowered hybrids. Only a single plant of it was originally collected and it is morphologically identical with typical *P. besseae*. It has been considered an albino variant. Seedlings have been raised from it and are now widely available in the trade. Counts of four colchicine-treated plants at the Eric Young Orchid Foundation revealed four different counts: 2n = 24, 25, 26, 36, the last being a triploid (Moon & Cribb, 1998).

Fig. 70. *Phragmipedium besseae f. flavum*. A flower from a plant cultivated at the Eric Young Orchid Foundation, Jersey. (EYOF)

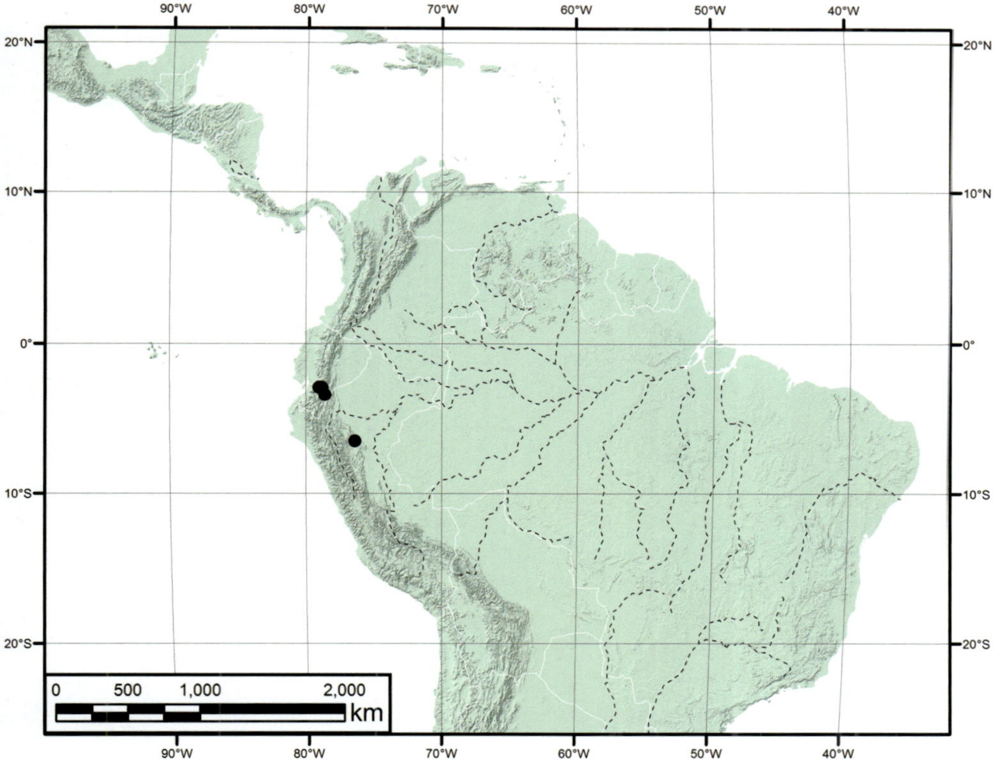

Map 15. Distribution of *Phragmipedium besseae* in the northern Andes of Ecuador and Peru.

3. PHRAGMIPEDIUM DALESSANDROI

The frequent taxonomic conflicts between botanists and horticulturists are well illustrated by this species, which is most satisfactorily treated botanically at varietal rank, whereas species rank is preferred by horticulturists. Calaway Dodson and Olaf Gruss (1996) based *P. dalessandroi* on a specimen collected by Denis D'Alessandro at between 900 and 1300 m elevation in the Cordillera del Condor on the upper Rio Bombuscara in Zamora-Chinchipe Province in southern Ecuador. They distinguished it from the now well-known *P. besseae* by its more compact non-climbing habit, more villose inflorescence, apricot-coloured flowers, narrow, drooping petals, rhombic staminode and different chromosome number (2n=28). *Phragmipedium besseae* has a pubescent but not villose flower stalk, red flowers, flat broad, spreading petals, and a chromosome number of 2n=24 (Wimber, 1994). At first glance these seem to be good distinguishing features. However, neither morphological nor chromosomal differences are perhaps quite as clear-cut as the above simple statement appears to show.

The status of *Phragmipedium dalessandroi* (Dodson & Gruss, 1996) was discussed in detail by Moon & Cribb (1998). The early hybrids that were registered by the Eric Young Orchid Foundation and others with *P. besseae* named as one parent were made using the plant now considered by Dodson and Gruss to be *P. dalessandroi*. New names were indeed published by the International Orchid Registrar based at the Royal Horticultural Society who accepted evidence from hybridisers that the hybrid progeny of *P. dalessandroi* looked different from those of *P. besseae*. Moon & Cribb (1998) suggested that before creating a lot of new hybrid grex names the evidence behind the establishment of *P. dalessandroi* as a species distinct from *P. besseae* should be examined, bearing in mind that growers were happy previously to accept that *P. besseae* occurred as two variants, respectively called the Ecuadorian and Peruvian forms. Indeed, we find Gruss (1996) referring to the Ecuadorian form as *P. besseae* var. *dalessandroi*, a then unpublished combination.

There are in cultivation at least four morphologically distinct red-flowered types of *Phragmipedium*: *P. besseae* from Peru and southern Ecuador; *P. dalessandroi* from southern Ecuador; *P. besseae* (including cultivars such as 'Wössen', 'Wössener Feuer' and 'Elisabeth') possibly from central Ecuador (Dalessandro, pers. comm.); and *P. besseae f. flavum*, a yellow-flowered variant said to have come from Peru but possibly from the same area as the 'Wössen' Type.

The third of these is of particular interest because it is morphologically intermediate between *P. besseae*, which has broad flat petals and a pubescent but not villose flower stalk, and *P. dalessandroi*. It has a chromosome number of 2n=26, that is likewise intermediate.

Variation in chromosome number has already been reported in a number of slipper orchids such as *Paphiopedilum venustum, P. bullenianum, P. javanicum* and *P. dayanum* (Karasawa 1979) and alone is insufficient to recognise the cytological races as distinct at specific level. Furthermore, the count of 2n=24 for *P. besseae* was based upon a single plant, that of 2n=28 for *P. dalessandroi* on a number of plants from a single pod sent by Calaway Dodson to Don Wimber (Wimber, pers. comm.).

On balance, it appears likely that the red-flowered phragmipediums represent a single species that is undergoing active evolution in the Andes. At one extreme the plants resemble *P. dalessandroi*, at the other *P. besseae*, but those in between, and there may be other populations in valleys that have yet to be explored, are intermediate. Therefore,

Plate 13. *Phragmipedium dalessandroi*. Drawn by Carol Woodin for *Curtis's Botanical Magazine*, n.s. t. 750 (2013) from a cultivated plant.

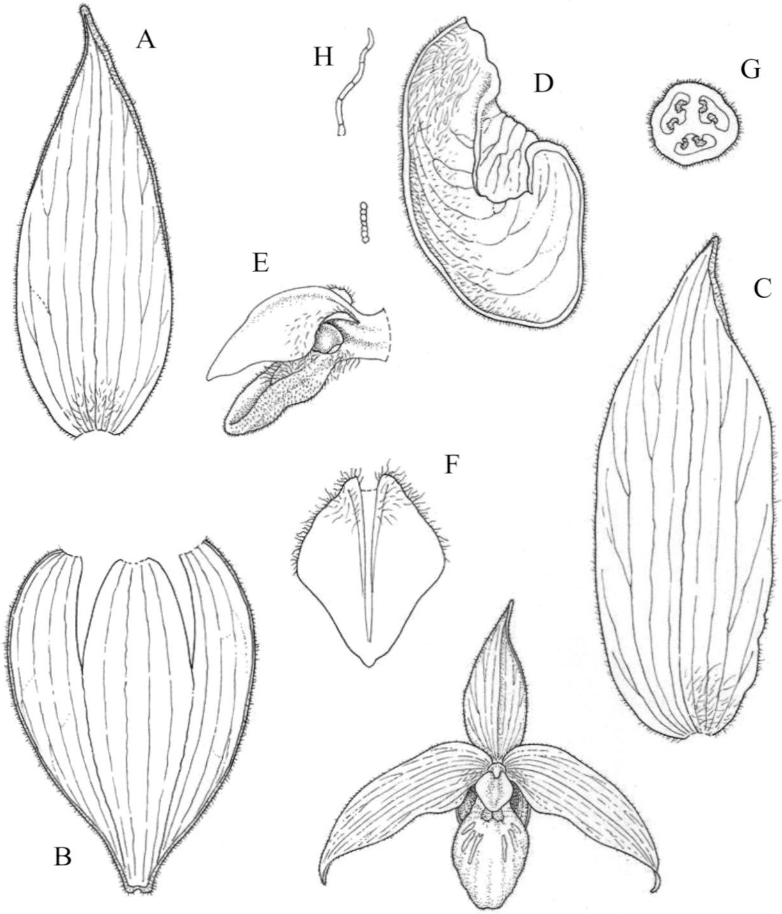

Fig. 71. ***Phragmipedium dalessandroi.*** **A**, dorsal sepal; **B**, synsepal; **C**, petal; **D**, lip, longitudinal section; **E**, column, side view; **F**, staminode; **G**, ovary, transverse section; **H**, hair on ovary. All drawn by Carol Woodin from cult. Kuhn. A–D, × 2.25; E, F, × 4.5; G, × 12; H, × 60.

botanically there seems to be little to be gained by recognising *P. dalessandroi* at species rank but horticulturally its status has been recognised and that decision is accepted here. Its pollinators are unknown, but are undoubtedly similar to those of *P. besseae*.

DESCRIPTION. Lithophytic perennial *herb* with closely spaced growths; rhizomes short; roots striate. *Leaves* elliptic-oblong, minutely trifid and oblique at acute apex, 13–30 × 2.2–5 cm, dull green, glabrous, midvein strongly keeled and pointed, margins revolute. *Inflorescence* racemose, usually branching, several-flowered, appearing in succession on each branch, up to 50 cm tall; peduncle brown, villose; sterile bracts 1–2, elliptic, apex minutely trifid, 4–5 × 1.6 cm, lowest 5 cm from plant base; floral bracts widely ovate, apex minutely bifid or trifid, 2–3 × 1.6–2 cm, shorter than ovary, glabrous, fused 0.7–1 cm from base. *Flowers* with orange-red or apricot-red sepals and petals, with yellow bases, and a yellow lip flushed red or orange in the apical part, the outer surfaces mostly velutinous, the inner surfaces mostly puberulous, the margins ciliate; ovary, including pedicel, 50–80 mm long, velutinous, green. *Dorsal sepal* ovate-elliptic, obtuse, 18–23 × 07–13 mm, 15-nerved, margins ciliate; synsepal elliptic, obtuse, 20–22 × 9–15 mm, yellow-green, margins ciliate. *Petals* decurved, oblong-elliptic, acuminate and usually recurved at tip, oblique, 22–30 × 10–15 mm, nerves branching, inner surface puberulent, basal hairs arachnoid, margins ciliate. *Lip* ovoid-ellipsoidal, tapering to a rounded tip, 20–22 × 10–14 mm at the widest point, translucent areas on sides and back of lip between the veins; lip opening ovoid, tomentose, the apical rim scalloped, shallowly incurved, the lateral rim glabrous to tomentose; claw interior pubescent. *Column* yellow, pubescent; staminode obovate-rhombic, bilobed at tip, 6–7 × 6–8 mm, entire surface puberulent, orange-red with yellow margins and mid-vein; apex acute, minutely emarginate; median stigmatic lobe conical. *Capsule* brown, velutinous, slightly curving apically. 2n = 28. **Plate 13. Figs. 71–74.**

Figs. 72–74. *Phragmipedium dalessandroi*, cultivated at the Eric Young Orchid Foundation, Jersey (EYOF, SD). **Fig. 75.** *Phragmipedium besseae*. Flower, side view. Cultivated plant (SD).

Phragmipedium dalessandroi Dodson & Gruss in Die Orchidee 47, 4: 217 (1996). Type: Ecuador, Zamora-Chinchipe Prov., *Dalessandro* 745A (holo. RPSC).

P. besseae Dodson & Kuhn var. *dalessandroi* (Dodson & Gruss) A.Moon & P.J.Cribb in Orchid Rev. 105 (1216): 229 (1997).

DISTRIBUTION. S Ecuador (Zamora-Chinchipe). **Map 16.**

HABITAT. On steep banks above rivers; 900–1500 m.

PHENOLOGY. Flowering January to July.

CONSERVATION STATUS. Critically Endangered (CR) A1acd; B1ab(i–v); C 1 + 2a(i–ii); D; E. It is a rare and very local species with a distribution restricted to S Ecuador. Although it is protected by its usually precipitous habitat, it is still threatened by factors such as collecting, anthropogenic habitat degradation and deforestation. The estimated extent of occurrence and estimated area of occupancy of the species are 20 km^2 and 5 km^2, respectively, with a continuing decline in the quality of habitats in all the locations. Its original habitat was rapidly cleared of plants, almost certainly before it was described. The assessment is made on the basis of its limited range and the considerable collection for horticulture that has already severely impacted its known populations.

ETYMOLOGY. Named for Denis D'Alessandro, an American orchid collector.

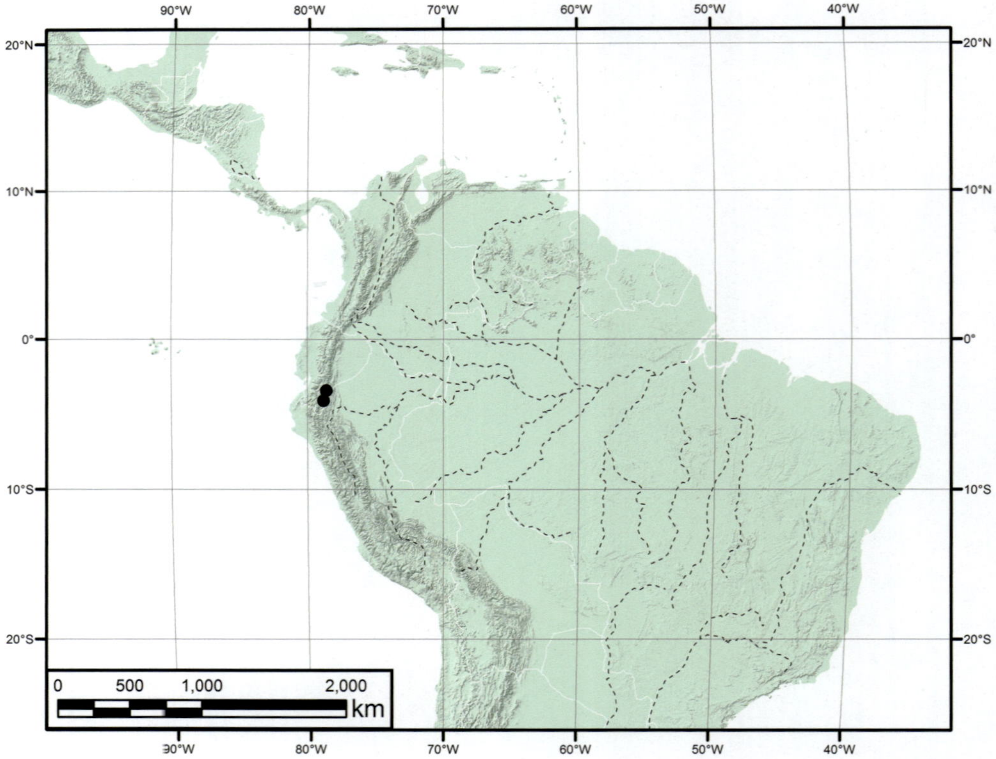

Map 16. Distribution of *Phragmipedium dalessandroi* in southern Ecuador.

4. PHRAGMIPEDIUM ANDREETTAE

A small-flowered species, allied to *P. schlimii* (Rchb.f.) Rolfe and *P. fischeri* Braem et H. Mohr, *P. andreettae* was collected in the early years of the 21st century in north-eastern Ecuador in the same region where *P. fischeri* was discovered a decade before. It had a similar habit and habitat to *P. fischeri* according to the collector, although exact details of the provenance are not available. It is similarly coloured to some pale forms of the Colombian *P. schlimii* but its lip and staminode shape are quite distinct, being differently coloured, longer than broad and distinctly, if minutely, bifid at the tip. The type plant was flowered in cultivation in the collection of Ecuagenera of Gualaceo, Ecuador and the type was deposited in the National Herbarium in Lima. It is named in honour of Padre Andreetta, a Salesian priest who had a life-long interest in orchids and who inspired the Portilla family in the foundation and development of Ecuagenera.

DESCRIPTION. A lithophytic *herb*. *Leaves* linear, acute, up to 15 × 1.3–2 cm, mid-green. *Inflorescence* erect to arching-suberect, 13–16 cm long, unbranched or with one branch, flowers produced in succession; peduncle purple, finely white-pubescent; sterile bract on peduncle acute, 2–3 cm long, green; fertile bracts conduplicate, ovate-lanceolate, acute, 2–2.5 × 1 cm, green with a purple base. *Flowers* with pale pink sepals, white petals flushed with pale pink on back, and white staminode with a yellow central spot; lip pink with purple and pink spots and with a yellow stripe along the back wall within; pedicel and ovary 55–65 × 2–3 mm, purple, finely white-pubescent. *Dorsal sepal* elliptic, obtuse, 21–22 × 12–13 mm, lateral margins reflexed, finely pubescent on both surfaces; synsepal similar in shape, shorter than the lip, 20–22 × 16–20 mm, bi-keeled on reverse, margins reflexed. *Petals* elliptic to obovate, rounded, 24 × 16 mm, margins reflexed, finely white pubescent on inner surface. *Lip* urceolate, 21 × 12–14 mm, finely pubescent on outside; side margins incurved, more so at the base, hairy within and without. *Column* 6 mm long; staminode obtrullate, 7 × 6 mm, bifid at tip; anthers small, bilocular; stigma 4.5–5 mm long, spatulate, hidden by staminode. **Plate 14. Figs. 76–78 & 82.**

Fig. 76. *Phragmipedium andreettae*. Flower of type plant. Cultivated plant, Ecuagenera, Gualaceo, Ecuador (PC).

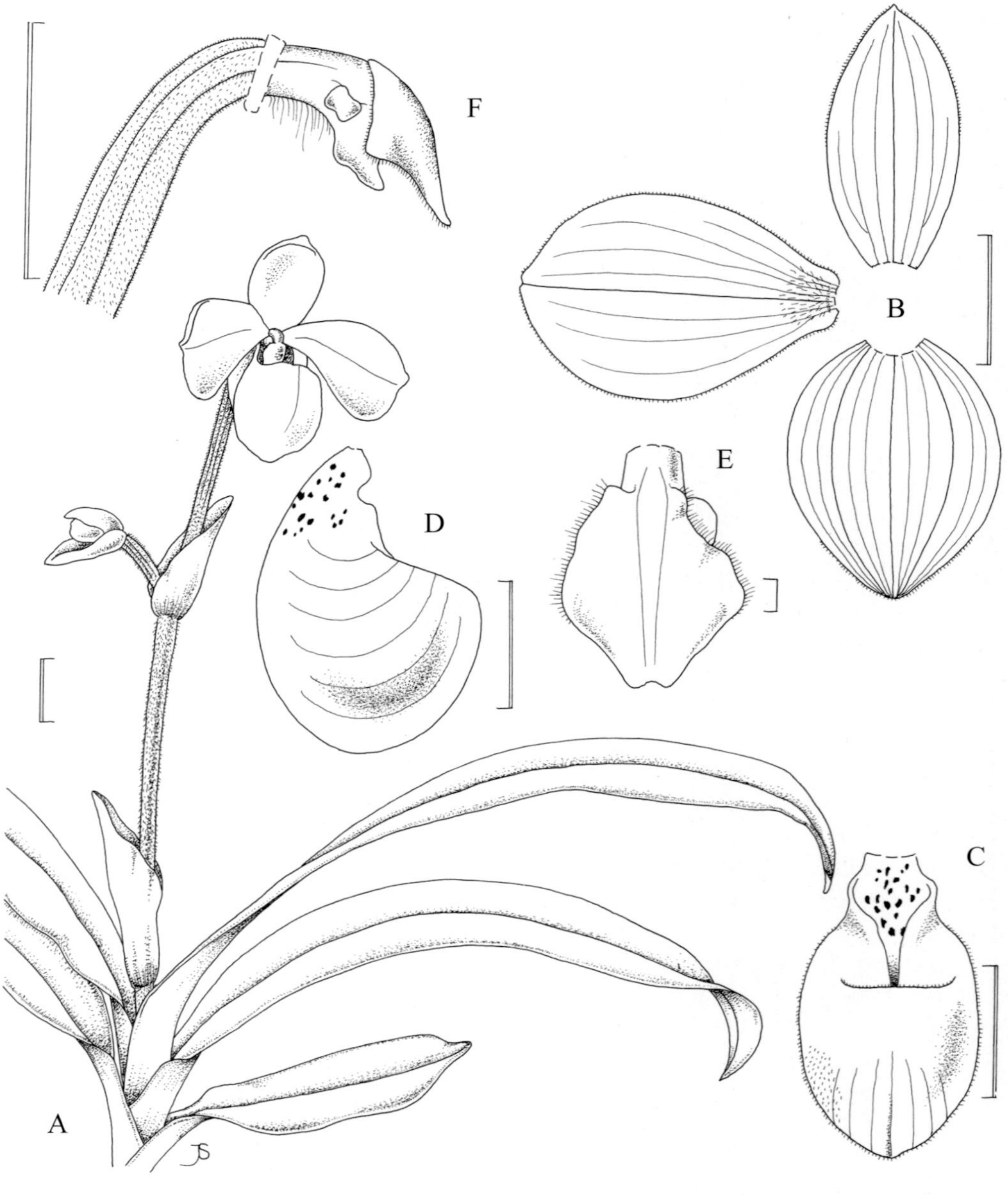

Fig. 77. ***Phragmipedium andreettae.*** **A**, habit; **B**, dorsal sepal, synsepal and petal from dissected flower; **C**, lip front view; **D**, lip, longitudinal section; **E**, staminode, front view, **F**, ovary and column, side view. All drawn by F. Pupulin from the type collection; inked and rendered by J. Stone. Double bar = 10 mm; single bar = 1 mm

Fig. 78. *Phragmipedium andreettae* in the type locality in the Ecuador/ Colombia border region. (PP).

Plate 14. *Phragmipedium andreettae*. Drawn by Carol Woodin for *Curtis's Botanical Magazine*, n.s. t. 697 (2010) from a plant cultivated by Ecuagenera, Gualaceo, Ecuador.

Phragmipedium andreettae P.J. Cribb & Pupulin in Lankesteriana 6(1): 1; figs. 1–2 (2006). Type: ?NW Ecuador, without exact prov., hort. Ecuagenera, November 2005, *Portillo s.n.* (holotypus QCA!).

DISTRIBUTION. Known only from NW Ecuador and SW Colombia. **Map 17**.

HABITAT. On steep banks along fast flowing rivers in montane forest; 500–1000 m.

PHENOLOGY. Flowering October to December.

CONSERVATION STATUS. *Phragmipedium andreettae* is assessed as Critically Endangered (CR): A3cd+4cd; B1ab(ii,iv,v)+2ab(ii,iv,v); C1+2a(i,ii); D. It is a rare and very local species, known only from a restricted area of NW Ecuador and SW Colombia. The trend of the population is decreasing, the number of mature individuals is low possibly less than 1000 and the abundance of the species is very low due to collection for the regional and international horticultural trade, human disturbance and habitat degradation. The estimated extent of occurrence and estimated area of occupancy of the species are 100 km^2 and 10 km^2, respectively, with an estimated continuing decline in the number of mature individuals and the quality of habitat. The population has declined by 80% over the last generation and is predicted to decline further in the next three generations if not protected.

ETYMOLOGY. Named for Padre Angel Andreetta (1913–2011), a Salesian priest and pioneer in the study of Ecuadorian orchids.

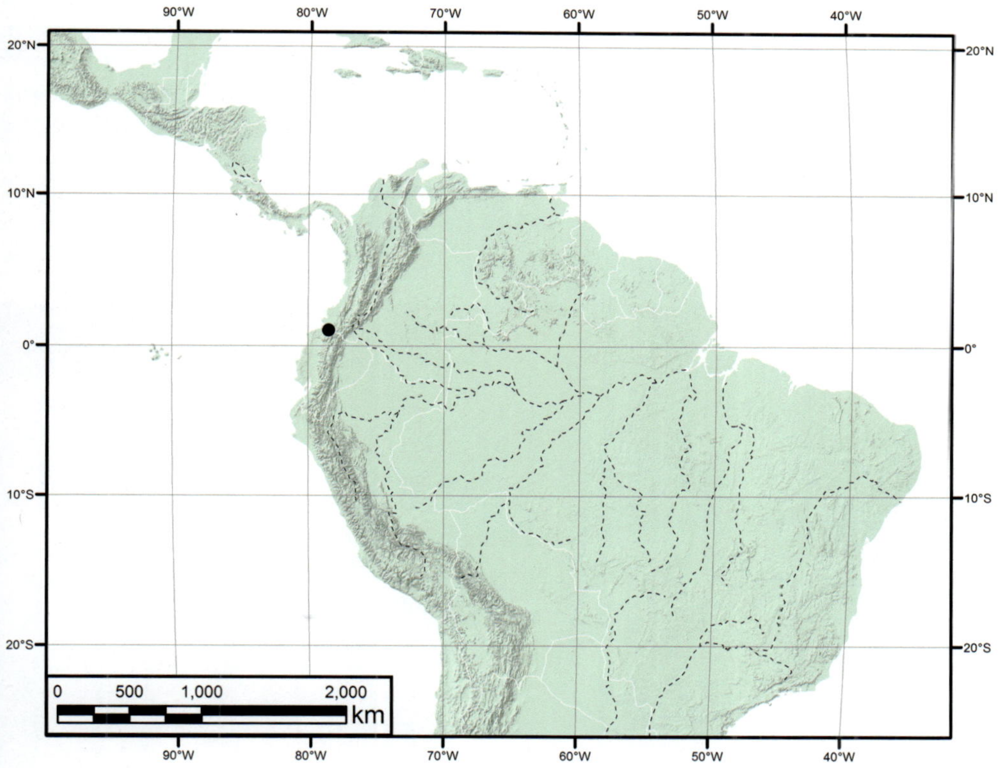

Map 17. Distribution of *Phragmipedium andreettae* in the northern Andes of Ecuador and Colombia.

5. PHRAGMIPEDIUM ANGULOI

This small-flowered species, closely allied to *P. andreettae* P.J. Cribb & Pupulin, was described by Braem, Teson and Macias (2014) based upon a collection by Roberto de Angulo Blum in southern Colombia. The type locality was given as the Patia-Timbio valley in the Department of Cauca on the western side of the Andes. Another unverified source gives its provenance as the Department of Putumayo on the eastern side of the Andes. It grows at 1600–1700 m elevation growing on rocky slopes in very wet montane forest. This contrasts with the habitat of *P. andreettae* and *P. fischeri* which grow at lower elevations on banks and cliffs above fast-flowing streams and rivers.

It has a similar habit to *P. andreettae*, but is a more compact plant with slightly broader leaves. Its flowers are white or very pale pink with a lilac margin to the front rim of the lip and a yellow spot in the middle of the broadly obovate staminode which is obtuse at the tip. The petals are elliptic and cute. The lip is ellipsoid-ovoid and has distinct and clearly visible window on the sides towards the synsepal. In contrast, *P. andreettae* has pink flowers with petals that appear obovate because of the recurved proximate basal margin, a lip lacking obvious side windows and a retuse rhombic staminode. It is named in honour of Roberto de Angulo Blum.

DESCRIPTION. A more or less lithophytic or terrestrial *herb* with somewhat clustered growths. *Leaves* linear, acute, 10–13 × up to 2.5 cm, green. *Inflorescence* erect to suberect, 9–12 cm long, unbranched, with flowers produced in succession; peduncle green to brown, finely white-pubescent; sterile bract on peduncle acute, 2–3 cm long, green; fertile bracts conduplicate, ovate-lanceolate, acute, 2–2.5 × 1 cm, green. *Flowers* with white to very pale pink sepals with greenish veins and greenish on the outer surface, white or very pale pink petals with pink at the base and tip, a white or pale pink lip with a rose-lilac rim to the orifice and a white staminode with a yellow central mark and pink at the tip; lip pink with purple and pink spots and with a yellow stripe along the back wall within; pedicel and ovary 36–55 × 2–3 mm, purple-brown, finely white-pubescent. *Dorsal sepal* concave,

Figs. 79 & 80. *Phragmipedium anguloi*. Plant and flower in type locality in Colombia. (RA)

elliptic, obtuse, 11–20 × 6–15 mm, finely pubescent on both surfaces; synsepal similar in shape, shorter than the lip, 12–20 × 9–15 mm, bi-keeled on reverse, margins reflexed. *Petals* elliptic to obovate, rounded, 19–23 × 15–19 mm, margins flat, finely white pubescent on inner surface. *Lip* ellipsoidal, 18–20 × 11–19 mm, with marked window on the lateral ventral sides towards the base, finely pubescent on outside; side margins incurved, more so at the base, hairy within and without. *Column* 4–5 mm long; staminode somewhat convex, broadly obovate to rhombic, 7 × 5–6 mm, obtuse at tip, pubescent; anthers small, bilocular; stigma 4.5–5 mm long, spatulate, hidden by staminode. **Figs. 79–81.**

Phragmipedium anguloi Braem, Teson & Manzur in Richardiana 14: 290; figs. 1–2 (2014). Type: SW Colombia, Dept. of Cauca, Patia-Timbio valley, *R. de Angulo Blum s.n.* (holo. FAUC).

DISTRIBUTION. Known only from S Colombia. **Map 18.**

HABITAT. In pockets of humus and leaf debris on steep rocky slopes in very wet montane forest; 1600–1700 m.

PHENOLOGY. Flowering September and October.

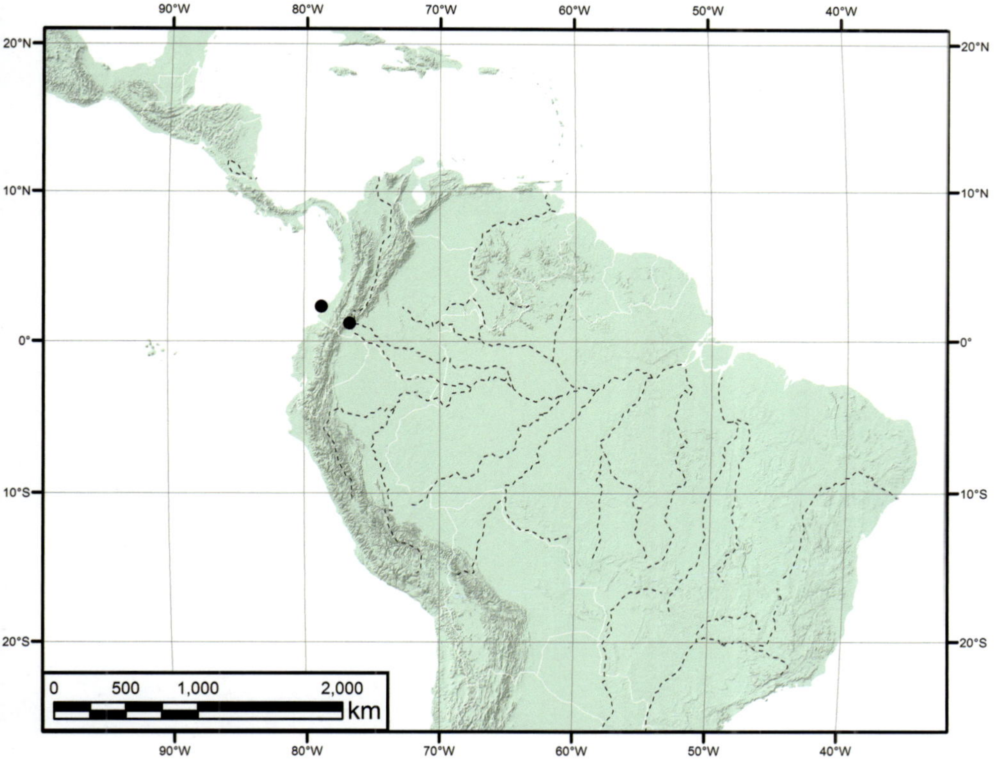

Map 18. Distribution of *Phragmipedium anguloi* in Colombia.

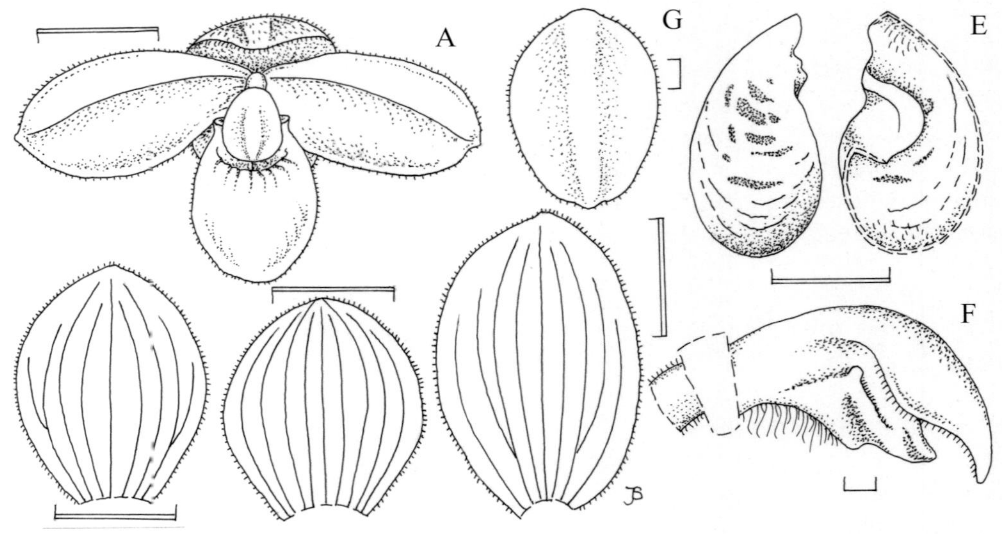

Fig. 81. ***Phragmipedium anguloi.*** **A**, flower; **B**, dorsal sepal; **C**, synsepal; **D**, petal; **E**, lip, longitudinal section; **F**, column, side view; **G**, staminode. All drawn by Judi Stone after drawings of type by E. Téson. Single bar = 1 mm; double bar = 10 mm.

CONSERVATION STATUS.Data Deficient (DD) but probably Critically Endangered (CR) A1acd; B1ab(i–v); C 1 + 2a(i–ii); D; E.. Its original habitat was rapidly cleared of plants, almost certainly before it was described. The above conservation rating is given on the basis of its very limited range and the collection for horticulture that has already severely reduced its known population.

ETYMOLOGY. Named for its discoverer.

6. PHRAGMIPEDIUM FISCHERI

Phragmipedium fischeri, closely allied to the Colombian *P. schlimii* and considered by some to be a variety of that species, was discovered in Ecuador in April 1996. Guido Braem and Hartmut Mohr described it in September of the same year based upon a cultivated plant flowered in the USA by Jerry Lee Fischer of Orchids Limited, after whom it was named They distinguished it from *P. schlimii* by its shorter, more horizontal leaves, more intensely coloured flowers with a solid red-purple lip, and, in particular, the morphology of its staminode which is subquadrate or transversely oblong with a longitudinal, central, raised ridge, rather than transversely elliptic-triangular and with an obscure central ridge, and distinctively coloured creamy white with a central yellow ridge terminating in a purple spot.

Gruss (2013) provided a detailed account of its morphological variation, habitat and climatic preferences. In Ecuador, it is found in the Maldonado region at about 1400 m elevation. In the wild it grows on steep banks and mossy cliffs in shaded but breezy places along montane rivers. The temperatures here range from 26–30°C and drop to 20°C at night.

Christenson (1997) indicated that Schlechter had reported that Hoppe (*Hoppe 163*) collected a similar plant with rose-coloured flowers in the same region. Unfortunately, that specimen in the Berlin (Dahlem) herbarium was destroyed in World War II. Its pollinators are unknown, but undoubtedly are similar to those of *P. schlimii*.

DESCRIPTION. Terrestrial perennial *herb*; rhizomes short, up to 3 cm long and 1 cm wide; roots up to 4 mm in diam. *Leaves* elliptic-lanceolate to lanceolate-ligulate, acute, 4–20 × 1.5–3.2 cm, often widening appreciably 3–5 cm from base, thin, glabrous, pale green; margins revolute. *Inflorescence* 2–10-flowered, racemose, 12–20 cm tall; peduncle purple,

Fig. 82. *Phragmipedium fischeri* and *P. andreettae* flowers (SD).

Plate 15. *Phragmipedium fischeri*. Drawn by Carol Woodin from a plant cultivated by Jerry Fischer.

Fig. 83. *Phragmipedium fischeri.* Plants growing on a steep bank above a river, N Ecuador. (SD).
Fig. 84 (inset).

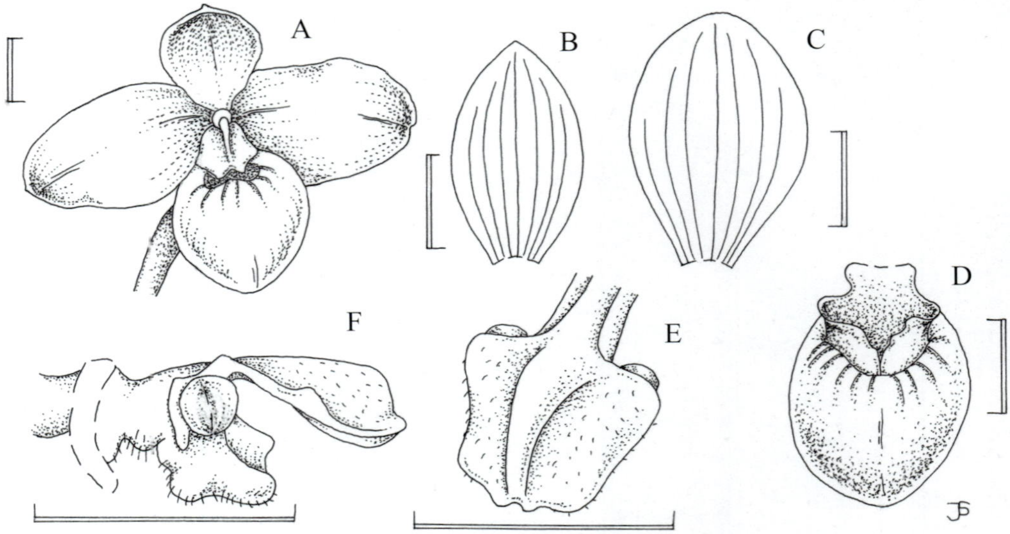

Fig. 85. ***Phragmipedium fischeri*****.** **A**, flower; **B**, dorsal sepal; **C**, petal; **D**, lip, from front; **E**, column, side view; **F**, staminode. All drawn by Judi Stone from cult. Fischer. Single bar = 1 mm; double bar = 10 mm.

Figs. 86 & 87. *Phragmipedium fischeri*. Ecuagenera in Gualaceo, Ecuador. Flower, oblique view (PC) and side view (SD). Cultivated by Ecuagenera, Gualaceo, Ecuador.

Fig. 88. *Phragmipedium fischeri*. Plant in the wild, NE Ecuador. (JH).

sparsely hairy; sterile lower bract ovate, acute, 3–4 × 1.4 cm, 6–9 cm from plant base; floral bracts conduplicate, ovate-lanceolate, acute, 2–4 × 1.4–2 cm, shorter than ovary, glabrous, strongly keeled, fused up to 0.6 cm from base on marginal side, green. *Flowers* produced in succession, 4.5–5 cm in diameter, rose-pink to rose-purple, the staminode creamy white with a yellow basal mark with a V-shaped purple spot at its lower end, outer surfaces mostly pilose-velutinous; inner surfaces mostly glabrous to sparsely pubescent, the margins glabrous to ciliate; ovary and pedicel 50–90 mm long, dark purple, white pubescent. *Dorsal sepal* elliptic, acute, 15–20 × 10–11 mm, 7-nerved, the outer surface pilose, the margins ciliate; synsepal elliptic, rounded at apex, 13–17 × 10 mm, smaller than lip, not strongly nerved, the margins ciliate. *Petals* somewhat recurved in apical half, obovate or oblong-obovate, obtuse, 15–23 × 11–16 mm, central nerve prominent, petals held at more or less 90° from lip, the inner surface sparsely puberulent; basal hairs white, villous-arachnoid, margins ciliate, slightly undulate. *Lip* round to slightly pointed, 16–21 × 12–17 mm wide at the widest point, pink to magenta, thin interveinar tissue forming window-like network on sides and back of lip; labellar opening oval to round, slightly scalloped on apical rim, apical margin shallow, incurved, lateral margins minutely pubescent, 5–8 mm wide. *Column* up to 3 mm long, whitish-pink, pilose; anther dark-pigmented; staminode somewhat flattened, subquadrate to transversely oblong, 6–10 × 5–9 mm wide in apical half, with a prominent central ridge, puberulous, the apex shallowly notched, the margins undulate, often white, finely ciliate; stigma up to 2 mm wide, papillose, the midlobe conical, up to 3 mm long, the lateral lobes rounded. *Capsule* dark purple-brown, hirsute, splitting in middle. 2n=36 (fide Fischer). **Plate 15. Figs. 82–88.**

Phragmipedium fischeri Braem & H.Mohr in Leaflets of the Schlechter Institute 3: 28 (1996); Gruss in Orchideenjournal 2013: 7–10 (2013). Type: Ecuador, Maldonado, 1400 m, April 1996, cult. Orchids Ltd., USA (holo. SCHL no. 96/0414).

Phragmipedium schlimii var. *fischeri* (Braem & H.Mohr) Gruss in Japan Orchid Society Bulletin 43: 34 (2000).

DISTRIBUTION. NW Ecuador (Carchi) and SW Colombia. **Map 19**.

HABITAT. Grows in shaded but breezy places along montane rivers; 1000–1500 m elevation.

PHENOLOGY. Flowering throughout the year.

CONSERVATION STATUS. Critically Endangered (CR): A3cd+4cd; B1ab(ii,iii,v)+2ab(ii,iii,v); C1+2a(i,ii). It is a very local, narrowly endemic species, restricted to very few localities. Although it is somewhat protected by the inaccessible habitat (steep river banks), the population is decreasing and the number of mature individuals is low (possibly less than 100). The abundance of the species is decreasing due to collection for regional and international horticultural trade, deforestation, human disturbance and habitat degradation. The estimated extent of occurrence and estimated area of occupancy of the species are less than 100 km^2 and 10 km^2, respectively with an estimated continuing decline in the number of mature individuals and the quality of habitats in the single location. The decline of the population size is estimated to be around 80% over the last two generations and is predicted to decline further in the next three.

ETYMOLOGY. Named for Jerry Fischer, nurseryman and owner of Orchids Limited.

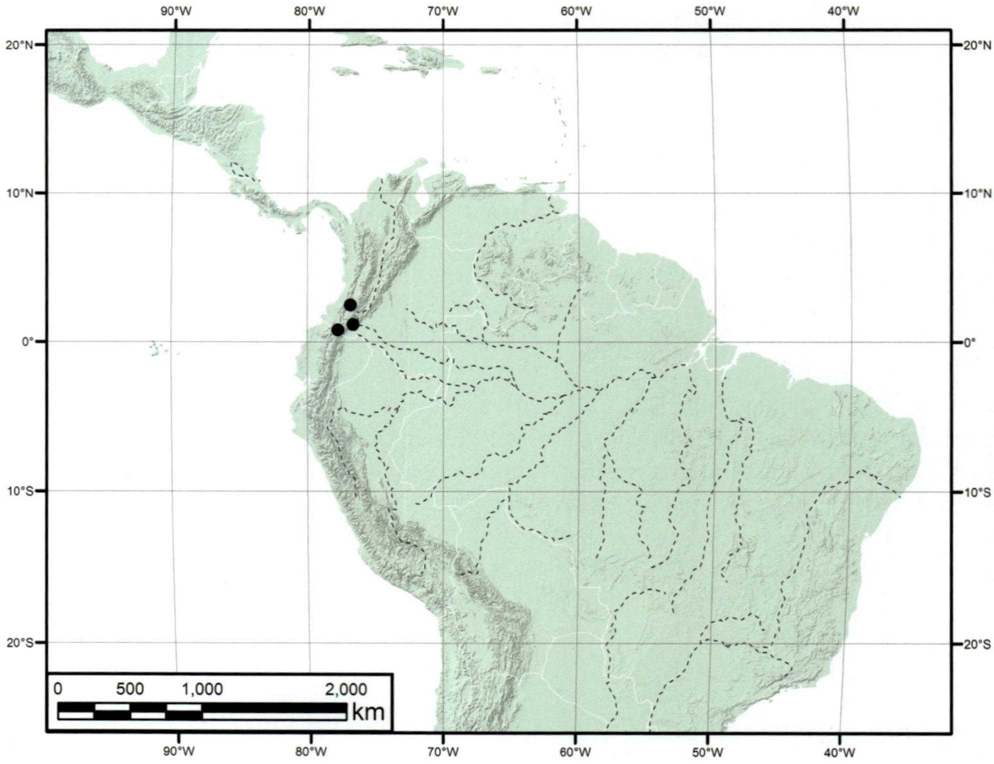

Map 19. Distribution of *Phragmipedium fischeri* in the northern Andes of Ecuador and Colombia.

7. PHRAGMIPEDIUM SCHLIMII

Louis Schlim, collecting for Jean Linden's nursery in the early 1850s, discovered this pretty species near Ocaña in Colombia. Jean Linden and H.G. Reichenbach (1854) described it as *Selenipedium schlimii* in his honour in *Bonplandia* and it was transferred to the present genus by Rolfe in 1896.

Fowlie (1970) provided an account of its habitat near Ocaña in Colombia where it grew in seepages on granite above a stream. Large clumps were found growing in permanently wet areas on the roots of small trees and shrubs that had caught algae, detritus and leaf litter. Below, he also found plants attached to crevices in the granite, also in seepages. The region otherwise was arid, protected from the rainclouds of the Caribbean by high mountains to the north-east.

Phragmipedium schlimii has an elongate, conically-shaped median staminode lobe, which may facilitate autogamy. It is reported that the sticky pollen mass has a thin envelope that dries, leaving the free granular pollen that may then contact the stigmatic surface and, thereby, effect self-pollination (Anon., 1922). Plants in cultivation vary in their flower colour from pure white, marked with purple only on the staminode, to pink with a deeper rose-coloured lip.

Phragmipedium schlimii caused something of a sensation when it was introduced into cultivation in the 1850s. Since the other tropical American slipper orchids in cultivation at that time were mostly yellow and green, the bright pinks and whites of this plant were immediately recognised for their horticultural value. It is relatively easy to cross, and is one of the parents of many early *Phragmipedium* hybrids (Stein, 1892). Its hybrids with *P. besseae* and *P. dalessandroi* are particularly pleasing and popular in cultivation. *P. schlimii* hybrids are small plants with pleasingly shaped and coloured flowers, suitable for a small greenhouse.

DESCRIPTION. Terrestrial perennial *herb*; rhizomes up to 3 cm long and 1 cm wide; roots up to 4 mm in diam, usually striate. *Leaves* elliptic to lanceolate-ligulate, minutely trifid

Figs. 89 & 90. *Phragmipedium schlimii*. Flowers from plants cultivated by the Eric Young Orchid Foundation, Jersey. (PC).

Plate 16. *Phragmipedium schlimii*. Drawn by Carol Woodin from a plant cultivated by Lynn and Bob Wellenstein, Candor, New York.

at acute apex, 4–30 × 1.5–3.6 cm, often widening appreciably 3–5 cm from base, thin, glabrous, pale green, often purple basally; margins revolute. *Inflorescence* 1–10-flowered, the flowers produced in succession, simple or branching, 12–27 cm tall; peduncle purple, sparsely velutinous; sterile lower bract ovate, acute, 3–4 × 1.4 cm, 6–9 cm from plant base; bracts ovate-triangular, acute, 2–4 × 1.4–2 cm, shorter than ovary, glabrous, strongly keeled, fused up to 6 mm from base on marginal side. *Flowers* whitish-pink to magenta, the lip pink to magenta with a dark pink, magenta, or often striped with magenta on rim, the staminode bright yellow with white lateral margins and 2 purple apical marks, pilose-velutinous on outer surface, glabrous to sparsely pubescent within, the margins glabrous to ciliate; pedicel and ovary, 5–9 cm long, velutinous to hirsute, dark purple. *Dorsal sepal* ovate-elliptic, acute, 15–20 × 10–12 mm, 7-nerved, pilose outside, the margins ciliate; synsepal elliptic, acute to minutely bifid, 13–20 × 10 mm, smaller than the lip, not strongly nerved, the margins ciliate. *Petals* borne at 90° from lip, obovoid-rhombic, obtuse, 15–24 × 11–14 mm, nerves branching, central nerve prominent, sparsely puberulent within, the basal hairs white, villous-arachnoid, the margins ciliate, slightly undulate. *Lip* round to slightly pointed, 16–25 × 14 mm wide at the widest point, with thin interveinar tissue forming a window-like network on sides and back of lip, the mouth oval to circular, slightly scalloped on apical rim, its apical lip rim, the apical lining incurved, not sharply folded, the lateral rim minutely pubescent, 5–8 mm wide, the claw face white or yellow with pink spots, tubercles absent, villous and dark red spotted within. *Column* up to 12 mm long, whitish-pink, pilose; stigma up to 2 mm wide, the median lobe conical, to 3 mm long, the lateral lobes rounded; anther dark-pigmented; staminode convex, transversely elliptic-triangular, 6–10 × 5–9 mm, central vein often prominent, puberulous, the lateral and basal margins often white, finely ciliate, the apex acute to shallowly bilobed. *Capsule* dark purple-brown, hirsute. 2n=30 (Karasawa, 1980). **Plate 16. Figs. 89–92.**

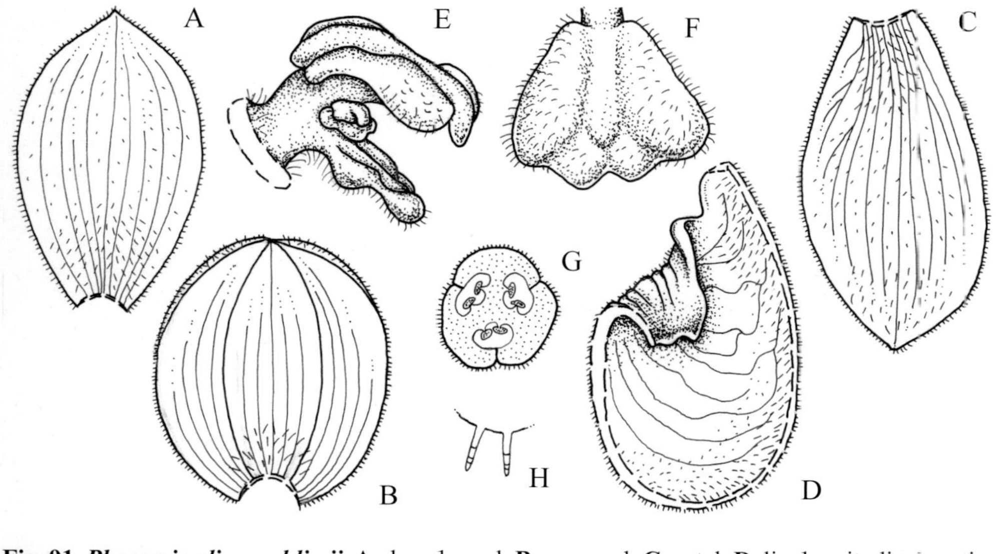

Fig. 91. ***Phragmipedium schlimii.*** **A**, dorsal sepal; **B**, synsepal; **C**, petal; **D**, lip, longitudinal section; **E**, column, side view; **F**, staminode; **G**, ovary, transverse section; **H**, hairs on ovary. All drawn by Judi Stone from cult. Wellenstein (Kew Spirit Collection no. 56856). A–D, × 3; E, F, × 4.5; G, ×9; H, × 48.

Phragmipedium schlimii (Linden & Rchb. f.) Rolfe in Orchid Rev. 4: 332 (1896). Type: Colombia, Ocaña, *Schlim* 407 (lecto. W!, K! ; syn. K!, G!).
Selenipedium schlimii Linden & Rchb.f. in Bonplandia 2: 277–278 (1854).
Cypripedium schlimii (Rchb.f.) Linden ex Bateman in Curtis's Bot. Mag. 22: 5614 (1866).
Paphiopedilum schlimii (Rchb.f.) Stein, Orchideenbuch, 483 (1892).
Phragmopedilum schlimii (Rchb.f.) Pfitzer in Bot. Jahrb. Syst. 25: 527 (1898).

DISTRIBUTION. Colombia, on both the eastern and western slopes of both Andean cordilleras. **Map 20**.

HABITAT. Found on wet slopes and road cuts, and rocky crevasses of wet forests. Its roots reach deep between rocks, making them difficult to dislodge (Fowlie, 1970); 1200–2000 m elevation.

PHENOLOGY. June to February.

CONSERVATION STATUS. Endangered (EN): A2cd+3cd+4cd; B2ab(ii,iii,v); C1. It is a local species with a broad distribution in Colombia but its populations are widely scattered. The trend of the population is decreasing, the number of mature individuals in many populations is low due to habitat change, deforestation and collection for regional and international horticulture trade by local collectors and orchid growers. The estimated area of occupancy of the species is less than 500 km^2 with an estimated continuing decline in the number of mature individuals and the quality of habitats in many locations. The population size has declined over the last three generations and is projected to decline further.

ETYMOLOGY. This species is named for its discoverer, Louis Schlim, the half-brother of Jean Linden.

Fig. 92. *Phragmipedium schlimii*. Flowers from plants grown at the Eric Young Orchid Foundation, Jersey. (PC).

7a. var. manzurii (W.E. Higgins & P. Viveros) P.J. Cribb **comb. et stat. nov.**
Phragmipedium manzurii W.E. Higgins & P. Viveros in Lankesteriana 8, 3: 89 (2008).
Type: Colombia, Santander, cult. June 2008, *D.A. Manzur* 1501 (holo. FAUC).

This pretty orchid was discovered by David Angel Manzur, a retired professor at the Faculty of Agronomy at the University of Caldas, Manzinales, in northern Colombia who first flowered it in his collection in June 2008. It is undoubtedly closely allied to the more widely distributed *P. schlimii*, being similar in habit and flower size and shape. It differs in flower colour, the sepals and petals being predominantly yellow-green, the elliptic, flat petals being violet-flushed, and in staminode shape and colour, which is subquadrate-subcircular rather than transversely oblong-elliptic and yellow-green rather than bright yellow with two apical purple blotches. Wesley Higgins and Paula Iveros described it in December 2008 based upon a plant in Manzur's private orchid collection. They also commented that the lip is white and purple-spotted within.

The description by Olaf Gruss (2011) of *P.* × *colombianum*, as a natural hybrid between *P. schlimii* and *P. manzurii*, casts doubt on the distinctiveness of the latter. The type and a second plant, both flowered in cultivation by Franz Glanz in Germany, are well illustrated by Gruss. They show flowers with a yellow-green dorsal sepal and elliptic white petals, scarcely marked with purple at the base, both characters considered typical in *P. manzurii*. However, its lip shape and colour and staminode are typical of *P. schlimii*. It seems probable that *P. manzurii* is no more than a variety of *P. schlimii*. The necessary new combination is made here.

DESCRIPTION. Plant similar to *P. schlimii*. *Flowers* with pale yellow-green sepals and petals, the petals flushed with purple at the base, the lip pale pink with magenta markings around the rim, the staminode bright yellow, minutely pilose-velutinous on outer surface, glabrous to sparsely pubescent within, the margins glabrous to ciliate; ovary and pedicel, 4.8–5.8 cm long, velutinous, green. *Dorsal sepal* elliptic, subacute, 20 × 10–15 mm, 7-nerved, pilose outside, the margins ciliate; synsepal oblong-elliptic, acute to minutely bifid, 18–20 × 14–17 mm, smaller than lip, pink to white infused with pink, not strongly nerved, the margins ciliate. *Petals* borne at 90° from lip, obovoid-rhombic, obtuse, 22–28

Fig. 93. *Phragmipedium schlimii* var. *manzurii*. Flower. (CA). **Fig. 94.** *Phragmipedium schlimii*, a flower intermediate between *var. manzurii* and *var. schlimii*. (CA)

× 11–20 mm, central nerve prominent, sparsely puberulent within, the basal hairs white, villous-arachnoid, the margins ciliate. *Lip* obovoid, slightly pointed, 18–24 × 14–17 mm wide at the widest point, the mouth oval to circular, slightly scalloped on apical rim, the apical lining incurved, not sharply folded, the lateral rim minutely pubescent, 5–8 mm wide, the claw face yellow-green with pink spots, tubercles absent, villous spotted within. *Column* 15–16 mm long, whitish-pink, pilose; stigma to 5 mm wide, the median lobe conical, to 3 mm long, the lateral lobes acute; staminode convex, subquadrate-subcircular, emarginate, 6–8 × 6–8 cm, central vein often prominent, puberulous, the lateral and basal margins often white, finely ciliate. Chromosome number unknown. **Figs. 93 & 94.**

DISTRIBUTION. Colombia: Santander Province only.

HABITAT. On damp slopes, seepages and rocky crevices in wet montane forest; 1300–2000 m elevation.

PHENOLOGY. Flowering in June and possibly in other months.

CONSERVATION STATUS. Critically Endangered (CR): B1ab(ii,iii,v)+2ab(ii,iii,v); C2a(i,ii); D. It is a newly discovered taxon that appears to be restricted to one site in Santander province in Colombia. Its population is likely to be decreasing rapidly as plants are collected for horticulture, the number of mature individuals is almost certainly low and is decreasing. The estimated extent of occurrence and estimated area of occupancy of the species are less than 100 km^2 and 10 km^2, respectively, with an estimated continuing decline in the number of mature individuals and the quality of habitats in the single location.

ETYMOLOGY. Named for the collector of the type specimen.

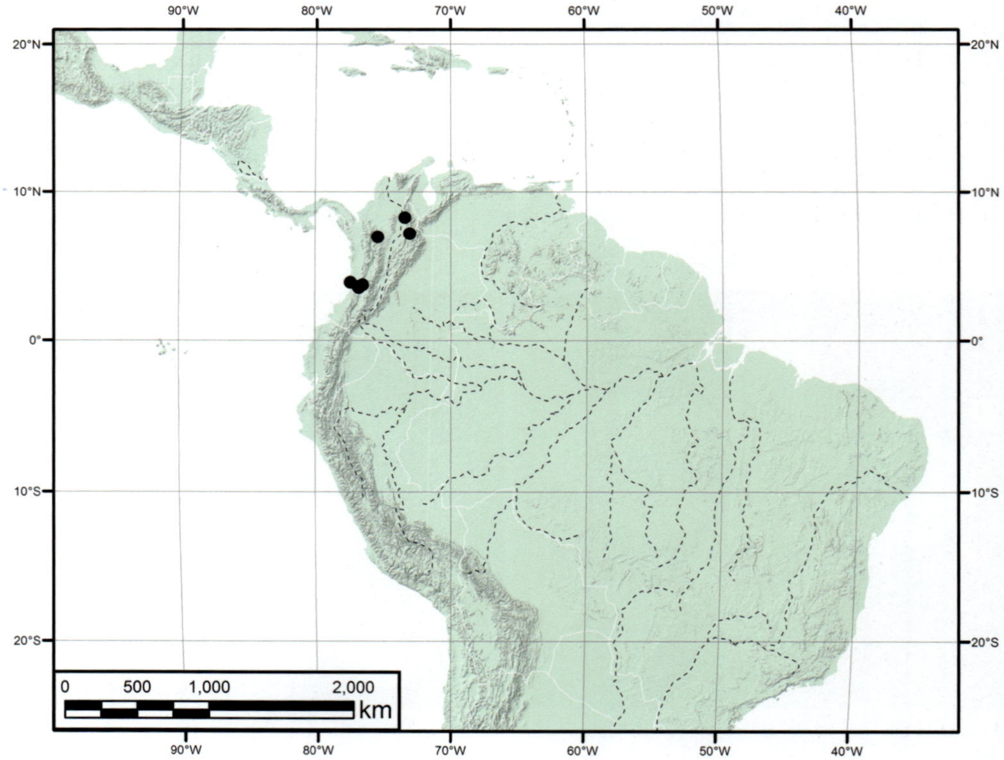

Map 20. Distribution of *Phragmipedium schlimii* in Colombia.

PHRAGMIPEDIUM
SECT. PLATYPETALUM

Phragmipedium section **Platypetalum** (Pfitzer) Garay in Orchid Dig. 43, 4: 133 (1979). Type: *Cypripedium lindleyanum* Lindl. (= *P. lindleyanum* (Schomb. ex Lindl.) Rolfe).
Phragmopedium sect. *Platypetalum* Pfitzer in Engler, Pflanzenr. 12: 43 (1903).
Paphiopedilum sect. *Platypetalum* (Pfitzer) V.A. Albert & Börge Pett. in Lindleyana 9, 2: 137 (1994).
Phragmipedium subgenus *Platypetalum* (Pfitzer) Braem in Richardiana 12, 1: 20 (2011), **syn. nov.**

A section of two large extra-Andean species ranging from Venezuela and the Guyanas to Brazil and lowland Bolivia. The species are characterised by their elongate multi-flowered inflorescences with bare scapes and bearing simultaneously several flowers with spreading spatulate petals.

8. PHRAGMIPEDIUM LINDLEYANUM

John Lindley described this species in 1840 as *Cypripedium lindleyanum*, a name given it by Robert Hermann Schomburgk (1804–1865) who discovered this large species on Mt Roraima in Guyana (then British Guiana) in the late 1830s. Schomburgk's original field sketch is attached to the type sheet in Lindley's herbarium at Kew. He found it growing amongst species of *Drosera*, *Heliamphora*, *Utricularia* and *Xyris*, the first three being carnivorous plants. In describing this species, Lindley commented "I unwillingly consent, at the particular instance of Mr Schomburgk, to allow this plant to bear my name, who have no title to the compliment."

N.E. Brown (1885) described *P. kaieteurum* (as *Selenipedium kaieteurum*) based on a specimen from the Kaieteur Falls in Guyana. He considered it distinct from the *P. lindleyanum* by reason of its leaf shape and colour, more glabrous sheaths and bracts and differently coloured flowers. Reichenbach (1886) treated it as conspecific with *P. lindleyanum*. However, Garay (1979) also considered it to be distinct from the sympatric *P. lindleyanum*, on the basis of its shorter ligulate leaves with yellow margins, the edge of the lip orifice being spotted in front and the triangular-rhombic staminode which is subtruncate-apiculate at the tip. Although sympatric with *P. lindleyanum* it appears to be far more restricted in its range. It is likely to be an ecological variant of P. *lindleyanum* and we are inclined to follow Cogniaux (1893) who considered it to be a variety of *P.* (as *Selenipedium*) *lindleyanum*. Both were illustrated by Dunsterville (K, illustrations collection).

Phragmipedium lindleyanum and *P. sargentianum* (Fowlie, 1986), are very closely related and have similar habit and floral morphology. McCook (1989a) argued that a range of slight morphological differences can be observed which grade between flowers

Plate 17. *Phragmipedium lindleyanum*. Drawn by Jeanne Koch for N.E. Brown from a plant cultivated at Kew (Kew collection).

and plants. She argued that the characters listed by Rolfe for *P. sargentianum* are not consistently different from those of *P. lindleyanum*. However, other characters have been found to delimit these taxa. This species is distinguished from *P. sargentianum* by its green leaves with yellow margins, and its flowers which have an elliptic dorsal sepal rather than a lanceolate one, a more bulbous lip with a subquadrate orifice and a deltoid rather than subquadrate staminode that is as long as broad and has an apical acumen. Dunsterville's unpublished illustration comparing *P. lindleyanum* var. *kaieteurum* with *P. sargentianum* demonstrates the minor differences between them.

Gorinsky (1973) described in detail an adventurous visit to the habitat of *P. lindleyanum* in the Kanuku Mountains of Guyana where he found plants growing in semi-shade on the ground and as low epiphytes in scrubby vegetation below bare black rock facing north-east areas at 1000 m where water seepages oozed through the fallen leaves of *Clusia rosea*. He noted that the main flowering period was from March to August.

DESCRIPTION. Terrestrial or lithophytic *herb* up to 2.5 m tall; rhizomes large, up to 2 cm in diam.; roots up to 5 mm in diam., densely woolly, striate. *Leaves* lorate to elliptic-lanceolate, acuminate, often minutely and roundly bilobed at apex, 27–110 × 3–7.3 cm, glabrous, brittle, base often red, the margins yellow; lower mid-vein enlarged and sharp; margins revolute, shiny and stiff, usually yellow, comprised of perpendicularly elongate ranks of cells with a heavy cuticle which are visible with a hand lens. *Inflorescence* racemose or a large panicle, 55–220 cm tall, 10–30-flowered; peduncle up to 1 cm in diam., reddish brown, hirsute to rusty hairy; branches of peduncle erect, ascending, often zigzag; floral bracts ovate to linearovate, 2–6 × 1.3–2.4 cm, shorter than ovary, green to rusty brown and often veined in red-brown, midvein scarcely more prominent than lateral veins, hirsute to villous on both surfaces; hairs may have bulbous terminal cell, or only doughnut-shaped base may remain; margins glabrous to ciliate, not fused basally; apex acute, mucronate; sterile bracts subtending panicle branches ovate, larger than floral bracts, but generally resembling them in pubescence and margin characters; sterile lower bracts 2–5, lanceolate to ovate-lanceolate, sharply keeled and generally resembling basal leaves, 6–15 × 2.5–5.5 cm, lowest 5–25 cm from plant base; margins as in basal leaves; apex acute to mucronate. *Flowers* yellow to olive green, suffused with reddish brown, the

Plate 18. *Phragmipedium lindleyanum.* Grown by W. Kimball of Rochester, New York. September 1888. Artist unknown (Kew collection)

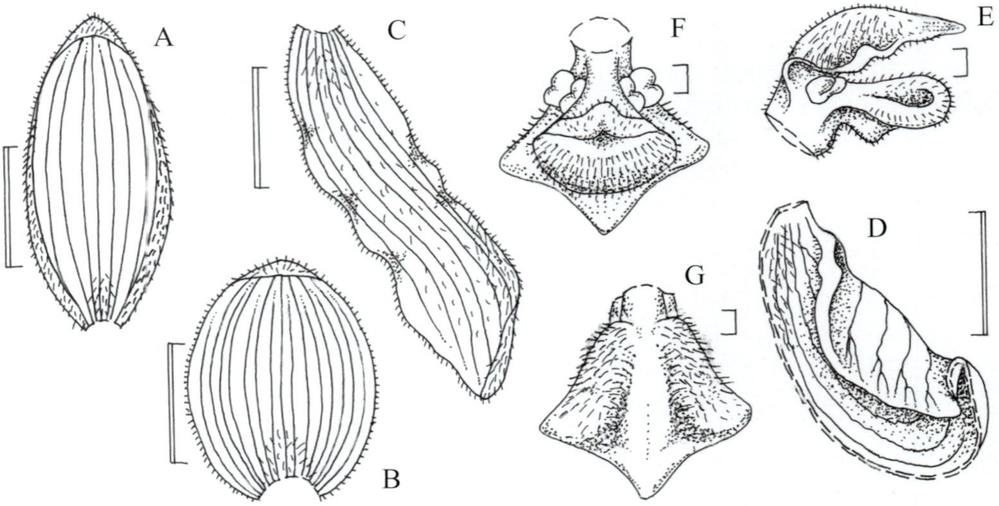

Fig. 95. ***Phragmipedium lindleyanum.*** **A**, dorsal sepal; **B**, synsepal; **C**, petal; **D**, lip, longitudinal section; **E**, column, side view; **F**, column from beneath; **G**, staminode. All drawn by Judi Stone from Kew Spirit Collection no. 48153. Single bar = 1 mm; double bar = 10 mm.

lip rim yellow-green with red-brown, the lateral rim olive-green to yellow, suffused with small brownish red spots and a white chevron-shaped spot at the marginal fusion, often green below the chevron, the claw interior blotched with reddish brown, the outer sepal surfaces mostly pubescent, the inner surfaces glabrous to finely pubescent, the margins glabrous to ciliate; ovary and pedicel 7–10 cm long, hirsute to densely villous, reddish brown. *Dorsal sepal* ovate to elliptic, obtuse, minutely hooded at apex, 28–40 × 12–22 mm, light green with 14–18 × 12–20 mm, with green or light green nerves; outer surface pubescent, enlarged glandular hair bases may remain after hairs are gone; margins often undulate, glabrous to apically ciliate; synsepal elliptic, often reflexed, obtuse, minutely hooded at apex, 24–32 × 16–26 mm, shorter than lip, green with 14–17 purple-brown veins; margins ciliate

Fig. 96. *Phragmipedium lindleyanum.* Flower of plant, growing on Auyantepui, Bolivar State, Venezuela (EK).'

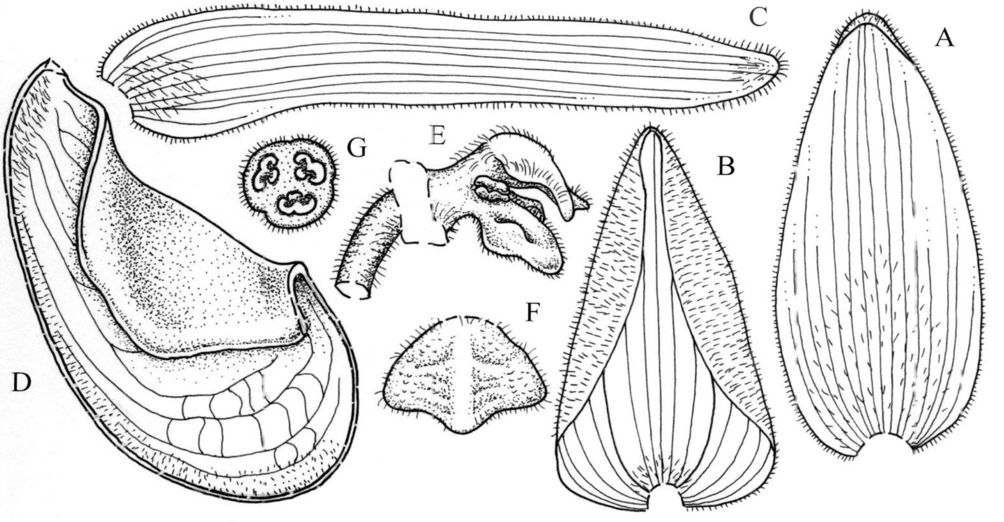

Fig. 97. ***Phragmipedium lindleyanum*** var. ***kaieteurum.*** **A**, dorsal sepal; **B**, synsepal; **C**, petal; **D**, lip, longitudinal section; **E**, column, side view; **F**, staminode; **G**, ovary, transverse section. All drawn by Paul Harwood from cult. Kunisch (Kew Spirit Collection no. 48153). A–D, × 3; E–G, × 5.

to glabrous, often undulate. *Petals* oblong to oblanceolate, obtuse to minutely retuse at apex, 30–50 × 5–8 mm, green or yellow-green basally, often becoming white apically with reddish brown longitudinal stripes, flat to twisting one revolution, held at 45° angle from lip, the inner surface mostly glabrous, basal hairs long, white to pink, the margins long-ciliate with dark multi-cellular hairs, more so apically, undulate in older flowers. *Lip* elliptic, round to pointed at the tip, 25–40 × 15–18 mm, mouth shallowly formed, round to slightly rhombic, the apical rim minutely pubescent, the apical lining 5 mm, the lateral rim minutely pubescent, the claw interior pubescent. *Column* 4 mm long, white to yellow-green, pubescent; staminode 3-lobed, rhombic to triangular, narrow basally, acute to truncate at apex, 4–5 × 5 mm, white to yellow, its pubescence pinkish-red, multi-cellular, long, scattered laterally and basally, but concentrated at margins; stigma rounded, its median lobe 5 mm wide. *Capsule* 5.5–6.5 cm long, with dark brown hairs, flaring at apex, edges curling outward. 2n=22 (Karasawa, 1980). **Plates 17 & 18. Figs. 95–99.**

Phragmipedium lindleyanum (R.H. Schomb. ex Lindl.) Rolfe in Orchid Rev. 4: 332 (1896). Type: Venezuela, Bolívar, Roraima, *Robert Schomburgk s.n.* (lecto. K!; syn. K!, L!).
Cypripedium lindleyanum R.H. Schomb. ex Lindl., Gen. Sp. Orch. Pl. 531 (1840).
Selenipedium lindleyanum (R.H. Schomb.) Rchb.f., Xenia Orch. 1: 8 (1854).

DISTRIBUTION. Venezuela, Guyana, French Guyana and Surinam. **Map 21.**

HABITAT. In wet meadows, scrub and forest along streams and below bare north-east facing rocks where water seepages occur; 600–2300 m.

PHENOLOGY. Flowering from January to August.

CONSERVATION STATUS. Vulnerable (VU): B2ab(ii,iii,v). It is local over a broad range in Venezuela, Guyana, Surinam, French Guyana and northern Brazil. The species is no longer collected very often as it has been in cultivation for many years and is not that popular in horticulture because it tends to grow into a large straggling plant. Its small and isolated populations are vulnerable to habitat change and destruction. The abundance of the species is likely to have been significantly reduced in recent decades due to deforestation, habitat degradation, human disturbance, agriculture expansion, the destruction of the watershed areas and the industrial activity. It is possible that the number of mature individuals is declining. The estimated area of occupancy of the species is less than 2000 km^2 with an estimated continuing decline in the number of mature individuals and the quality of habitats in all the locations.

ETYMOLOGY. Named for John Lindley (1799–1865), the father of orchid science.

Fig. 98. *Phragmipedium lindleyanum*. Auyantepui, Venezuela (EK)

Fig. 99. *Phragmipedium lindleyanum*. Cultivated at RBG Kew. (PD)

8a. var. kaieteurum (N.E. Br.) Pfitzer in Engler, Pflanzenr., 45 (1903), as '*P. kajeteurum*'.
Selenipedium kaieteurum N.E. Br. in Gard. Chron. n.s. 24: 262 (1885). Type: Guyana, Essequibo, Potaro River below Kaieteur Falls, *Jenman* 879 (lecto. K!; syn., AMES!, K!, W!).
S. lindleyanum (Lindl.) Rchb.f. var. *kaieteurum* (N.E. Br.) Cogn., Fl. Bras., Orchid. 1, 3: 14 (1893).
Phragmipedium kaieteurum (N.E. Br.) Garay in Orchid Dig. 43: 136 (1979).

Differs from the typical variety by the colour and shape of its leaves which are much shorter and have a yellow margin, slightly more reflexed and twisted petals, blunter lip with an obtuse rather than acute apex and a truncate shortly acuminate apex. **Plate 19.**

DISTRIBUTION. Venezuela (Gran Sabana, Cerro Venamo).

HABITAT. In scrub and rocks, often in the spray-zone of waterfalls; 400–600 m.

CONSERVATION STATUS. Vulnerable (VU): B2ab(ii,iii,v). It is known from a limited region in Venezuela. It is not that popular in horticulture. Its small and isolated population is vulnerable to habitat change and destruction.

ETYMOLOGY. Named for Kaieteur Falls, the type locality.

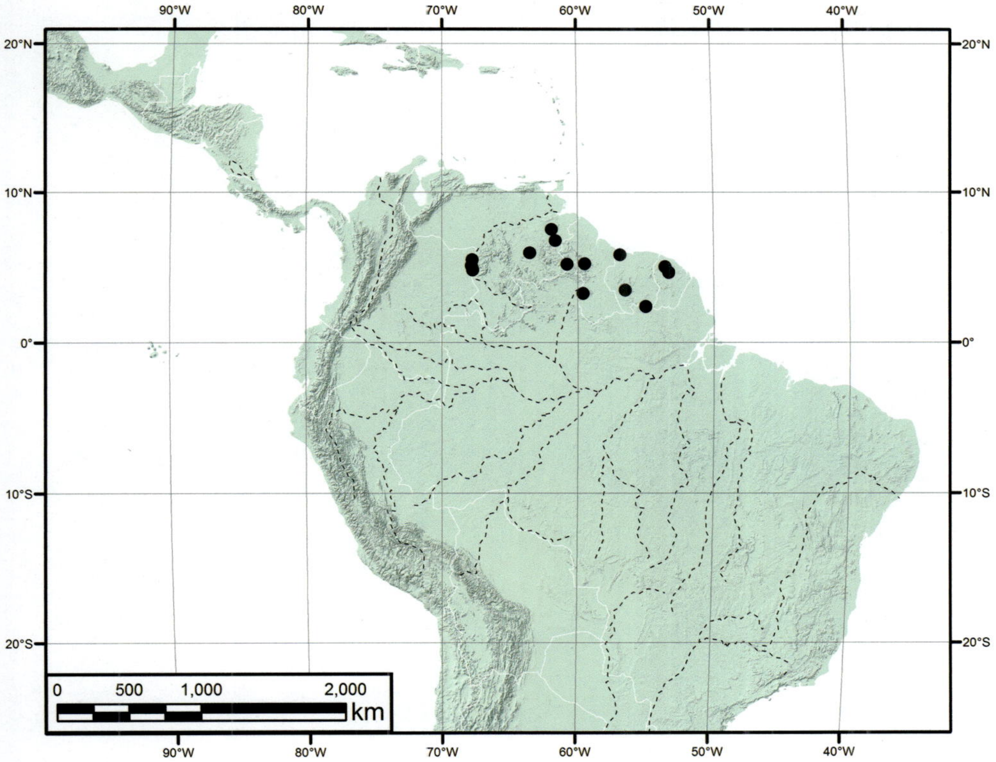

Map 21. Distribution of *Phragmipedium lindleyanum* in northern South America.

Plate 19. *Phragmipedium lindleyanum* var. *kaieteurum*. Drawn by Carol Woodin from a plant cultivated by Lou Hegedus, Fort Collins, Colorado.

9. PHRAGMIPEDIUM SARGENTIANUM

In 1893, Robert Rolfe described this magnificent plant as *Selenipedium sargentianum* after Charles Sprague Sargent (1841–1927), Director of the Arnold Arboretum at Harvard University, and an expert on temperate trees and shrubs. The type specimen was collected in Pernambuco in Brazil by one of Frederick Sander's collectors. Rolfe transferred it to the present genus the following year.

Phragmipedium sargentianum is very closely allied to *P. lindleyanum* and has variously been treated as conspecific with it (McCook, 1989 a & b, 1990, 1998) or as a distinct species (Garay, 1979; see comments under previous species). However, its populations are very remote from those of *P. lindleyanum*, by a distance of about 2500 km. It has entirely green leaves, rather than ones with yellow margins, and its flowers have an elliptic rather than a lanceolate dorsal sepal, a less bulbous lip and a subquadrate rather than deltoid staminode that is as broad as long and lacks an apical acumen.

This species exists in two discrete areas in north-eastern Brazil: in Pernambuco, where the type collection originated, and adjacent Algoas and in central and southern Bahia, some 600 km to the south. The Bahian plants have broader red petals than those from further north.

Ghillany (1972) provided a detailed and entertaining account of a locality near São Caetano do Norte in Pernambuco where a population of large plants grew at 850–950 m in marshy seepages in forest on a steep east-facing ridge.

DESCRIPTION. Terrestrial or lithophytic *herb* up to 2.5 m tall; rhizomes large, up to 2 cm in diam.; roots up to 5 mm in diam., densely woolly, striate. *Leaves* lorate to elliptic-lanceolate, 27–110 × 5–7.3 cm glabrous, brittle, uniformly green; lower mid-vein enlarged and sharp; margins revolute, shiny and stiff, usually yellow, comprised of perpendicularly elongate ranks of cells with a heavy cuticle which are visible with a hand lens; apex acuminate, but often minutely and roundly bilobed. *Inflorescence* a tall raceme to a large panicle, 27–110 × 5–7.3 cm tall, 10–30-flowered; peduncle often sturdy, branching, up to 1 cm in diam., reddish brown, hirsute to rusty hairy, the branches erect, ascending, often zig-zag; floral bracts ovate to linear-ovate, acute, mucronate, 2–6 × 1.3–2.4 cm, shorter than ovary, green to rusty brown and often veined in red-brown, mid-vein scarcely more prominent than lateral veins, hirsute to villous on both surfaces; hairs may have bulbous terminal cell, or only doughnut-shaped base may remain; margins glabrous to ciliate, not fused basally; sterile bracts subtending panicle branches ovate, larger than floral bracts, but generally resembling them in pubescence and margin characters; sterile lower bracts 25, lanceolate to ovate lanceolate, acute to mucronate, sharply keeled and generally resembling basal leaves, 6–15 × 1.5–5.5 cm, lowest 5–25 cm from plant base; margins as in basal leaves. *Flowers* yellow to olive-green, suffused with reddish brown, the dorsal sepal green with 14–17 purple-brown veins, the petals green or yellow-green basally, often becoming white apically with reddish brown longitudinal stripes, the lip yellow-green with a red-brown mouth, the apical rim often spotted, the lateral rim olive-green to yellow, suffused with small brownish red spots, white chevron-shaped spot at margin fusion, often green below chevron, the claw blotched with reddish brown; outer calyx surfaces mostly pubescent; inner surfaces glabrous to finely pubescent, floral vestiture ranges from 2-celled and pointed to 8-celled with a globose terminal cell; margins glabrous to ciliate; ovary and pedicel 70–100 mm long, hirsute to densely villous, reddish brown. *Dorsal sepal* ovate to elliptic, obtuse, minutely hooded at apex, 28–40 × 12–20 mm, light green with 14–18 green or reddish brown nerves; outer surface pubescent, enlarged glandular hair bases may remain after hairs are gone; margins often undulate, glabrous to apically ciliate; synsepal elliptic, often reflexed, obtuse, minutely hooded at apex, 24–32 × 10–20 mm, shorter than lip; margins ciliate to glabrous, often undulate. *Petals* oblong to oblanceolate, obtuse to minutely retuse at apex, 30–50 × 5–8 mm, flat to twisting one revolution, held at 45° angle from lip, the inner surface mostly glabrous, basal hairs long, white to pink, the margins with long ciliae ciliate with dark multi-cellular hairs, more so apically, undulate in older. *Lip* elliptic, round to pointed at the tip, 25–40 × 15–18 mm, yellow-green with red-brown mouth shallowly formed, round to slightly rhombic, the apical rim minutely pubescent, often spotted, the apical lining 5 mm, the lateral rim minutely pubescent, olive-green to yellow, suffused with small brownish red spots white chevron-shaped spot at margin fusion, often green below chevron, the claw interior pubescent, blotched with reddish brown. *Column* 4 mm long, white to yellow-green, pubescent; staminode 3-lobed, subquadrate, narrow basally, umbonate at tip, 4–5 × 5 mm, white with a green apical two-thirds, with red multi-cellular hairs, long, scattered laterally and basally, but concentrated at margins; stigma rounded, its median lobe 5 mm wide. *Capsule* 5.5–6.5 cm long, dark brown-pilose, point of floral attachment flaring, suture edges curling outward. 2n=22 (Karasawa, 1980). **Plates 20 & 21. Figs 100–103.**

Phragmipedium sargentianum (Rolfe) Rolfe in Orchid Rev. 4: 332 (1896). Type: Brazil, Pernambuco, cult. Sander, 1892 (lecto. K; syn. AMES!).
Selenipedium sargentianum Rolfe in Orchid Rev. 1: 239 (1893).

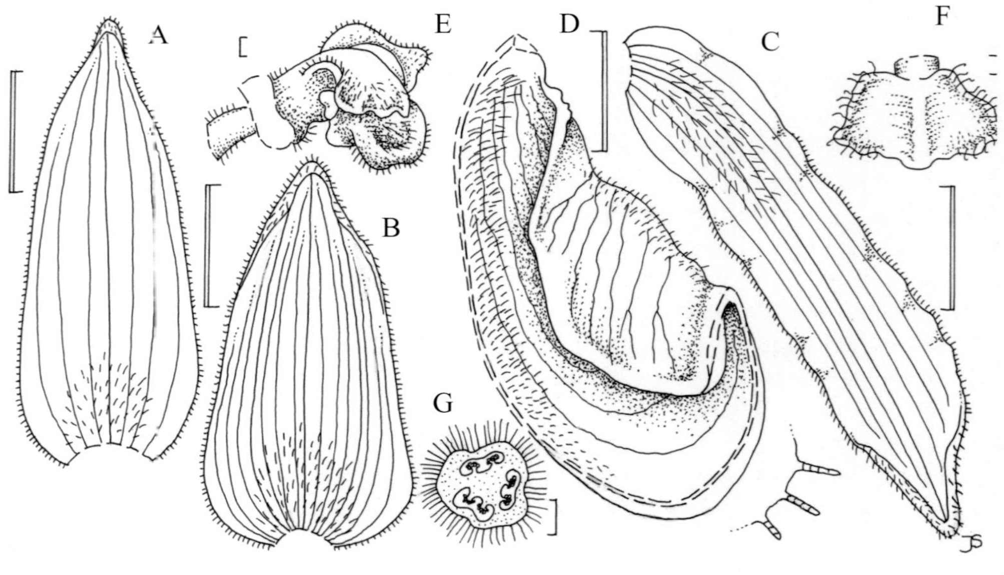

Fig. 100. ***Phragmipedium sargentianum.*** **A**, dorsal sepal; **B**, synsepal; **C**, petal; **D**, lip, longitudinal section; **E**, column, side view; **F**, staminode; **G**, ovary, transverse section. All drawn by Judi Stone from cult. E. Ross (Kew Spirit Collection no. 48112). Single bar = 1 mm; double bar = 10 mm.

Plate 20. *Phragmipedium sargentianum*. Drawn by Carol Woodin from a plant cultivated by Lynn and Bob Wellenstein, Candor, New York.

Plate 21. *Phragmipedium sargentianum*. Drawn by Matilda Smith for *Curtis's Botanical Magazine* t. 7446 (1895) from a plant cultivated at Kew.

Fig. 101 (above). *Phragmipedium sargentianum*. Inflorescence. Bahia State, Brazil. (CB).
Figs. 102 & 103 (below left & right). *Phragmipedium sargentianum.* Flowers. Bahia State, Brazil. (CB).

Paphiopedilum sargentianum (Rolfe) Hallier in Ann. Jard. Bot. Buitenzorg 14: 46 (1897).
Cypripedium sargentianum (Rolfe) Kraenzl., Orchid Gen. Sp. 1: 38 (1897).
Phragmopedilum sargentianum (Rolfe) Pfitzer in Bot. Jahrb. Syst. 25: 527 (1898).

DISTRIBUTION. NE Brazil: Pernambuco, Alagoas and Bahia states. **Map 22.**

HABITAT. On rocks in open areas on mountain tops, in damp humus in light shade in areas of constant seepage, in swampy seepages in forest on steep ridges; 850–950 m.

PHENOLOGY. Flowering over a long period.

CONSERVATION STATUS. Endangered (EN): B1ab(ii,iii,v)+2ab(ii,iii,v). It is a very local and rare species with a restricted distribution in north-eastern Brazil. Its known populations are widely separated. Much of its original habitat has been cleared for agriculture, mining and other activities. The overall population appears to be decreasing, the number of mature individuals is fairly low but some subpopulations still have hundreds of plants. It is threatened by habitat change, erosion, deforestation, mining activity and human disturbance. The estimated extent of occurrence and the estimated area of occupancy of the species are less than 5000 km^2 and 500 km^2, respectively, with an estimated continuing decline in the number of mature individuals and the quality of habitats.

ETYMOLOGY. Named for Professor Charles Sprague Sargent (1841–1927), Director of the Harvard University Arboretum.

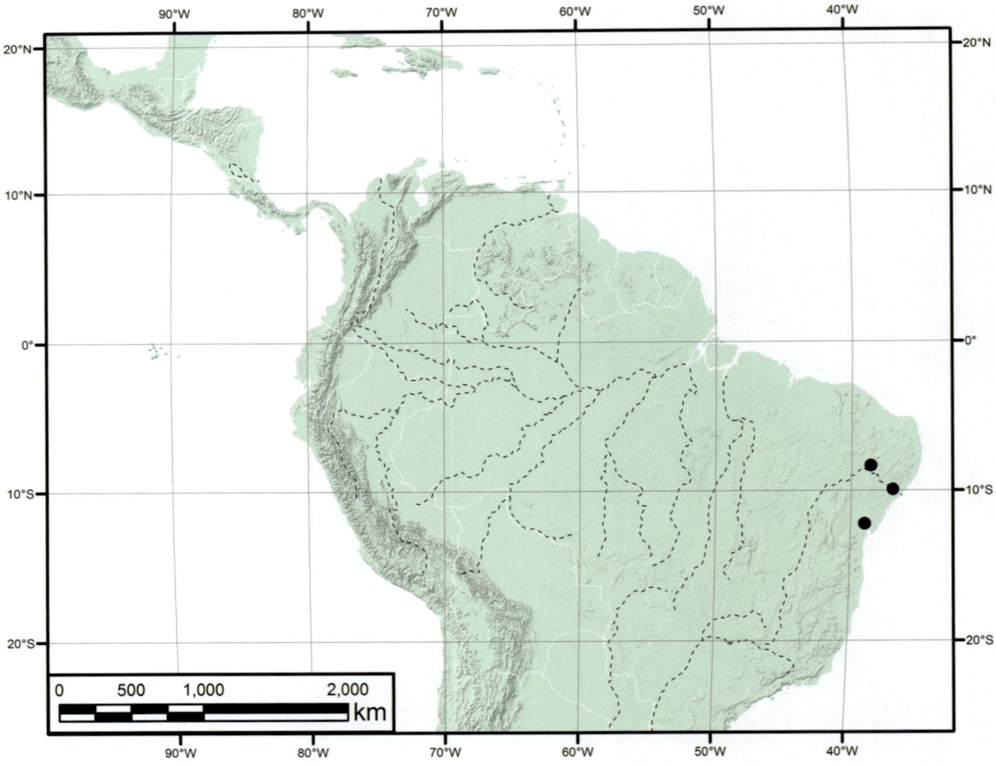

Map 22. Distribution of *Phragmipedium sargentianum* in Brazil.

PHRAGMIPEDIUM SECTION LORIFOLIA

Phragmipedium section **Lorifolia** (Kraenzl.) Garay in Orchid Dig. 43, 4: 141 (1979). Type: *Cypripedium longifolium* Rchb.f. & Warsz. (=*Phragmipedium longifolium* (Rchb.f. & Warsz.) Rolfe).

Phragmopedilum section *Himantopedilum* Pfitzer in Engler, Pflanzenr. 12, 43 (1896), nom. superfl.

Cypripedium section *Lorifolia* Kraenzl., Orchid. Gen. Sp. 1: 37 (1897).

Phragmipedium section *Himantopetalum* (Hallier) Garay in Orchid Dig. 140 (1979), in part. Lectotype (invalid): *Cypripedium caricinum* Lindl. (=*Phragmipedium caricinum* (Lindl.) Rolfe, selected by Garay (1979)).

Paphiopedilum section *Himantopetalum* Hallier in Jard. Bot. Buit. 14: 44 (1896).

Phragmipedium subgenus *Himantopetalum* (Hallier) Braem in Richardiana 21, 1: 22 (2011).

P. subgenus *Longifolium* Braem in Richardiana 21, 1: 22 (2011), **syn. nov.** Type: *P. longifolium* (Rchb.f. & Warsz.) Rolfe.

P. section *Viridiflorae* Braem in Richardiana 21, 1: 22 (2011), **syn. nov.** Type: *Phragmipedium boissierianum* (Rchb. f.) Rolfe.

McCook (1989a) included section *Himantopetalum* within the typical section, noting that *P. klotzschianum* shares dark pigmented staminode lobes and a rounded lip apex with *P. caudatum* and its allies. She averred that characters had not been found to support an independent clade for the species of sect. *Himantopetalum* and that all of its species possessed a well-defined rim of multi-cellular hairs along the sides of the lip orifice that borders the ribbed area on the claw face like that found in species of the typical section (McCook, 1989a). However, the DNA analyses of Cox *et al.* (1997), based upon nuclear ribosomal ITS nucleotide data, and Rodriguez Salas *et al.* (2007) who used Amplified Fragment Length Polymorphisms (SFLP), both show that the species of sections *Lorifolia* and *Himantopetalum* that they sampled form a single clade, with the exception of *P. caricinum* (Cox *et al.*, 1997). Although *Himantopetalum* is the earlier name, its original circumscription included *P. caudatum*, the type of the genus, making it a superfluous name. Thus, its later lectotypification by Garay (1979), using *P. caricinum* Lindl., cannot be upheld.

We agree with McCook's analysis of *P. klotzschianum* and include it here in the typical section but have chosen to include the remaining species of section *Himantopetalum* species within this section since the DNA-based analyses consistently include *P. pearcei* in the same clade as *P. longifolium* and *P. boissierianum*. A more detailed analysis that includes all the species of the genus is needed to resolve this question.

This section comprises a group of closely allied species characterised by lax leaves, unbranched inflorescences in which the flowers are produced in succession, greenish flowers, often marked with purple or brown, tapering spreading petals, and a staminode that is transversely rhombic or very broadly ovate, usually with an upper margin of short dark hairs. The species cluster in two groups: those with a lip with a horn-like protuberance on each side of the basal part and lateral lobes that are not connate, and a group in which such horns are absent and the side lobes are connate.

10. PHRAGMIPEDIUM BOISSIERIANUM

H.G. Reichenbach described this attractive species, as *Selenipedium boissierianum*, in 1854 based upon a specimen collected in 1787 by José Pavon in Huancayo, Peru. Edmond Boissier (1810–1885), after whom this species is named, was a Swiss botanist, explorer and mathematician who travelled widely in Europe, North Africa and the Middle East, specialising in the study of their plants. He and his daughter and son-in-law accumulated a substantial herbarium that is now housed in the herbarium in Geneva. Robert Rolfe transferred it to the present genus in 1896. A beautiful watercolour illustration of the species by Francisco Pulgar, artist on the Ruiz and Pavon expedition, exists in the archives of the Jardín Botaníco de Madrid (Plate 1). Although McCook (1989) considered the two herbarium specimens of this species preserved in the Madrid herbarium (MA810806 & 810805) as isotypes of *S. boissierianum*, Pupulin (2012) noted that they were collected along the Maraynioc River by Vitoc seven years after the type in 1794.

Restricted to Andean Peru and southern Ecuador, this species varies somewhat throughout its range, probably because its populations are rather scattered and disjunct. It varies in flower colour, the degree of crimping and reflexion of the sepals and petals and size, lip shape, and shape and size of the staminode. Taxonomically, it has had a chequered career, having been described several times under different names from various parts of its range in Peru and Ecuador. In the same year as he described *S. boissierianum*, Reichenbach described *Selenipedium czerwiakowianum* based upon a Warzscewicz collection from Peru. Twenty years later he described *S. reticulatum* based upon a Smith collection from Ecuador. Finally, in 1921, Rudolf Schlechter described *P. cajamarcae* Schltr. based upon a Weberbauer collection made in April 1912 from Tal des Flusses, St Tabaconas, Cajamarca, in Peru. Most authors had accepted the conspecificity of all three with *P. boissierianum* until Garay (1979) resurrected *P. reticulatum* and *P. czerwiakowianum* in his illustrated synopsis of the genus, placing *P. cajamarcae* in the synonymy of the last.

Garay (1979) distinguished the three species as follows:

1. Synsepal apex, and often the crispate margins, well recurved; staminode wider than the base of the lip .... ***P. czerwiakowianum***
   Synsepal apex and margins not recurved; staminode narrower than the base of the lip 2

2. Staminode rhombic-triangular to cordate-triangular in outline, acuminate; lip without a central groove, rounded in front .... ***P. boissierianum***
   Staminode transversely elliptic-reniform in outline, pentagonal or hexagonal, obtuse; lip with a central groove, shallowly bilobed at front .... ***P. reticulatum***

Bennett & Christenson (1993) provided drawings of both *P. boissierianum* and *P. reticulatum* from Peru. They distinguished the former by its glabrous column, subrhombic-triangular

Plate 22. *Phragmipedium boissierianum*. Drawn by Carol Woodin from a plant cultivated by Terry Partin, Chicago.

staminode, thickened and raised down the centre and with an acute apex, and the undulate synsepal margins. The latter was distinguished by its crisped petal margins, the horn-like protuberances on the sides of the lip, the pentagonal staminode and the raised veins on the inner surface of the lip. However, even using these features and the photographs that accompany Garay's key, it is possible to see that characters such as staminode width and shape vary more or less continuously across the three taxa. Similarly, the shape and attitude of the apex and sides of the synsepal do not seem to be significant characters in delimitation of the species. We prefer to treat *P. boissierianum* as a widespread and variable species because the supposed distinctions merge into one another when sufficient specimens are examined.

Capsules are common on herbarium specimens, and in wild populations. However, pollinators are unknown. Atwood said that flowers grown in the greenhouse at the Marie Selby Botanical Gardens were autogamous (Atwood quoted by McCook, 1989a). Koopowitz (2008) confirmed that Ecuadorian populations of *P. boissierianum* are autogamous.

DESCRIPTION. Terrestrial or lithophytic *herb* up to more than 1 m tall; rhizomes short and thick; plants closely clumped; roots thick and woolly, up to 2 mm in diam. *Leaves* linear, 40–100 × 2–4.5 cm, glabrous, bases often purple; margins revolute; apex finely bifid. *Inflorescence* 3–10-flowered, racemose, occasionally branching, 35–208 cm tall; peduncle to 1 cm in diam., straight to zig-zag, brown, pilose, somewhat more so just below floral bracts, with uniseriate and multicellular hairs; floral bracts lanceolate to ovate-lanceolate, 6.5–10 × 2–3.6 cm, longer than ovary, strongly reticulately veined, the margins hyaline, veins not running to the margins, the apex mid-vein shorter than blade; sterile lower bracts 3–4, linear-lanceolate, 7–25 × 2–4 cm, lowest 5–11 cm from plant base; margins hyaline,

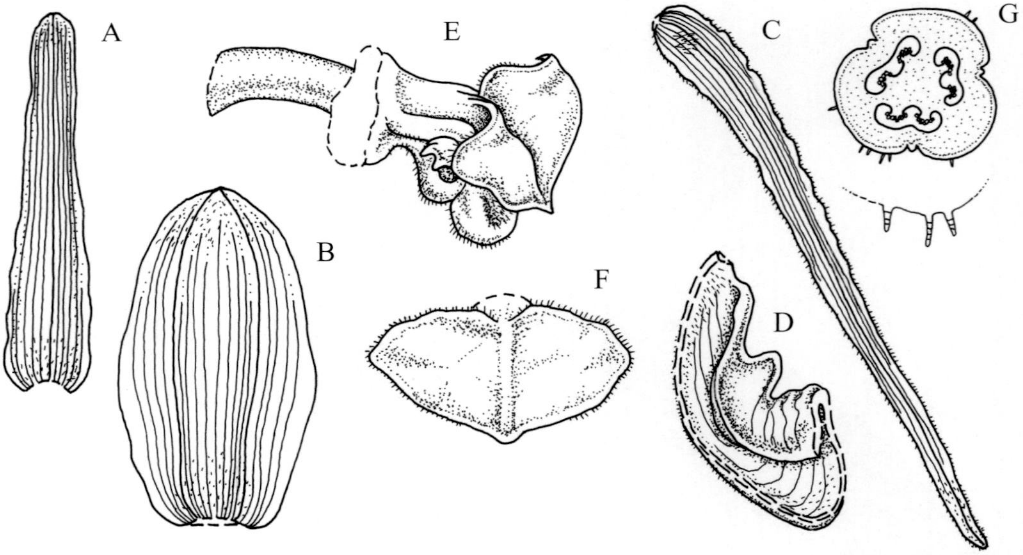

Fig. 104. ***Phragmipedium boissierianum.*** **A**, dorsal sepal; **B**, synsepal; **C**, petal; **D**, lip, longitudinal section; **E**, column, side view; **F**, staminode; **G**, ovary, transverse section; **H**, hairs on ovary. All drawn by Judi Stone from cult. Partin (Kew Spirit Collection no. 62738). A-D, × 1; E, F, × 3; G, × 6; H, × 48.

Figs. 105 & 106. *Phragmipedium boissierianum*. Plant growing on roadside bank, Moyombamba road, N Peru (PC).

apex minutely 3-lobed. *Flowers* green to greenish-yellow, the margins lighter coloured, entire and smooth to strongly undulate, varying with flower age, the outer surfaces mostly glabrous, the inner surfaces glabrous to pubescent; pedicel and ovary 40–55 mm long, light green-brown with short white hairs. *Dorsal sepal* oblong, broadly acute, 35–65 × 12–18 mm, yellow-green, 14–16 nerved, reticulately veined, sparsely glandular inside, the margins ciliate; synsepal ovate, obtuse, 35–65 × 20–28 mm, exceeding lip, yellow-green, 23–28-nerved, reticulately veined, sparsely glandular pubescent inside. *Petals* borne at 75° angle from lip, twisting with age, linear, acute to minutely retuse, 75–100 × 5–8 mm, green tinged with brownish purple, the inner surface shortly pilose, hairs uniseriate; basal hairs long, multicellular, uniseriate, to 0.5 mm long; margins strongly undulating with age, the apex covered in short, congested uniseriate hairs. *Lip* ovoid, 25–50 × 17–20 mm, green tinged with purple, mostly glabrous, the lip opening squared apically and notched, the notch shallow to 5 mm deep, the apical rim 8–10 mm deep, edged in minute hairs, the lateral rim glabrous; spurs present on lateral rim, rounded or pointed toward base, 6–10 mm high; lateral lip lining and 7–8 mm deep; claw face yellow-green to green with 8–12 brown, irregularly shaped spots; tubercles green; claw interior spotted, sparsely pilose. *Column* 5–6 mm long, green, covered in sparse, short, glandular hairs; staminode transversely rhombic to reniform, 5–7 × 7–12 mm, as wide to wider than base of lip, green, the basal margin often undulate or folded, lined with short, light coloured marks concentrated near the sides at the column, but continuing past the side lobes; apex apiculate or acuminate; stigma 0.5 cm wide. Capsule triangular in cross-section, 4.5–6.5 cm long, dark brown, glabrous to sparsely shortly white pubescent, curving 90° at apical end, dehiscing at the centre along 6 sutures, the two ends remaining intact. 2n=18 (Karasawa, 1980). **Plate 22. Figs 104–106.**

Phragmipedium boissierianum (Rchb.f.) Rolfe in Orchid Rev. 4: 332 (1896). Type: Peru, Huancayo, Pillao, 1787, *Ruiz & Pavon s.n.* (holo. G!; iso. BM!).

Cypripedium grandiflorum Pavon *in sched.*

Selenipedium boissierianum Rchb. f., Xenia Orch. 1: 3, 176 (1854).

S. czerwiakowianum Rchb. f. in Bonplandia 2: 116 (1854). Type: Peru, *Warszewicz s.n.* (holo. W!).

S. reticulatum Rchb.f., Xenia Orch. 2: 223 (1874). Type: Ecuador, *Smith s.n.* (lecto. W); Ecuador, along Rio Zamora, *Wallis s.n.* (para. W!, illustration).

Cypripedium boissierianum Rchb.f. in Gard. Chron. n.s. 18: 521 (1882).

C. czerwiakowianum (Rchb.f.) Rchb.f., Gard. Chron. n.s. 18: 521 (1882).

C. reticulatum (Rchb.f.) Rchb.f. in Gard. Chron. n.s. 18: 521 (1882).

Paphiopedilum boissierianum (Rchb.f.) Stein, Orchideenbuch: 456 (1892).

P. czerwiakowianum (Rchb.f.) Pfitzer in Bot. Jahrb. Syst. 19: 41 (1894).

P. reticulatum (Rchb.f.) Pfitzer in Bot. Jahrb. Syst.19: 41 (1894).

Phragmipedium czerwiakowianum (Rchb.f.) Rolfe in Orchid Rev. 4: 332 (1896); Pfitzer, Pflanzenr. 4(50): 50 (1903).

P. boissierianum (Rchb.f.) Rolfe var. *reticulatum* (Rchb.f.) Veitch in Pfitzer, Pflanzenr. 4, 50: 50 (1903).

P. reticulatum (Rchb.f.) Schltr. in Repert. Spec. Nov. Regni Veg. Beih. 8: 111 (1921); Garay, Flora Ecuador Orchid. 9: 22 (1979).

P. cajamarcae Schltr. in Repert. Spec. Nov. Regni Veg. Beih. 9: 41 (1921). Types: Peru, Cajamarca, Tal des Flusses, St Tabaconas, April 1912, *Weberbauer 6158* (holo. B, destroyed). Peru, San Martin: Tarapoto to Yurimaguas, 12 Oct. 1981, *Besse, Halton & Luther* 787 (neo. SEL!).

DISTRIBUTION. Southern Ecuador and Peru, along the eastern cordillera of the Andes. **Map 23**.

HABITAT. Usually in open sun on road cuts or rocky slopes, or on the edge of wet forest along roads or streams; 400 and 1800 m.

PHENOLOGY. In most months of the year.

CONSERVATION STATUS. Least Concern (LC). It is a local species with a broad distribution area along the eastern cordillera of the Andes in southern Ecuador and Peru. The trend of the population is stable, the number of mature individuals in some populations is large and the estimated extent of occurrence is more than 20000 km^2. The species is nowadays of limited horticultural interest but populations can be threatened by anthropogenic habitat degradation, deforestation, destruction of watershed areas and mountain erosion.

ETYMOLOGY. Named after the Swiss botanist Pierre Edmond Boissier (1810–1885), who held the holotype in his herbarium. Reichenbach (1882) called him "the modern father of Spanish and oriental botany".

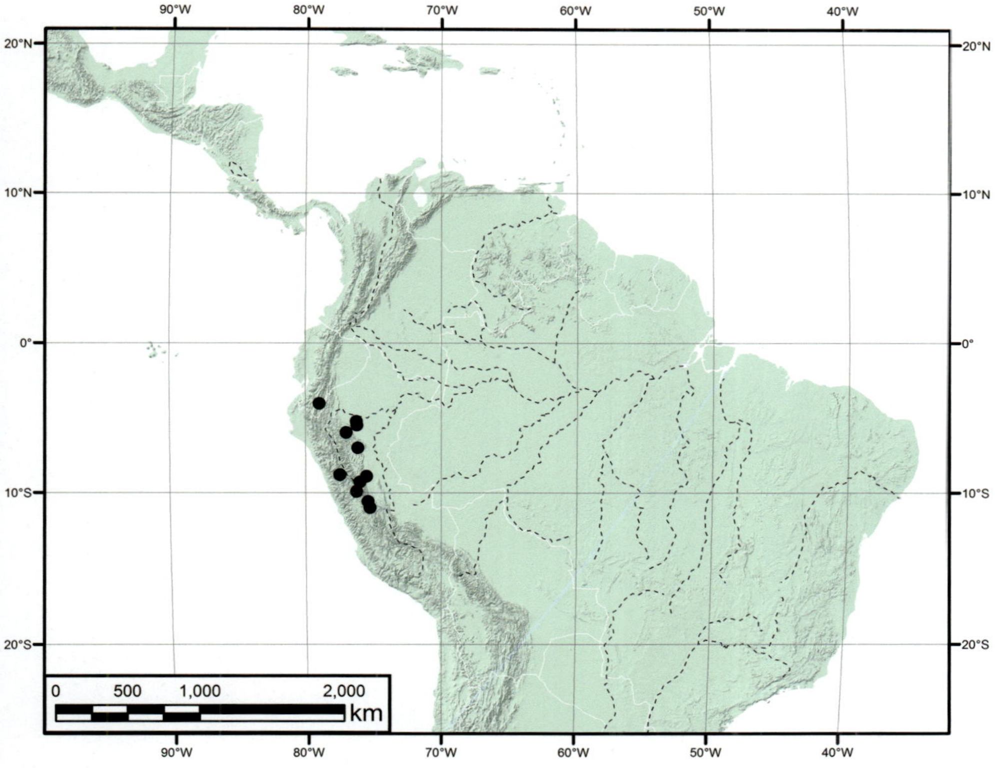

Map 23. Distribution of *Phragmipedium boissierianum* in the Andes of Ecuador and Peru.

11. PHRAGMIPEDIUM HIRTZII

Calaway Dodson described *Phragmipedium hirtzii* in 1987 based upon a plant found by the prolific Ecuadorian collector Alex Hirtz in NW Ecuador. He recognised its close affinity to *P. longifolium* which grows commonly nearby, especially on roadside verges, but it differs in several respects. Notably it is a smaller plant with grass-like thin leaves, up to 25 cm long and scarcely 10 mm broad, an inflorescence up to 40 cm long, almost pendent 7–8 times twisted petals, curled dorsal sepal, ivory lip face, and a glabrous staminode.

The type locality, a waterfall in northern Ecuador, is readily accessible and over-collected but plants still survived when one of us (PC) visited the site in 2007. The plants perched on large boulders at the top of a cascade in a fast-flowing stream above the road and are constantly wetted by the spray.

Gruss (2012) described forma *luteo-viride* (as *luteo-viridis*) which lacks any red-purple pigments, based on a plant cultivated in Germany.

Phragmipedium × *roethianum* O. Gruss & Kalina (1998) is now considered to be its natural hybrid with *P. longifolium*, a species which grows nearby in some quantity.

DESCRIPTION. Lithophytic *herbs* up to 40 cm tall; rhizome caulescent, up to 8 mm in diam., up to 12 cm between growths, covered with non-leafy sheaths, rooting at the nodes. *Leaves* distichous, leathery, linear, acuminate, up to 25 × 1 cm, carinate on the lower surface, sheathing and imbricate at the base. Inflorescence erect, up to 40 cm long; peduncle terete, bearing two or three distant appressed narrowly ovate, acuminate sheaths

Figs. 107 & 108. *Phragmipedium hirtzii.* Flower, two views, cultivated plant (HO, JL).

Plate 23. *Phragmipedium hirtzii*. Drawn by Carol Woodin from a plant cultivated by Joe Kunisch, Pittsford, New York.

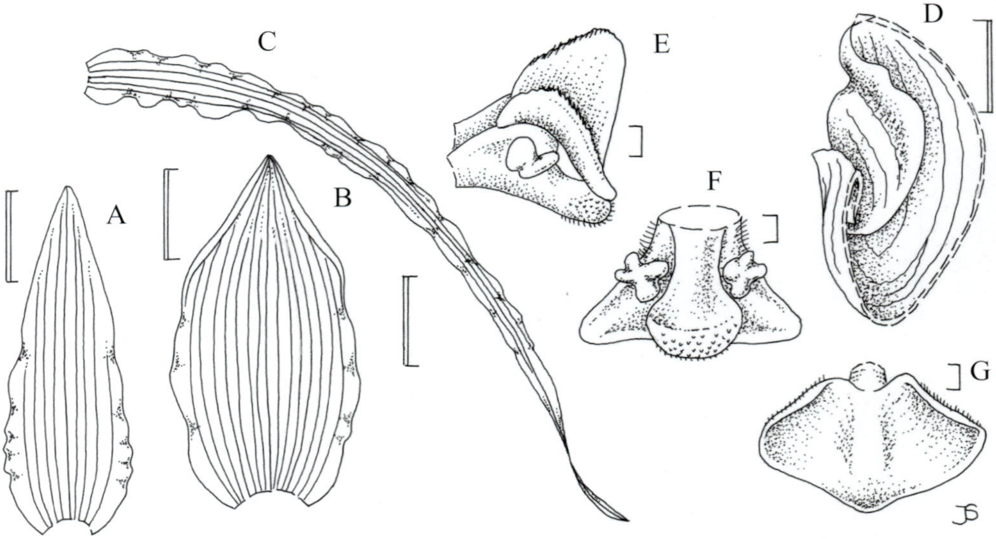

Fig. 109. ***Phragmipedium hirtzii.*** **A**, dorsal sepal; **B**, synsepal; **C**, petal; **D**, lip, longitudinal section; **E**, column, side view; **F**, column, from beneath; **G**, staminode. All drawn by Judi Stone from Kew Spirit Collection no. 62736. Single bar = 1 mm; double bar = 10 mm.

Fig. 110. Type locality of *Phragmipedium hirtzii* on Lita road, N Ecuador. (PC).

along its length. *Flowers* 5–7, produced in succession; dorsal sepal, synsepal and petals yellow-green to green with darker veins and usually white margins, tinged with purple towards the margins; lip green to purple with white side lobes spotted with green and purple; staminode pale to bright green; ovary 4–5 cm long, very shortly pubescent. *Dorsal sepal* suberect, ovate, acute, up to 38 × 12 mm, twisted towards the apex and with more or less undulate margins; *synsepal* concave, broadly ovate, obtuse, up to 32 × 22 mm. *Petals* pendent–spreading at about 60 degrees below the horizontal, linear-tapering to the apex, up to 70 × 6 mm, twisted spirally 7–8 times along length. *Lip* urceolate, oblong-obovoid, 35–40 × 15–17 mm, the side lobes with a blunt horn on each side, the margins involute. *Staminode* subcordate, lacking hairs and ciliae, up to 4 × 4 mm. **Plate 23. Figs 107–110.**

Phragmipedium hirtzii Dodson in Orchis 58: 129 (1988) & in Icones Platarum Tropicarum ser. 2, t. 0568 (1989). Type: Ecuador, Esmeraldas Prov., Lita, road from Ibarra to San Lorenzo, 600 m, 15 Aug. 1986, *Hirtz 2919* (holo. OCNE, iso. K!, MO!, RPSC).

DISTRIBUTION. NW Ecuador (Esmeraldas, Carchi, Imbabura) and SW Colombia only.

HABITAT. On large boulders in bed of montane streams and by waterfalls; 500–900 m.

PHENOLOGY. Flowering throughout the year. **Map 24.**

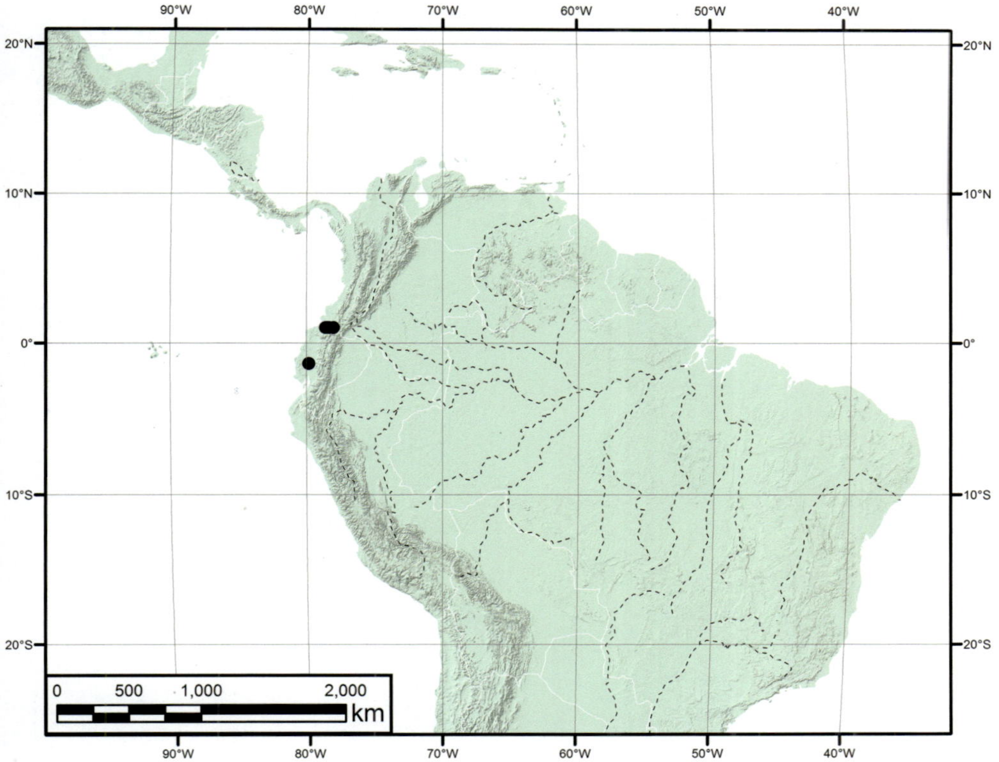

Map 24. Distribution of *Phragmipedium hirtzii* in Ecuador.

CONSERVATION STATUS. Endangered (EN): B1ab(ii,iii,v)+2ab(ii,iii,v). It is a rare and very local species with a distribution restricted to north-western Ecuador and possibly adjacent Colombia. The trend of the population is decreasing, the number of mature individuals is low and the abundance of the species is decreasing due to over-collection by local collectors and orchid vendors (especially in those populations near roads) and by anthropogenic changes to the rivers where it occurs. The estimated extent of occurrence and estimated area of occupancy of the species are less than 5000 km^2 and 500 km^2, respectively, with an estimated continuing decline in the number of mature individuals and the quality of habitats in its habitats.

ETYMOLOGY. Named after Alex Hirtz (born 1945), one of the most productive and perspicacious of all collectors of Ecuadorian orchids.

Phragmipedium hirtzii var. **anchicayense** (Braem, Tesón & Faust) P.J. Cribb **comb. et stat. nov.**

Phragmipedium anchicayense Braem, Tesón & Faust in Austr. Orchid Rev. 81, 5: 10 (2016). Type: Colombia, Anchicaya, Aug. 2016, *Tesón s.n.* (holo. COL).

DESCRIPTION. Leaves, linear, acute, up to 35 × 1–1.3 cm. Inflorescence 25–35 cm tall, usually carrying a single flower in succession; bracts up to 50 × 5 mm. Flowers 16.5–17.5 cm across, pale yellow-green with purple margins to the petals and a purple-flushed lip, heavily purple-spotted on the inturned side lobes. Dorsal sepal narrowly oblong-ovate, acute, up to 62 × 15 mm, with recurved sides; synsepal oblong-ovate, obtuse, up to 70 ×

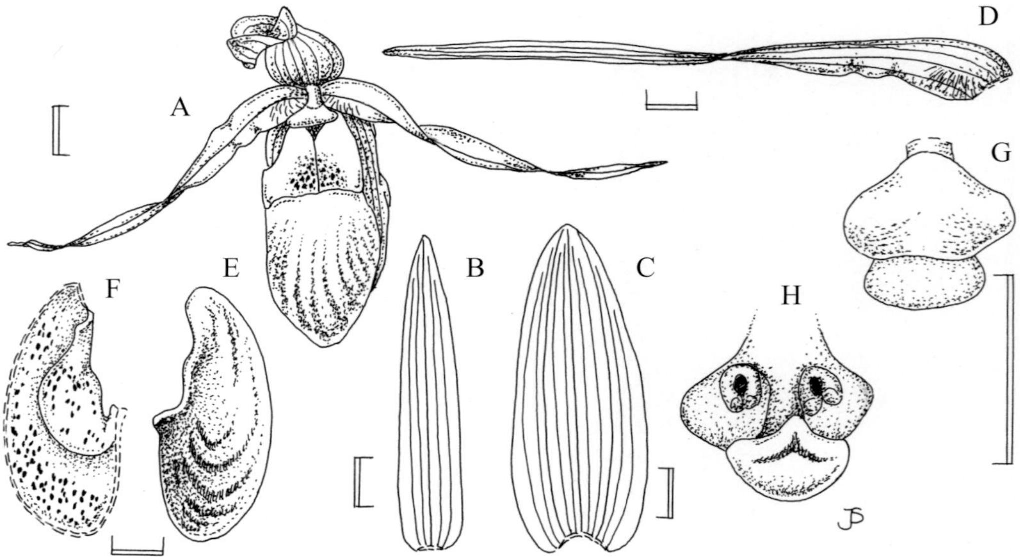

Fig. 111. ***Phragmipedium hirtzii*** var. ***anchicayense***. **A**, flower; **B**, dorsal sepal; **C**, lateral sepals; **D**, petals; **E**,**F**, lip, side and longitudinal section; **G**,**H**, column, from above and below. All drawn by Judi Stone: A after photograph by D. Mejia; B–H after drawings by D. Téson. Scale bar = 10 mm.

27 mm. Petals spreading widely, linear-tapering rounded at tip, up to 115 × 12 mm, with purple hairs at base. Lip deeply saccate, up to 55 × 24 mm; side lobes obliquely elliptic, strongly incurved, touching under staminode; apex straight, outcurved. Column 10–12 mm long; staminode transversely elliptic, 5 × 9 mm, glabrous on the basal margin. **Fig. 111.**

DISTRIBUTION. SW Colombia, Choco region.

HABITAT. Grows on rocky river banks and on boulders nearby in hill rain forest; c. 500 m.

This recently described taxon is close to *Phragmipedium hirtzii*, differing mainly in the less twisted, more widely spreading petals with scarcely crispate margins and the less jagged lip side lobes. The petals bear a similarity to those of the common *P. longifolium*, a species that is widespread in Colombia. The lip apex is also contrasted with that of *P. hirtzii* but, as acknowledged, it varies in typical *P. hirtzii* and cannot be used to distinguish the former. Bearing in mind its many similarities to the Ecuadorian species, we consider it reasonable to treat it as a variety of that species.

Braem *et al.* (2016) based their description on a plant collected by Eliseo Tesón in the Anchicaya canyon in the Choco region of south-western Colombia. This is somewhat to the north of the known distribution of typical *P. hirtzii*, but its habitat is similar. He reported that its population is quite large.

12. PHRAGMIPEDIUM LONGIFOLIUM

Phragmipedium longifolium is the most widespread and variable species in the genus. The type specimen was discovered in 1848–1849 by Josef Ritter von Warszewicz, the prolific German orchid collector, in wet and marshy places in the Cordillera Chiriqui in Panama, an area rich in orchids. Warszewicz and H.G. Reichenbach described it in 1852 as *Cypripedium longifolium*. It was eventually transferred to the present genus by Robert Rolfe in 1896. Further collections, made by A. Endres in Panama and Berthold Seeman in the Darien area of northern Colombia, were respectively named as *C. reichenbachii* and *C. dariense*. The former was never formally described.

In 1871, Reichenbach described plants, collected by Benedict Roezl in the Cauca Valley in Colombia for Messrs Sander & Sons, as *Selenipedium roezlii* which he distinguished by its more robust habit. Garay (1979) distinguished it by the anterior margin of the lip being erose-denticulate to prominently rugose-tuberculate.

Reichenbach (1878) also described *Cypripedium hincksianum*, based upon a plant collected by Gustav Wallis at Cape Darien in northern Colombia and flowered in England by T.C. Hincks.

Campacci and Takase (2000) described *P. chapadense* based upon a specimen from the Chapada dos Veadeiros in Goias State in Brazil, a locality far removed from the main populations of *P. longifolium*. The authors distinguished *P. chapadense* by its brightly coloured flowers, staminode with somewhat recurved sides and habitat. However, it is clearly very close to *P. longifolium* and falls within the morphological range of that species.

Garay (1979) recognised several closely allied and narrowly defined species and provided the following key to distinguish them.

Plate 24. *Phragmipedium longifolium*. Drawn by Carol Woodin from a plant cultivated by Lynn and Bob Wellenstein, Candor, New York.

Key to the segregates of *P. longifolium* (after Garay, 1979)

1. Opening of lip in front of staminode with a pair of triangular-falcate processes originating from infolded margins .... ***P. dariense***
   Opening of lip in front of staminode without triangular-falcate processes 2

2. Staminode transversely elliptic, 3-lobulate in front, especially in dried condition, basally muriculate-puberulent .... ***P. hartwegii***
   Staminode somewhat rhombic to triangular, entire or obscurely lobulate in front, especially in dried condition, basally prominently hispid 3

3. Lip conical, its anterior margin of opening with a notch in middle, the infolded parts do not touch one another, but form an open channel in living specimens .... ***P. hincksianum***
   Lip ovoid, its anterior margin of opening emarginated to erose-denticulate, the infolded parts are tightly approximate, hence do not form an open channel in living specimens 4

4. Anterior margin of lip emarginated to distinctly excised in a broad V-shape .... ***P. longifolium***
   Anterior margin of lip erose-denticulate to prominently rugose-tuberculate ... ***P. roezlii***

Plate 25. *Phragmipedium longifolium*. Drawn by Walter Hood Fitch for *Curtis's Botanical Magazine* t. 5970 (1872) from a plant cultivated at Kew (Kew collection).

Atwood (1984) demonstrated that there was considerable infra-populational variation in the floral morphological characters used to define the above taxa. Based upon isoenzyme studies, McCook (1979) concurred that there was considerable morphological and genetic diversity in the six populations that she sampled and that these were compatible with those of other out-crossing species. She accepted Atwood's circumscription of *P. longifolium* but included *P. hartwegii* in her circumscription. Based upon this evidence and examination of herbarium and plants in living populations, we recognise a broadly defined and rather variable species which includes all of these taxa. Gruss (2003) recognised two species, namely *P. longifolium* and *P. hartwegii*, and three varieties and two forms of the former: var. *longifolium* (with forma *gracile*), var. *roezlii* (with forma *minutum*) and var. *chapadense*. Var. *chapadense* (Campacci & Takase) Gruss, from the State of Goias in Brazil, differs from typical *P. longifolium* in having a broader chestnut brown lip; forma *gracile* (Veitch ex Rolfe) Gruss has plants that are typical of the species but flowers that are only 12 cm across; and forma *minutum* Gruss differs in having shorter inflorescences but otherwise normal flowers. All these seem to be no more than minor variations on a theme. This

Fig. 112. *Phragmipedium longifolium*, growing on a roadside bank on the Lita road in N Ecuador. (PC).

has been recently acknowledged by Braem (2016) who has treated *chapadense*, *gracile*, *hartwegii* and *roezlii* at the rank of form.

Fowlie (1970) provided a description of its habitats in Costa Rica, Panama, Colombia and Ecuador where it grows in rocky and sandy areas in stream beds, in clefts and crevices in the rock where permanent water seepages occur. In such places plants can be inundated during periods of heavy rain or flooding. He also described large colonies inhabiting 'tinajas' or pools that provide fresh water in arid regions in Costa Rica during the dry season. Here they grow in fine sand and moist soil where they are protected from fire during the dry season which keep the roots moist. Man-made habitats can also be colonised as evidenced by their frequent occurrence on steep damp roadside cuttings.

DESCRIPTION. Terrestrial *herb* to 1 m tall; rhizomes up to 1 cm in diam.; roots striate, to 0.4 cm in diam., pubescent. *Leaves* linear, oblique and minutely trilobed and apiculate at apex, up to 80 × 0.6–4.5 cm, glabrous, dark green, the margins finely revolute. *Inflorescence* racemose or with one branch, 0.2–1 m tall, 5–14-flowered, elongate to congested; peduncle brownish-purple, straight to zig-zag, glabrous to sparsely and finely pubescent with short, light-coloured hairs from a glandular base, glandular base persisting if hairs deciduous; floral bracts lanceolate to lanceolate-ovate, finely trilobed, acute at apex, 2.5–11 × 1–2.5 cm at base, about as long as, or longer than ovary, green tinged with purple, the margins somewhat hyaline, the apex mid-vein slightly longer than blade; sterile lower bracts 2–4, linear-lanceolate, apex finely trilobed, 4.1–7 × 1–1.8 cm, the lowest 5–27 cm from plant base, the margins thin. *Flowers* green to yellow tinged with maroon, the nerves green or maroon, the outer surfaces mostly glabrous, shiny, the inner surfaces mostly glabrous, but petals pubescent, the margins white; ovary and pedicel 3–7 cm long, glabrous to finely rugose, pubescent and often deciduous, green to dark purplish brown. *Dorsal sepal* ovate-lanceolate, acute, 20–56 × 7–20 mm, green to yellow with purple nerves, 14–21-nerved, nerves straight to somewhat reticulate (especially on large flowers), glabrous, the margins slightly undulate; synsepal ovate, obtuse to acute, 20–50 × 15–32 mm, shorter than or equal to lip, green with 20–30 purple nerves, the margins entire to undulate. *Petals* borne at an angle of 45–75° from the lip, twisted up to once, linear, retuse to finely bilobed, oblique at apex, 85–130 × 4–13 mm, yellowish green suffused with purple, the central nerve green, purple inside the margins, the inner surface mostly glabrous with scattered pubescence, the basal hairs uniseriate, multi-cellular, to 2 mm long, the margins white, finely ciliate apically, slightly undulate, the petals covered with short red-brown multicellular glandular hairs. *Lip* pointed, 20–25 mm long, 10–18 mm wide at the widest point, yellow or green tinged with brown-purple, the mouth square in front, the apical rim jagged to undulating, the apical lining to 1 cm deep, the lateral rim glabrous, rounded spurs to 5–7 mm long, the lateral lining to 1 cm wide, the claw face glabrous, smooth, yellowish green, covered with irregular, small purple spots, 2–3 green tubercles per margin, with horizontally oblong purple spots and light-coloured pilose hairs within. *Column* green with patches of red hairs corresponding to lip base opening; staminode more or less broadly obtriangular or transversely rhombic, obtuse to acute, 3–6 × 6–10 mm, green to yellowish green, the basal margin lined with purple-black short hairs of equal length throughout, sometimes continuing onto column; stigma 3 mm long and wide, the median stigmatic lobe largest and rounded. *Capsule* dark purplish brown, sparsely rugose. 2n=20 (Karasawa, 1980). **Plates 24 & 25. Figs. 111–115.**

Fig. 113 (above). *Phragmipedium longifolium.* Costa Rica, cult. Lankester Botanic Garden (PC).
Fig. 114 (below left). *Phragmipedium longifolium.* Growing on a roadside bank in C Ecuador (PC).
Fig. 115 (below right). *Phragmipedium longifolium* habitat in C Ecuador (PC).

Phragmipedium longifolium (Rchb.f. & Warsz.) Rolfe in Orchid Rev. 4: 332 (1896). Type: Panama, Chiriqui, Cordillera Chiriqui, *Warszcewicz* 40 (holo. W!).

Cypripedium longifolium Rchb.f. & Warsz. in Bot. Zeitung 10, 40: 690 (1852).

Selenipedium longifolium (Rchb. f. & Warsz.) Rchb.f., Xenia Orch. 1: 3 (1854).

Selenipedium hartwegii sensu Rchb.f., Xenia Orch. 1: 3 (1854) & in Seemann, Bot. Voy. Herald, 6: 214215 (1854).

S. reichenbachii Endres, *in sched.* (1867). Based upon herbarium sheets W15652 & 40322 (both W).

S. dariense Rchb. f. in Gard. Chron. 1869, 47: 1206 (1869). Type: Colombia, Choco, Darien, *Seeman s.n.* (holo. W!).

Selenipedium roezlii Rchb.f. in Gartenflora 20: 164 (1871). Type: Colombia, El Valle del Cauca, Rio Dagua, *Roezl s.n.* (lecto. W!).

Cypripedium roezlii (Rchb. f.) Regel in Gartenflora 22: 97 (1873).

C. hincksianum Rchb.f. in Gard. Chron. n.s. 9: 202 (1878). Type: Colombia, Cape Darien, 1876, cult., *Wallis s.n.* (holo. W!; iso. K!, W!).

C. dariense Rchb.f. in Gard. Chron. n.s. 18: 521 (1882).

Selenipedium hincksianum (Rchb.f.) Desbois, Monog. Cyp. 136 (1888).

S. gracile Desbois, Monog. Cyp. 136 (1888), *nomen nudum*.

Cypripedium longifolium (Rchb.f.) var. *gracile* A.H. Kent, Man. Orch. Pl. 4: 65 (1889).

C. longifolium var. *hincksianum* (Rchb.f.) H.J. Veitch, Man. Orch. Pl. 4: 66 (1889).

Paphiopedilum longifolium (Rchb.f.) Stein, Orchideenbuch 475 (1892).

P. longifolium var. *hincksianum* (Rchb.f.) Stein, Orchideenbuch, 475 (1892).

P. hincksianum (Rchb.f.) Pfitzer in Bot. Jahrb. Syst. 19: 41 (1894).

P. longifolium (Rchb.f.) Kerchove, Livre Orchid. 32 (1894).

P. hartwegii (Rchb.f.) Pfitzer in Bot. Jahrb. Syst.19: 41 (1894).

P. roezlii (Rchb.f.) Pfitzer in Bot. Jahrb. Syst. 19: 42 (1894).

Selenipedium longifolium var. *roezlii* (Rchb.f.) Autran & Durand, Hort. Boiss., 364 (1896).

Paphiopedilum longifolium var. *dariense* (Rchb. f.) Hallier in Ann. Jard. Bot. Buitenzorg 14: 45 (1897).

Phragmopedilum longifolium (Rchb.f.) Rolfe emend. Pfitzer in Bot. Jahrb. Syst. 25: 518, 527 (1898).

P. longifolium var. *dariense* (Rchb.f.) Pfitzer in Engler, Pflanzenr. 4 (50): 50 (1903).

P. longifolium var. *roezlii* (Rchb.f.) Pfitzer in Engler, Pflanzenr. 4 (50): 50 (1903).

P. longifolium var. *hincksianum* (Rchb.f.) Pfitzer in Engler, Pflanzenr. 4 (50): 49 (1903).

P. longifolium (Rchb. f.) Rolfe var. *gracile* in Engler, Pflanzenr. 4 (50): 49 (1903).

P. longifolium var. *roezlii* (Rchb.f.) Hallier in Ann. Jard. Bot. Buitenzorg 14: 45 (1911).

Phragmipedium dariense (Rchb.f.) Garay in Orchid Dig. 43 (4): 141 (1979).

P. roezlii (Rchb.f.) Garay in Orchid Dig. 43 (4): 145 (1979).

P. hincksianum (Rchb.f.) Garay in Orchid Dig. 43 (4): 144 (1979).

P. longifolium (Rchb. f.) Rolfe subsp. *gracile* (A.H. Kent) Atwood & Dressler, *in sched.* Based upon *Atwood 77177* (SEL!).

P. chapadense Campacci & Takase in J. Hokkaido Orchid Soc. 28, Suppl. 1 (2000). Type: Brazil, Goias State, Chapada do Veadeiros, 700–900 m, *Campacci s.n.* (holo. not seen).

P. longifolium (Rchb.f.) Rolfe forma *gracile* (A.H. Kent) Gruss in Caesiana 16: 40 (2001).

P. longifolium (Rchb.f.) Rolfe var. *chapadense* (Campacci & Takase) Gruss in Orchid Digest 67, 4: 232 (2003).

P. longifolium (Rchb.f.) Rolfe forma *chapadense* (Campacci & Takase) Braem in Orchid Digest 80, 1: 40 (2016).

P. longifolium (Rchb.f.) Rolfe forma *roezlii* (Rchb.f.) Braem in Orchid Digest 80, 1: 42 (2016).

DISTRIBUTION. Widely distributed species, ranging through Costa Rica, Panama, Colombia, and Ecuador. Also very rare in Brazil. **Map 25**.

HABITAT. It has weedy tendencies, and is often found in disturbed places such as road cuts (Atwood, 1984). Terrestrial or lithophytic, growing on and among rocks along rivers and spray basins of waterfalls, sometimes below highwater mark; from sea level to 1600 m in elevation. In wet forests, it may be found on slopes, disturbed wet road embankments, vertical wet escarpments, and volcanic ash cliffs by rivers. Coastal populations in Colombia occur on rocks above high tide and sea spray. In Ecuador, populations have been found in pastures.

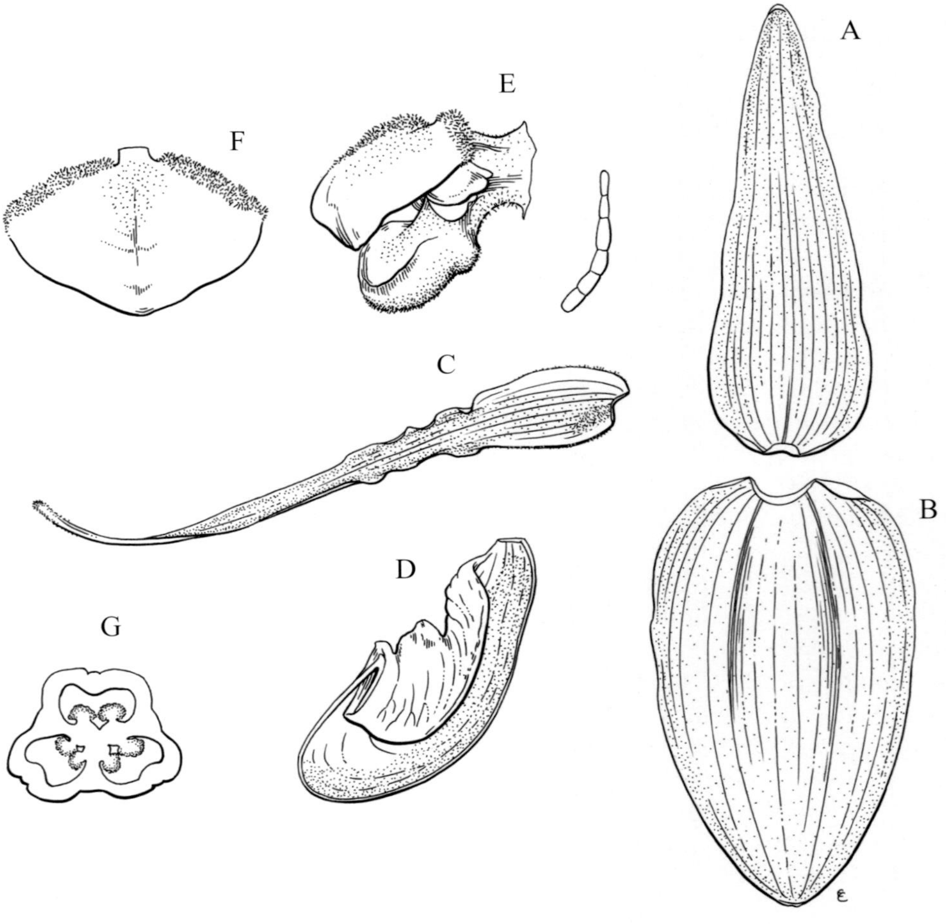

Fig. 116. *Phragmipedium longifolium*. A, dorsal sepal; **B**, synsepal; **C**, petal; **D**, lip, longitudinal section; **E**, column, side view; **F**, staminode; **G**, ovary, transverse section. All drawn by Paul Harwood from cult. J. Kunisch. A-D, × 1.5; E-F, × 3; G, × 6.

PHENOLOGY. Flowers are very long-lived in this species, many lasting for more than a month. The combination of long flower life, flowering in succession, and an elongate raceme of up to 14 flowers, results in a year-long flowering season for very large plants. In the wild it can flower in most months of the year, and large cultivated plants will continue to flower regardless of the season. Occasionally, a faint, sweet odour of daffodils can be detected.

CONSERVATION STATUS. Least Concern (LC). It is a widely distributed species with weedy tendencies, being a primary coloniser of bare soil and rock, for example, on the banks of newly constructed roads. Population numbers fluctuate with the age of the habitat and, as competition increases, seedling establishment is much reduced. However, the number of mature individuals can be high in any single population. Nowadays, it is not collected very often as the plant is common in cultivation and readily available in trade from nursery-raised material. In addition, no major threats affecting the species are likely to drive it to extinction in the near future.

ETYMOLOGY. Named for its long leaves.

12a. var. hartwegii (Rchb.f.) Stein, Orchideenbuch, 475 (1892).
Phragmipedium hartwegii (Rchb.f.) Pfitzer emend. L.O. Williams in Ann. Miss. Bot. Gard. 26: 280 (1939). Type: Ecuador, Pichincha, Nanegal, *Hartweg s.n.* (holo. K!, iso. W!).
Cypripedium macranthum Hartweg in Trans. Roy. Soc. London 2, 3: 155 (1848), *nom. nud., non C. macranthos* Sw.
Selenipedium hartwegii (Rchb.f.) Rchb.f., Xenia Orch. 1: 3 (1854), *nom. illeg.*
Cypripedium hartwegii Rchb. f. in Bot. Zeitung 10, 41: 714 (1852).
Selenipedium hartwegianum (Rchb.f.) Desbois, Monog. Cyp. 136 (1888), *orth. var.*
Cypripedium longifolium (Rchb.f.) Rolfe var. *hartwegii* (Rchb.f.) H.J. Veitch, Man. Orch. Pl. 4: 65 (1889).
Phragmipedium hartwegii (Rchb.f.) Pfitzer in Bot. Jahrb. Syst.19: 41 (1894).
Phragmopedilum hartwegii (Rchb.f.) Pfitzer in Engler, Pflanzenr. 4(50): 48 (1903).
Phragmipedium longifolium (Rchb.f.) Rolfe forma *hartwegii* (Rchb.f.) Braem in Orchid Digest 80, 1: 41 (2016).

DESCRIPTION. Similar in habit to *P. longifolium* but with dark green leaves, the margins green and finely revolute. *Flowers* with white to pale green sepals and petals, often suffused with pink, the nerves green, the petals with purple submarginal lines, the lip yellow-green, the outer surfaces mostly glabrous, shiny, the inner surfaces mostly glabrous, but petals pubescent, the margins white. *Dorsal sepal* ovate-lanceolate, acute, 20–56 × 7–20 mm; synsepal ovate, obtuse to acute, 20–50 × 15–32 mm. *Petals* linear, retuse to finely bilobed, oblique at apex, 35–130 × 4–13 mm. *Lip* 20 × 10–18 mm, the apical rim jagged to undulating, the apical margin repand. *Column* green with red hairs; staminode transversely oblong-elliptic, obtuse, 3–6 × 6–10 mm, green to yellow-green, the basal margin sparsely white-pubescent. **Fig. 117.**

DISTRIBUTION. Panama, Colombia and Ecuador.

HABITAT. Terrestrial or lithophytic, growing in swampy areas in wet montane forests; from sea level to 1600 m elevation.

PHENOLOGY. In the wild it can flower in most months of the year, and large cultivated plants will continue to flower all year.

CONSERVATION STATUS. Endangered (EN): B2ab(ii,iii,v). It is local, with scattered populations in three countries. It appears to be much rarer than the typical variety. The number of mature individuals and the abundance of the species are low. It grows on roadside banks and similar places as a primary coloniser. Changes brought about by increasing competition from other plants, habitat degradation, some collection for horticulture can affect populations. The estimated area of occupancy of the species is less than 500 km^2 with an estimated continuing decline in the number of mature individuals and the quality of habitats in its range.

ETYMOLOGY. Named after Theodore Hartweg (1812–1871), a German plant collector of the Horticultural Society of London.

H.G. Reichenbach described *Cypripedium hartwegii* in 1852 based upon a plant collected by Theodore Hartweg at Nanegal in Ecuador. Hartweg collected plants in Mexico, Guatemala, Ecuador, Jamaica, Madeira and Peru from 1836 until 1854 for the Horticultural Society of London, now the Royal Horticultural Society. He named it in 1848, as *Cypripedium macranthum*, without validating the name which is, anyway, a later homonym of an Asian

Fig. 117. *Phragmipedium longifolium* var. *hartwegii.* Flower, cult. RBG Kew. (Kew collection).

species described by Olof Swartz in 1800. *Cypripedium hartwegii* was transferred to the present genus by Ernst Pfitzer in 1903, under the generic name *Phragmopedilum*, later amended to *Phragmipedium*.

It is very closely allied to the widespread *P. longifolium*, superficially resembling it closely in its floral morphology. However, var. *hartwegii* consistently has a transversely elliptic staminode with a basal margin of sparse white hairs rather than a very dense margin of short blackish purple hairs. Other features that are said to distinguish it include its uniformly green leaves, a lip with a repand apical margin and its paler sepals and petals, often suffused with rose-pink. Reichenbach (*Xenia Orchiadacea* 1: 3. 1854) transferred it to *Selenipedium* but without citing the basionym reference. He then identified a Seeman collection from Panama as *Selenipedium hartwegii* (in Seeman, *Botany of the Voyage of HMS Herald*) but the accompanying Fitch illustration quite clearly belongs to typical *P. longifolium*.

In 1889, James Veitch considered it no more than a variety of *P. longifolium* (as *Cypripedium longifolium*). McCook (1979) included it in the synonymy of *P. longifolium*, without any recognition at formal taxonomic rank. Gruss (2003) resurrected it and we agree that it is distinguishable but only on minor characters, notably the staminode ornamentation. Treatment of it as a variety of *P. longifolium* seems reasonable as it occurs sporadically over much the same range as the typical variety.

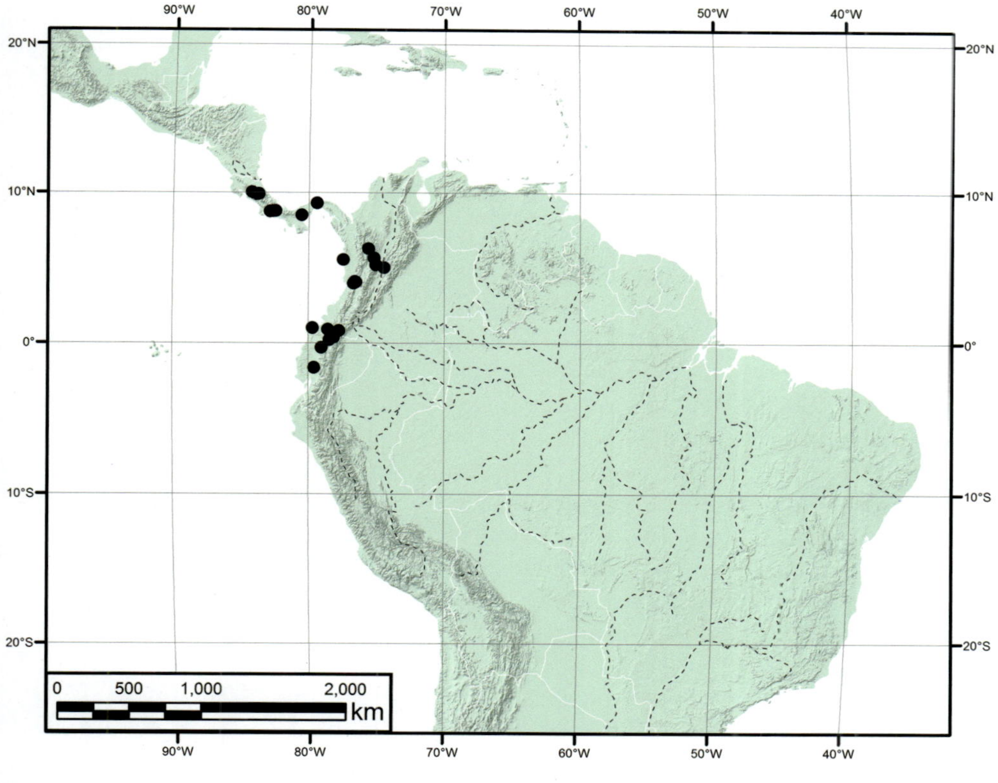

Map 25. Distribution of *Phragmipedium longifolium* in South America.

13. PHRAGMIPEDIUM CARICINUM

John Lindley described *Cypripedium caricinum* in 1846 in Paxton's *Flower Garden* based upon an unlocalised herbarium collection made by Bridges in Bolivia. Robert Rolfe transferred it to the present genus in 1896.

It is closely allied to, and has been confused with, *P. pearcei*, another delicate slipper orchid species found growing on rocks in mountain streams and rivers in northern Peru and Ecuador. However, it differs in having shorter 2 cm long rhizomes, a simple or branching inflorescence with a densely hairy, tomentose stalk, petals with densely hairy tips, a brownish-orange lip with a glabrous margin to its mouth and a white or pale brown claw with red-brown spotting and a transversely broadly cordate staminode with densely hairy basal margins but only sparsely hairy lateral ones.

It is endemic to the eastern flanks of the Andes in Bolivia, where it grows on rocks in fast-flowing streams and rivers in dense lower montane forest between 600 and 1000 m elevation.

Its pollinators are unknown, but flowers that are artificially self-pollinated in the glasshouse set fruit and produce viable seed.

DESCRIPTION. Terrestrial or lithophytic *herb*, clinging tightly to rocky stream banks, forming mats with mosses and grasses; rhizome brown, up to 2 cm long, plants congested. *Leaves* sedge-like, arcuate, linear, acute, finely bilobed or trilobed, oblique at apex, 22–55 × 0.6–2 cm, up to 3 cm wide at base, wiry, strongly keeled, glabrous, glossy, dark green, lighter green below, the margins revolute. *Inflorescence* 3–6-flowered, when branching to 10-flowered, simple or branched, up to 55 cm tall, but often shorter, erect; bracts sheathing axils of panicle; peduncle purple-brown, tomentose, the hairs uniseriate, multicellular, glandular, basal cell often swollen; floral bracts longer than the ovary, ovate to triangular lanceolate, acuminate, minutely bifid at apex, 2.5–5 × 0.9–1.8 cm, greenish yellow, often tinged with dark red, the margins hyaline, overlapping just above the base; sterile lower bracts lanceolate, acute, minutely bifid at apex, 70–250 × 8–14 mm, lowest 8–12 cm from plant base, the margins thin. *Flowers* yellow-green suffused with rose-tan to brownish orange, the nerves reddish brown, the outer surfaces mostly glabrous, the inner surfaces mostly puberulent to tomentose; margins white; ovary and pedicel, 48–55 mm long, glabrous to sparsely puberulent; the hairs uniseriate, 1–4-celled, glandular, often deciduous at maturity. *Dorsal sepal* elliptic, acute, minutely hooded at apex, 25–40 × 8–14 mm, white with 12–15 red-brown and green nerves, the margins to 1 mm wide, slightly undulate; synsepal ovate, obtuse, minutely hooded at apex, 2–30 ×

Fig. 118. *Phragmipedium caricinum*. Flower of cultivated plant, RBG Kew. (PC).

Fig. 119. *Phragmipedium caricinum*. Flower of cultivated plant (JL).

Plate 26. *Phragmipedium caricinum*. Drawn by Carol Woodin from a plant cultivated by Lynn & Bob Wellenstein, Candor, New York.

18–25 mm, equal to or shorter than lip, 18–22-nerved, the margins white. *Petals* spirally twisted, borne at 25–45° angle from lip, twisting 1–2 turns, linear-lanceolate, acute, 50–83 × 4–7 mm, inner surface densely shortly puberulent, more so at apex, the hairs reddish gold, multicellular, glandular, the basal hairs to 2 mm long, the margins white, sparsely ciliate, somewhat undulate, the apex densely covered in glandular hairs. *Lip* ovoid, 25–36 × 15–19 mm, brownish orange outside, whitish yellow with brown-red spots inside, the mouth opening widely ovate, slightly flared; apical lip rim entire, glabrous, edged with a thin brown line, the apical lining to 7 mm wide, the lateral rim lacking spurs, edged with multicellular hairs, the lateral lining to 7 mm wide, area of margin fusion to 8 mm long, the claw face beige with brown-red spots that grade in size and clarity from small and distinct above to larger and poorly defined below, the claw interior pilose, spotted. *Column* 3 mm long, yellow with sparse, fine red hairs; staminode transversely rhombic, widest at the centre, acute, 4–5 × 6–8 mm, light yellow to green, glabrous, the basal margins sparsely to densely edged with short, dark purple multicellular hairs; stigma 4 mm wide. *Capsule* dark brown to brown with maroon splotches, glabrous to roughly glaucous, floral end strongly curved. 2n=21 and 22 (Karasawa, 1980). **Plate 26. Figs 118–120.**

Phragmipedium caricinum (Lindl.) Rolfe in Orchid Rev. 4: 332 (1896). Type: Bolivia, 1846, *Bridges s.n.* (lecto. K!; syn. BM).
Cypripedium caricinum Lindl. in Paxton's Fl. Gard. 1: 39 (1850).
Selenipedium caricinum (Lindl.) Rchb.f., Xenia Orch. 1: 3 (1854).

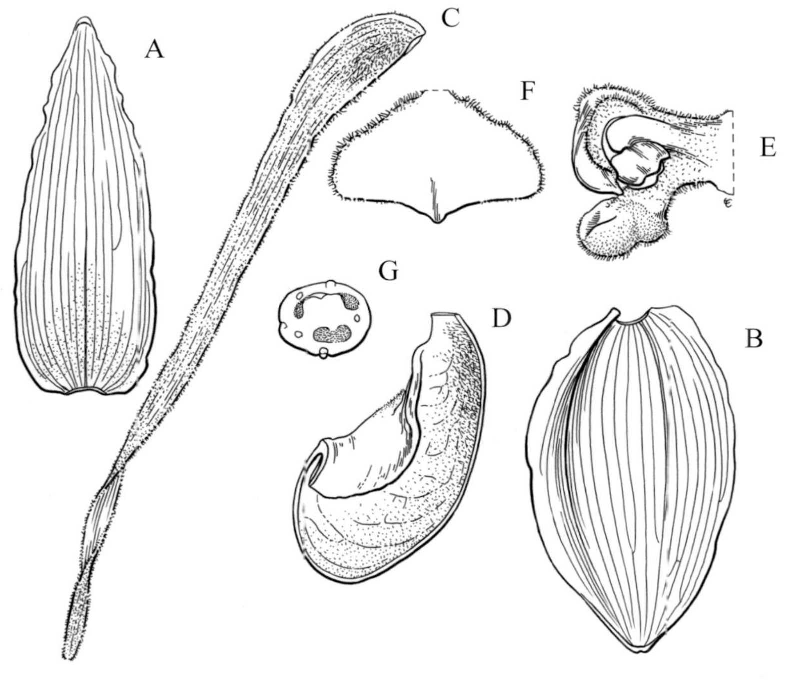

Fig. 120. ***Phragmipedium caricinum.*** **A**, dorsal sepal; **B**, synsepal; **C**, petal; **D**, lip, longitudinal section; **E**, column, side view; **F**, staminode; **G**, ovary, transverse section; **H**, hairs on ovary. All drawn by Paul Harwood from cult. J. Kunisch. A–D, × 1.5; E-F × 4.5; G, × 6.

Paphiopedilum caricinum (Lindl.) Stein, Orchideenbuch 458 (1892).
Phragmopedilum caricinum (Lindl.) Rolfe *emend.* Pfitzer in Engl., Pflanzenr. 4, H.12, 46 (1903). As *P. carycinum* (dePuydt, 1880).

DISTRIBUTION Bolivia. **Map 26**.

HABITAT. A terrestrial or lithophyte, clinging to rocks along streams and wet limestone banks, sandstone rocks, or in red clay among mosses and grasses; from 500–1500 m elevation. Often found growing below the high water mark.

PHENOLOGY. Flowers from May through November. In the greenhouse, flowers last up to two weeks.

CONSERVATION STATUS. Near Threatened (NT): B2ab(ii,iii,v); D1. It is a local species with distribution restricted to the eastern flanks of the Andes in Bolivia. The trend of the population is stable or decreasing, but we still can find large sub-populations. *Phragmipedium caricinum* is of little interest to horticulture and occurs in the little disturbed areas but the species is still threatened by anthropogenic habitat degradation, destruction of watershed areas and deforestation. The estimated area of occupancy of the species is less than 500 km^2 with an estimated continuing decline in the number of mature individuals and the quality of habitats in all the locations.

ETYMOLOGY. The root of the specific epithet, *carex*, is Latin for sedge, referring to its thin, sedgelike leaves.

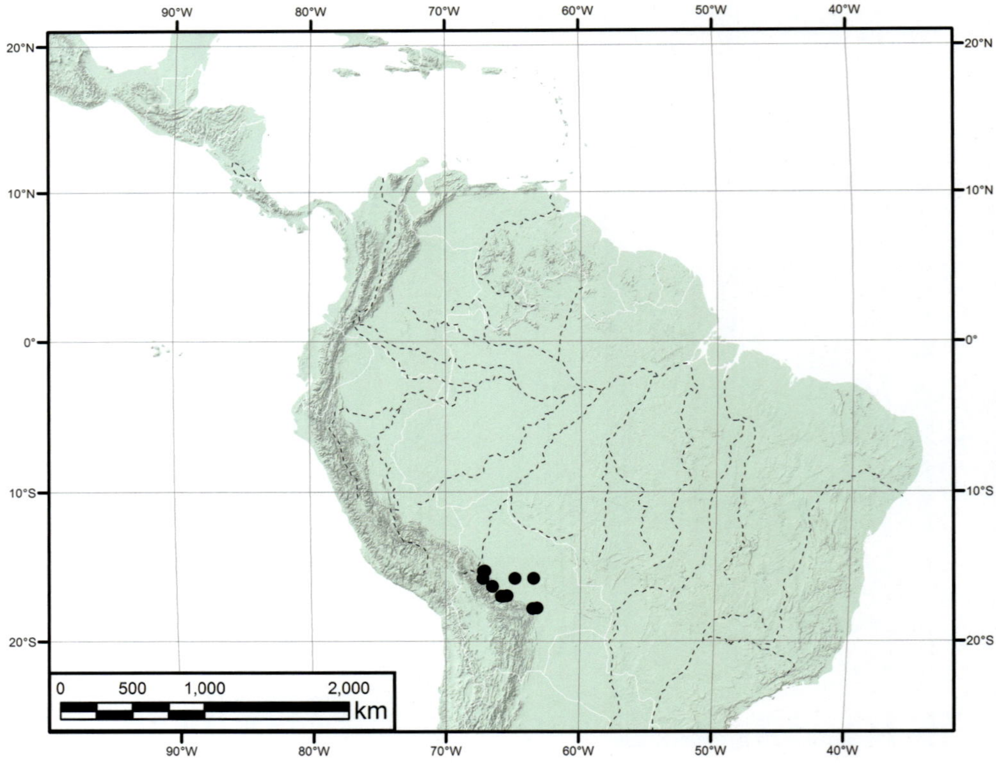

Map 26. Distribution of *Phragmipedium caricinum* in Peru and Bolivia.

14. PHRAGMIPEDIUM PEARCEI

H.G. Reichenbach described this small species, as *Selenipedium pearcei*, in 1865, based the name upon three specimens: a collection from Peru made by Richard Pearce, collecting orchids for Messrs Veitch & Sons of Chelsea and flowered by van Houtte in July 1864, a Richard Spruce collection from Tarapoto in northern Peru and a flower from a plant cultivated by Kramer. McCook (1989) selected the Pearce collection as the lectotype. Werner Rauh and Karl-Heinz Senghas transferred it to the present genus in 1975.

In 1978, Leslie Garay described *P. ecuadorense* based upon a Gunnar Harling collection from the Rio Napo in Ecuador. He distinguished it from *P. pearcei* by its umbonate, cordate, acute staminode and sparsely pubescent, lightly twisted and apically bifid petals. However, all of these characters fall within the range of variation found in populations on the Rio Napo (McCook, 1989a), and recent authors have consistently considered *P. ecuadorense* as conspecific with *P. pearcei*.

Bennett & Christenson (1993) illustrated a plant (*Schunke 14191* at USM) as a hybrid of *P. pearcei* and *P. ecuadorense*. It can equally well be interpreted as being part of the variation within the former.

Fig. 121 (left). *Phragmipedium pearcei.* Flower, C Ecuador (IO). **Fig. 122** (right). *Phragmipedium pearcei.* Flower. C Ecuador (PC).

Plate 27. ***Phragmipedium pearcei*** (as *Selenipedium pearcei*). Cult. van Houtte, from *Flore des Serres* 7: t. 9 (1867).

Phragmipedium pearcei is closely allied to *P. caricinum* but differs consistently in its stoloniferous rhizomes up to 5 cm in length, its inflorescence being unbranched with a sparsely pubescent stalk, petals with sparsely pubescent tips, and a green lip with a minutely hairy tip to the opening, and a white claw with small green spots basally and larger red-brown spots apically.

This is a variable species, particularly in plant and flower size and morphology. It is distinguished by its small stature, less than 25 cm tall, elongate rhizomes, very narrow sedge-like leaves, ovary pubescence, differently coloured flowers and the band of hairs along the front rim of the lip mouth. It can be confused with *P. caricinum*, another species with sedge-like leaves, but that species is a taller plant with broader leaves, a simple inflorescence, shorter petals with pubescent tips, a lip lacking hairs on the front rim, and a transversely elliptic staminode with dense marginal hairs on basal and side margins.

Large populations of this orchid grow on rocks by and in the river in fast-flowing water in a montane river between Topo and Mera in Pastaza province in central Ecuador. The plants are frequently inundated when the rivers are in spate.

In the NW Amazon basin, an infusion of the leaves has been reported as being used by the Kofan tribe to cure stomach problems (McCook, 1989a).

DESCRIPTION. Small clump-forming terrestrial or rarely epiphytic *herb*; rhizomes stoloniferous, up to 12 cm long, covered with sheaths; roots wiry, striate, to 3 mm in diam. *Leaves* wiry, arcuate, linear, acute to apiculate, minutely tri-lobed and oblique at apex, 17–40 × 0.6–2 cm, strongly keeled, glabrous, medium to dark green, lighter green below, the margins revolute. *Inflorescence* simple, 4- to 9-flowered in succession, simple, 10–45

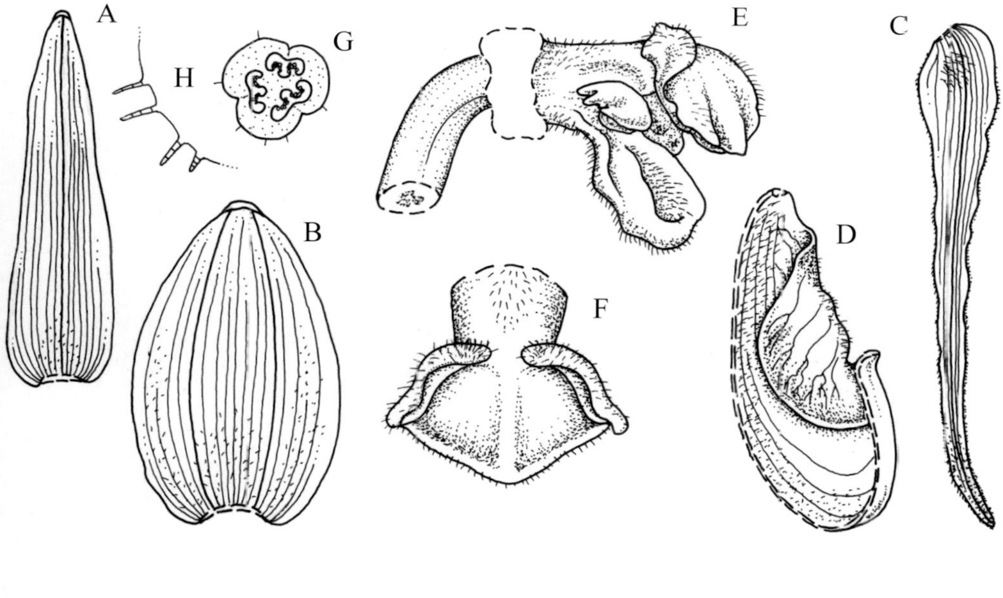

Fig. 123. *Phragmipedium pearcei.* **A**, dorsal sepal; **B**, synsepal; **C**, petal; **D**, lip, longitudinal section; **E**, column, from beneath; **F**, staminode; **G**, ovary, transverse section; **H**, hairs on ovary. All drawn by Judi Stone from Kew Spirit Collection no. 52509. A–D, × 1.5; E–F, × 4.5; G, × 6; H, × 48.

cm tall; peduncle dark brown to maroon, pilose with uniseriate multicellular hairs; floral bracts ovate, acute, minutely bi- or tri-lobed at apex, 1.5–3 × 1–1.6 cm, green with maroon veins, the margins hyaline; sterile lower bracts 2–3, linear to linearovate, acute, minutely bilobed at apex, 2.5–10 × 0.8–1.2 cm, strongly keeled, lowest 3–12 cm from plant base, the margins thin, often purple. *Flowers* with green to yellowish green sepals, with green or maroon-brown veins, similarly coloured petals with a dark red streak, and a greenish yellow lip with small dark red spots on the infolded white margins, the column green, with dark maroon red short hairs on the upper margin; outer surfaces mostly glabrous, the inner surfaces mostly glabrous, but with some short, glandular pubescence; margins white, slightly to strongly undulate; pedicel and ovary 4–5 cm long, glabrous to sparsely pilose, green to maroon-brown. *Dorsal sepal* ovate-lanceolate to elliptic, acute, minutely hooded at apex, 27–35 × 6–14 mm, white with 13–15 nerves, glabrous, with an occasional glandular hair; margins white, undulate; synsepal ovate, obtuse, 24–29 × 10–19 mm, shorter than lip, white, with 20–21 nerves; margins white. *Petals* helically twisted, borne at an angle of 10–45° from lip, linear-lanceolate, acute, often minutely bilobed at apex, 55–80 × 4–10 mm, white, the inner surface glabrous with sparse, short multicellular glandular hairs; basal hairs velutinous, dark red, the margins sparsely ciliate, undulate, with short glandular hairs at the apex, but not densely congested. *Lip* ovate, 20–38 × 12–18 mm, green to yellow-green with darker green veins, the mouth entire, rounded, flared, the apical rim minutely pubescent, edged with a thin brown line; apical lip lining to 6 mm deep, the lateral rim flaring and sometimes slightly bulbous, spurs absent, edged with multicellular hairs, the lateral lining to 7 mm wide; claw face white interior pilose. *Column* 3–5 mm long, green, with sparse short, pale hairs; staminode broadly cordate, acute to broadly acute, 3–5 × 4–8 mm, green, the basal margin and lateral lobes densely edged in shaggy maroon to black hairs; stigma 4–5 mm wide. *Capsule* brown, glabrous, apex straight to slightly curved; seeds to 0.5 mm long. 2n=22 (Karasawa, 1980). **Plate 27. Figs. 121–124.**

Phragmipedium pearcei (Rchb.f.) Rauh & Senghas in Orchidee (Hamburg) 26, 2: 62 (1975). Type: Peru, cult. Van Houtte, 17 July 1864, *Pearce s.n.* (lecto. W!; syn. K!).

Cypripedium caricinum Bateman in Curtis's Bot. Mag. 90: t. 5466 (1864).

Selenipedium pearcei Rchb.f. in Hamb. Gart.-Blumenz. 21: 298 (1865).

Phragmipedium ecuadorense Garay, Fl. Ecuador 9: 1517 (1978). Type: Ecuador, Napo, Pastaza, Rio Napo, *Harling* 3624 (holo. S!).

DISTRIBUTION. Eastern edge of the Andes of Ecuador and N Peru. **Map 27.**

Fig. 124. Habitat on the Rio Tigre, C Ecuador (PC).

HABITAT. On sandy banks and rocks along small submontane rivers in wet tropical forest, the plants frequently splashed by the river water, in marshes with horsetails (*Equisetum*) and occasionally epiphytic on tree on the margins of wet forest; 600–1200 m.

PHENOLOGY. Flowering mostly from September to February, but flowering specimens have been collected in almost every month of the year. The pollinators are unknown, but fruit set is common in the field.

CONSERVATION STATUS. Least Concern. It may be locally abundant. The extent of occurrence for this species is greater than that needed for a threatened category.

ETYMOLOGY. Named for Richard Pearce (1835–1868) who collected the type material.

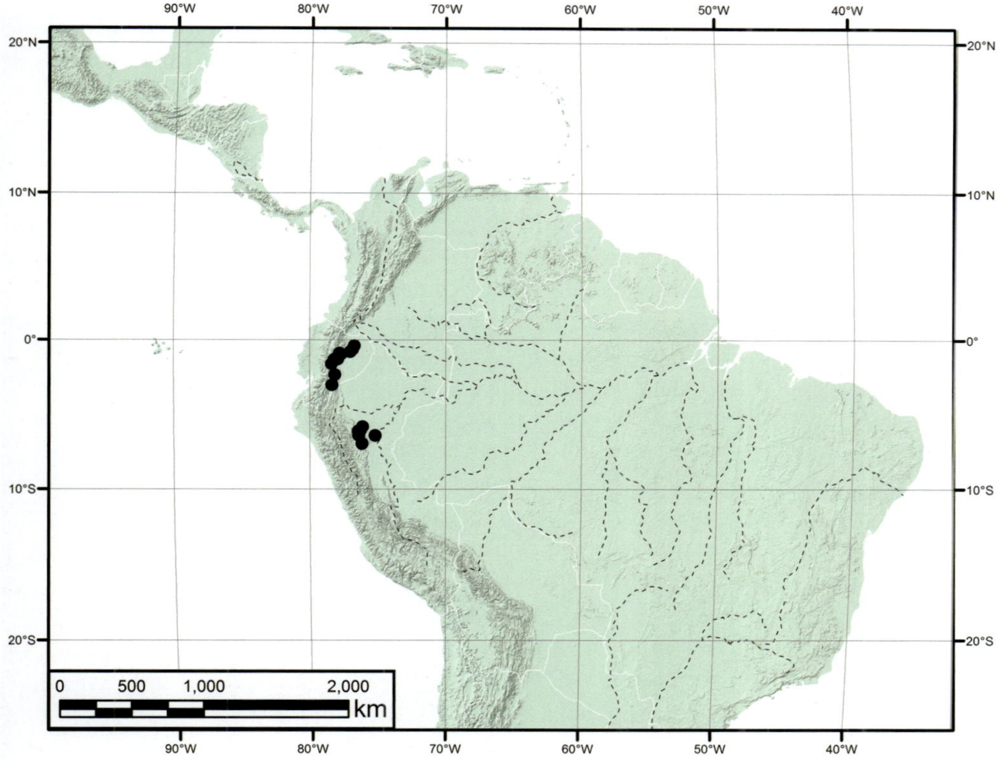

Map 27. Distribution of *Phragmipedium pearcei* in Ecuador.

15. PHRAGMIPEDIUM VITTATUM

The Brazilian botanist José Vellozo's (1742–1811) description of *Cypripedium vittatum* was published posthumously in 1832, based upon his own collection from near Rio de Janeiro in Brazil. Robert Rolfe transferred it to the present genus in 1896. Vellozo's specimen has, apparently, not survived and McCook (1989a) selected a collection made by Anders Regnell (1807–1884) from near Sao Paulo as a neotype.

Phragmipedium vittatum is readily distinguished by its yellow margined leaves, sepals with entire margins, a proximally rounded purple lip, with a yellow margined mouth and lacking a central groove, and a cordate-triangular staminode.

Barbosa Rodrigues (1877) described *Cypripedium paulistanum* based upon a Regnell collection from Cajurú in Sao Paulo State, distinguishing it by its free rather than fused lip, synsepal as long as the dorsal sepal and its lanceolate staminode. McCook (1989) examined the type collection and duplicates of it and found that the lip side lobes were fused, the staminode was transversely rhombic and the synsepal to dorsal sepal ratio varied with the age of the flowers. She, therefore, consigned Barbosa's species to synonymy.

Pabst and Dungs (1975) published an illustration (no. 1) of a flower of it under the name *P. caricinum*. Their illustration (no. 4) labelled *P. vittatum* is of *P. longifolium*. This species displays a great deal of variation in flower colour and size (McCook, 1989a).

Fowlie (1973) reported that it grew in boggy places in association with the purple-flowered *Bletia catenulata* in acidic conditions. Menezes and Braem (1993) provided a

Fig. 125. Habitat of *Phragmipedium vittatum* in Rio de Janeiro state. (RW)

Plate 28. *Phragmipedium vittatum* (as *Selenipedium vittatum*). Illustrated by P. De Pannemaeker, cult. Linden, Brussels, in *L'Illustration Horticole* 23: t. 218 (1876).

detailed account of its ecology and, in particular, the effect of fire in its habitat. They conclude that fire during the dry season, July to September, does not have an adverse effect upon *P. vittatum*. The main effect of fire appears to be an increased alkalinity with the soil pH raised from about 5.6–6.8 to 7.2–7.8.

It is native to marshy elevated areas of the 'Cerrado' of the south-eastern and central-western regions of Brazil. It is said to be restricted to the Brazilian central west (Menezes 1989). It appears to have had a large geographic range but to be much localised in distribution, being only found in very specific peaty habitats, where it may be locally abundant, e.g., 500 to 1000 plants in 1 square metre (Fowlie, 1973, Menezes, 2004). For decades, it remained little known and only small specimens were found in odd habitats presumed to be the remains of original degraded ones. When Brasilia was founded it was found in several areas of the Federal District. Intensive collections were made in the course of various landscaping operations. Much of the native wetland habitat was also lost in the construction of Brasilia and rice cultivation, pastures and coffee plantations (Fowlie, 1973, Menezes, 2004, 2014). The Cerrado is one of the most threatened biomes of South America. New subpopulations have been located, e.g., in Goias, where only four clumps were found (Menezes, 1989). Miller *et al.* (2002) and Miller (2008) reported that it was found in only one zone of the eight that cover the Organ Mountains in the state of Rio de Janeiro, and to be rare within it. The current size and dynamics of the metapopulation were unknown but they found one colony of some 300 plants on a wet rock face where the roots were attached to the rock some 8 cm below a turf cover.

Populations may be largely clonal and seed recruitment may be low (Menezes, 1989) but Miller *et al.* (2002) reported copious vegetative reproduction from the roots.

Menezes (1989, 2014) considered it difficult to grow in cultivation due to the extreme specificity of the habitats in which it thrives and Koopowitz (2007) reported that it was

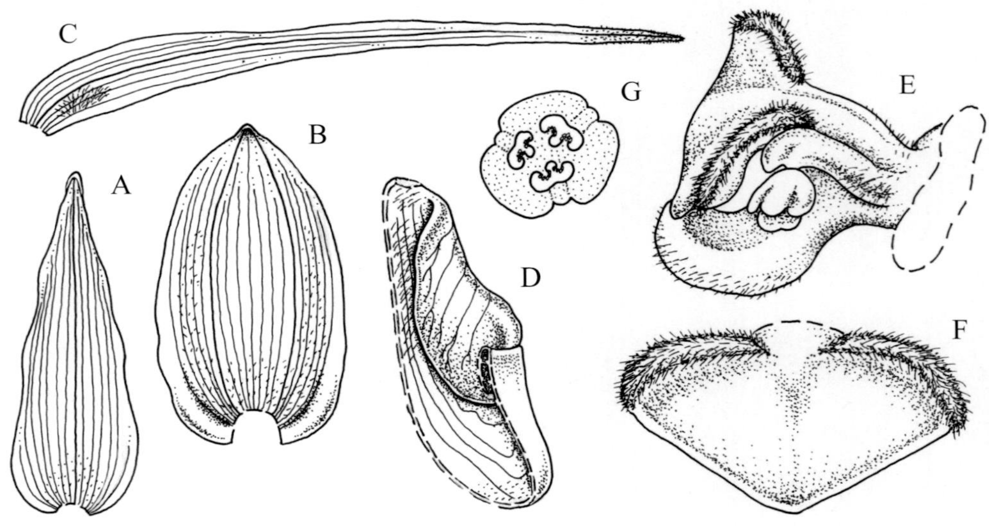

Fig. 126. *Phragmipedium vittatum*. A, dorsal sepal; **B**, synsepal; **C**, petal; **D**, lip, longitudinal section; **E**, column, side view; **F**, staminode; **G**, ovary, transverse section. All drawn by Judi Stone from cult. Wellenstein (Kew Spirit Collection no. 56851). A-D, × 1.5; E–F, × 6; G, × 9.

difficult to obtain, which may increase incentive to collect wild specimens. Collecting from natural sources is still very common in Brazil in spite of prohibitions of international commerce of species collected in the wild (Saddi *et al.,* 2005).

DESCRIPTION. Terrestrial *herb* up to 1 m tall, forming clumps; rhizome short and stout, up to 1.4 cm in diam.; roots striate, woolly, up to 4 mm in diam. *Leaves* linear, oblique, minutely 3-lobed at apex, 30–55 × 2.3–3.4 cm, glabrous, the margins yellow, shiny, to over 1 mm wide. *Inflorescence* 3–6-flowered, racemose, 50–80 cm tall; peduncle dark brown to purple, light brown to green basally, sparsely hirsute, more so basally and below floral bracts, the hairs short, 1–2-celled from a glandular base, often deciduous, then with the glandular base remaining; floral bracts broadly lanceolate, 7–9.5 × 2.3–2.8 cm, almost twice the length of the enclosed ovaries, tinged with purple, reticulately veined apically, the margins yellow, lacking veins, to more than 1 mm wide; apex acute, yellow; sterile lower bracts 4–5, lanceolate, acute, minutely 3-lobed and rounded at apex, 70–120 mm long, up to 40 mm wide, the lowest bract 10–29 cm from plant base, the margins same as leaves. *Flowers* greenish purple to pale yellow-brown, the margins white tinged with purple, the nerves green, the outer surfaces mostly glabrous, the inner surfaces mostly glabrous; ovary and pedicel 45–50 mm long, shorter than subtending bract, glabrous apically, the rest glabrous to sparsely puberulous, brown, curving apically to 90°. *Dorsal sepal* linear-elliptic, acute, minutely hooded at apex, 43–50 × 15–17 mm, green, 12–15-nerved, somewhat reticulate, glabrous, the margins entire to slightly undulate; synsepal elliptic-ovate, obtuse, minutely hooded at apex, 40–48 × 21–32 mm, about equal to lip in length, green, 13–17-nerved, somewhat reticulate, the margins entire. *Petals* held at 20–30° angle from lip, twisting only slightly, linear, 55–80 × 5–8 mm, green with purple margins,; inner surface glandular pubescent apically, basal hairs pilose, multi-cellular, the margins undulate and on basal half of petals, the apex glandular pubescent on inner surface. *Lip* ovoid, 30–40 mm long, 15–17 mm wide, green inside, purple outside, the mouth square in front, flaring slightly, the apical lip rim undulate to jagged, glabrous, the apical lining to 8 mm deep, the lateral rim spurred, the spurs rounded, to 6 mm high, the lateral lining 11–15 mm wide, the claw face glabrous, ribbed, beige with yellow and green spots centrally located, often in vertical rows; green tubercles present at lower lateral lobe margins, green, often obscured on herbarium specimens; claw interior villous. *Column* to 3 mm long, green, puberulent; staminode broadly rhombic, 4–5 × 6–8 mm, green, the basal margins lined with long brown to dark red multi-cellular hairs; hairs often continue onto column, shorter toward lobes; apex acute to roundly acuminate; stigma to 6 mm wide. *Capsule* dark purple-brown, glabrous to sparsely hirsute, curving sharply at apical end. 2n=18 (Karasawa, 1980). **Plate 28. Figs. 125–128.**

Phragmipedium vittatum (Vell) Rolfe in Orchid Rev. 4: 332 (1896). Type: Brazil. Rio de Janeiro, Alps around Rio de Janeiro in the direction of Cunha, *Vellozo s.n.* (holo. not found); Brazil, Sao Paulo: Cajurú, *Regnell* 1206 (neo. US!; isoneo. US! S!).
Cypripedium vittatum Vell, F1. Flum. 9: 62 (1831).
C. paulistanum Barb. Rodr., Gen. Sp. Orch. 1: 203 (1877). Type: Brazil, Sao Paulo, Cajurú, *Regnell* 111 1206 (lecto. S!; syn. S!, US!).
C. binotii Hort. in Ill. Hort. 34 (1887) & 116 (1890).
Selenipedium vittatum (Vell.) Rchb.f. in L'Ill. Hort. 23: 57, t. 238 (1876).
S. paulistanum (Barb. Rodr.) Pucci, Les Cypripedium, 208 (1891).

Paphiopedilum vittatum (Vell) Stein, Orchideenbuch, 491 (1892).
Phragmopedilum vittatum (Vell) Rolfe *emend.* Pfitzer in Engler, Pflanzenr. 4, 50: 47 (1903).

DISTRIBUTION. Brazil (Goias, Mato Grosso del Sur, Federal District, Minas Gerais, Rio de Janeiro); 800–1400 m. **Map 28.**

HABITAT. It grows in clumps in bright sunlight, scattered by sedges and grasses, to partial shade, in areas only a few metres wide on the very edge of gallery forests (e.g., elfin forest), which occur in peat bogs, on creek margins, on rocky wet slopes, in Campo limpo, swamps and in other riparian habitats, in the 'Panalto' of the barren 'Cerrado'; from 900 to 1600 m elevation. It almost always grows with species of the orchid genera *Epidendrum, Habenaria* and *Cyrtopodium* and *Bletia catenulata* and with grasses (*Panicum* spp.). The association with grasses may protect the orchid from damage by direct sunlight and the grass roots may help to keep the soil porous. Roots extend widely from the base of the rhizome and hold tenaciously to the peat where they are kept moist all year round. It is adapted to survive the grass fires, which sweep the surrounding dry plains in the dry seasons and extend to the surface of the peat bog although not to the gallery forest (Fowlie, 1973; Menezes, 1989, 2004; Manza, 2006; Koopowitz, 2007).

PHENOLOGY. Flowering December to April (Manza, 2006).

Fig 127. *Phragmipedium vittatum*. Flower, Rio de Janeiro State. (RW). **Fig. 128.** *P. vittatum*. Flower. Goias State, Chapada dos Veideiros. (RH).

CONSERVATION STATUS. Least Concern. It is native to marshy elevated areas of the 'Cerrado', of the south-eastern and central western regions of Brazil where it is highly localised in distribution, being only found in very specific peaty and grassy wet habitats. It may be locally abundant, although populations are largely clonal. Intensive and indiscriminate collections of this slipper orchid made in the course of various landscaping operations and habitat loss are said to have virtually eradicated this orchid from some parts of the Brazilian countryside in the past (Menezes, 1989). Threats to terrestrial orchids in the 'Cerrado' are numerous and ongoing. However, the taxon is found within protected areas. The extent of occurrence for this species is greater than that needed for a threatened category (Menezes, 2004) and Rio de Janeiro (Miller, 2008). Herbarium records found did not cover all this cited range but still give a range which exceeds the threshold. Menezes (1989) raised concerns about its decline and rarity near Brasilia, but it was reported as abundant on the Brazilian Central Plateau in 2004, which infers a population, an area of occupancy that also exceed the thresholds. However, it is threatened by collection, expanding human populations and pollution of its marshy habitat. Surveys are needed to confirm its current geographic range and habitat status. Monitoring is needed to detect any declines in view of habitat threats and possible collection pressures.

ETYMOLOGY. From the Latin *vittatus*, meaning longitudinally striped.

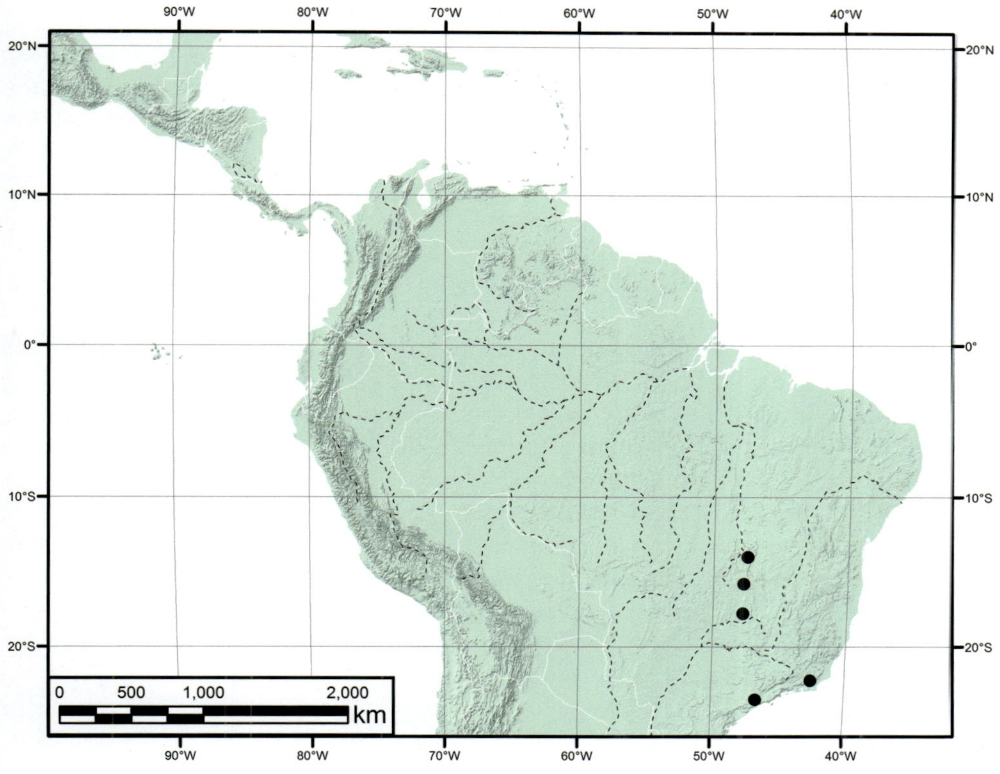

Map 28. Distribution of *Phragmipedium vittatum* in Brazil.

PHRAGMIPEDIUM SECTION PHRAGMIPEDIUM

Phragmipedium sect. **Phragmipedium**

Cypripedium sect. *Caudata* Kraenzl., Orch. Gen. Sp. 1: 13 (1897). Type species: *P. caudatum* (Lindl.) Rolfe.

Cypripedium sect. *Uropedium* Kuntze, Lex. Gen. Phan. 159 (1903).

Paphiopedilum sect. *Phragmipedium* (Rolfe) V.A. Albert & Börge Pett. in Lindleyana 9, 2: 137 (1994).

The species traditionally included in this section are characterised by their mainly epiphytic habit, few-flowered inflorescences and flowers that open more or less simultaneously, elongate sepals, long pendent petals that are much longer than the sepals and a transversely rhombic, rather thin-textured staminode with purple tips and bearing sparse, long hairs on the upper margin. Four species are closely related, enough to have been treated by some authors as conspecific. However, they are distinct morphologically and have discrete geographical ranges. *P. lindenii* and possibly *P. humboldtii* var. *exstaminodium* are self-pollinating forms allied, respectively, to *P. wallisii* and *P. humboldtii*. In *P. lindenii*, the lip is petaloid, whereas in *P. exstaminodium* the staminode is absent.

Using isoenzymes, McCook (1988) investigated the genetic diversity of *P. caudatum*, *P. warszewiczianum* (as *wallisii*), *P. exstaminodium* and the self-pollinating *P. lindenii*. Despite their morphological distinctiveness, she found them to be genetically identical using these loci, suggesting their recent origin without significant concomitant genetic change.

16. PHRAGMIPEDIUM KLOTZSCHIANUM

Robert Schomburgk discovered this pretty orchid in 1842 or 1843 on the first expedition to climb Mt Roraima (2810 m). Schomburgk (1848) gave the provenance of the type collection as Ruimeru Falls, according to his manuscript, but the type material only provides the information as 'Roraima, British Guiana, November 1842–43'. Steyermark (1981) suggested that this locality is probably located in Venezuela.

In 1848, H.G. Reichenbach named it, as *Cypripedium klotzscheanum*, for Dr. Johann Klotzsch, curator of the herbarium at Berlin from 1833 until 1860. Rolfe (1896) subsequently corrected the epithet to '*klotzschianum'*.

It is one of only five species of the section not found in the Andes. It is distinctive in the section by reason of its densely hairy peduncle and ovaries and because of it ellipsoidal, rather than urn-shaped, lip.

Pabst and Dungs' (1975) illustration of a flower of *P. klotzschianum* clearly shows its distinctive ciliate dorsal sepal and rhombic sparsely pubescent staminode. Its staminode more closely resembles that of *P. caudatum* and its allies in its shape, attitude and sparsely ciliate margins, while its stigma is also similar. These features suggest that *P. klotzschianum*

Plate 29. *Phragmipedium klotzschianum*. Drawn by Matilda Smith for *Curtis's Botanical Magazine* t. 7178 (1895) from a plant cultivated at Kew.

might be better placed in the typical section of the genus, rather than in section *Lorifolium*.

Dunsterville (1970) gave a vivid description of its habitat on Cerro Venamo on the Venezuelan-Guyanan border. He found it growing in masses with *Sobralia stenophylla* on pink sand on the banks of a small river at about 1000 m, where it would be frequently inundated in the wet season.

Phragmipedium klotzschianum was brought into cultivation by Sander in the late 1800s, but is not commonly cultivated today.

In the literature, the spelling of the specific epithet has varied from '*klotzscheanum*' to '*klotzschianum*'. The former was used by Reichenbach when he described *Cypripedium klotzscheanum*. However, he used both when he transferred it to the genus *Selenipedium* in 1854, using the former spelling in *Xenia Orchidacea* and the latter in *Bonplandia*. The recommended formation of a Latin epithet for Klotzsch would be the latter which Rolfe used when he transferred it to the present genus and we follow him here.

Although this species has traditionally been placed in section *Himantopetalum*, it bears features that place it close to *P. caudatum* in the typical section, notably its staminode with sparse marginal hairs and purple lateral tips and it acute lip apex (McCook, 1989a). Based upon a morphological cladistics analysis, McCook suggested that section *Himantopetalum*

Fig. 129. *Phragmipedium klotzschianum* growing on sand by a river, Gran Sabana, Venezuela (GR).

should be subsumed within the typical section, but DNA-based analyses, including *P. pearcei*, suggest that at least part of section *Himantopetalum* belongs in the same clade as the species of Section *Lorifolia* (e.g. Cox *et al.*, 1997; Yap, pers. comm.).

DESCRIPTION. Terrestrial grass-like *herb* up to 50 cm tall; rhizomes stoloniferous, up to 0.3 cm in diam., and up to 7 cm long, often with persistent leaves; roots wiry, up to 2 mm or more in diam. *Leaves* 4–8, linear, minutely trifid at apex, 10–39 × 0.3–1.2 cm, glabrous, wiry, the margins revolute. *Inflorescence* simple, 2–4 (5-) flowered, 22–60 cm tall; peduncle flattened in cross-section, brownish red, pilose, the multicellular hairs having a bulbous, glandular base; floral bracts ovate, apex minutely 3-lobed, 1.8–2.1 cm long, 0.5–0.8 cm wide, shorter than ovary, pilose in bud and the base of mature bracts; bracts linear-lanceolate, acute or subacute, 2–4.5 × 0.8–1 cm, the margins revolute. *Flowers* yellow with reddish brown markings, claw face white, the outer surfaces mostly pilose, the inner surfaces mostly glabrous, the margins white; pedicel and ovary 35–64 mm long, brown-velutinous. *Dorsal sepal* lanceolate-elliptic, minutely hooded at the apex, 27–43 × 7–12 mm, whitish yellow with 7–11 red-brown nerves, glabrous inside, with glandular hairs basally, the margins white, somewhat undulate, glabrous, pubescent on outside; synsepal ovate, minutely hooded at the apex, 20–36 mm long, 13–22 mm wide, roughly equalling lip, whitish yellow with 9–11 red-brown nerves, the margins whitish yellow. *Petals* borne more or less parallel to the lip, twisting 1–3 times, linear, minutely bilobed and oblique

Figs. 130 (left). *Phragmipedium klotzschianum* flower, Gran Sabana, Venezuela (CB). **Fig. 131** (right). *Phragmipedium klotzschianum* in its natural habitat, Canaima National Park, Venezuela. (CL)

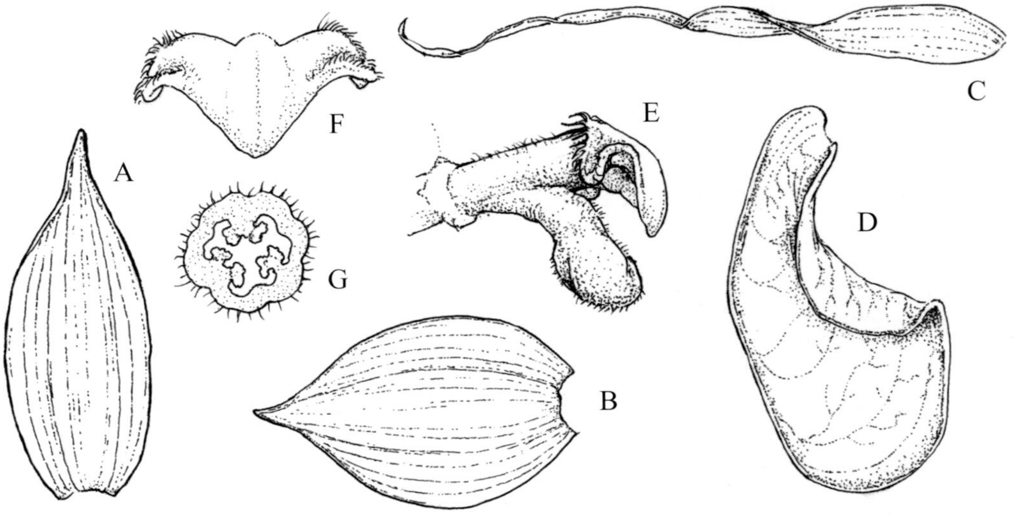

Fig. 132. ***Phragmipedium klotzschianum.*** **A**, dorsal sepal; **B**, synsepal; **C**, petal; **D**, lip, longitudinal section; **E**, column, side view; **F**, staminode; **G**, ovary, transverse section; **H**, hairs on ovary. All drawn by Paul Harwood from *Da Silva s.n.* (Kew Spirit Collection). A–C, × 2.5; D, × 12; E–F, × 6; G, × 12.

at the apex, 47–100 × 4–7 mm, white with reddish purple veins, mostly glabrous within, becoming glandular-pubescent apically, the basal hairs velutinous, the margins only slightly undulate. *Lip* obovate, 22–38 × 15–20 mm, yellow to pale green, the lip opening oval, entire, the apical rim glabrous, slightly emarginate in middle, the apical lining 4–5 mm deep, the lateral rim lacking nectar-less spurs, edged with well-defined area of multicellular velutinous hairs, the lateral lining 4–5 mm wide, the claw face white with green or reddish green spots at interface with yellow, the claw interior spotted, sparsely pilose. *Column* up to 5 mm long, yellow-green, sparsely pubescent; staminode triangular, lateral lobes falcate, 3–4 × 6–9 mm, yellow-green, lateral lobes reddish black; apex obtuse; basal margins and lateral lobes edged with dark purple multi-cellular hairs of 1 mm or less long; stigma 4 mm wide. *Capsule* 35–40 mm long, dark purple-brown, sparsely velutinous. Chromosome number unknown. **Plate 29. Figs. 129–132.**

Phragmipedium klotzschianum (Rchb.f.) Rolfe in Orchid Rev. 4: 332 (1896). Type: British Guiana (Guyana), Roraima, 1842–43, *Schomburgk* 718 (lecto. W!; syn. B †, BW, K!).

Cypripedium klotzschianum Rchb.f. ex Schomburgk, Vers. Flor. Br. Guiana, 1069 (1848), as '*klotzscheanum'*.

C. schomburgkianum Klotzsch, *nomen nudum,* Schomburgk, Vers. Flor. Br. Guiana, 1669 (1848).

C. klotzscheanum Rchb. f. in Linnaea 22: 811 (1849).

Selenipedium klotzschianum Rchb.f., Xenia Orch. 1: 3 (1854).

Paphiopedilum klotzschianum (Rchb.f.) Stein, Orchideenbuch, 473 (1892).

Phragmopedilum klotzschianum (Rchb.f.) Rolfe *emend.* Pfitzer in Engler, Pflanzenr. 4, 12: 47 (1903).

DISTRIBUTION. W Guyana, W Venezuela (Gran Sabana) and NW Brazil (Roraima State). **Map 29.**

HABITAT. Grows in sandy areas and on often inundated rocks in streams and rivers in savanna and forest; 415–1200 m.

PHENOLOGY. Flowering from October to May.

CONSERVATION STATUS. Endangered (EN): B1ab(ii,iii,v)+2ab(ii,iii,v). It is a very local species with a restricted distribution. Although the species occurs in very remote places it is not over-collected. The trend of the population is probably decreasing; the number of mature individuals and its abundance are affected by habitat changes, deforestation, destruction of watersheds, water quality and mining activities. One location lies close to the Orinoco industrial mining belt and the construction of a new road through the Gran Sabana towards the Brazilian border might impact it. The estimated extent of occurrence and estimated area of occupancy of the species are less than 5000 km^2 and 500 km^2, respectively, with an estimated continuing decline in the number of mature individuals and the quality of habitats in its locations.

ETYMOLOGY. For Dr. Johann Klotzsch (1805–1860), curator of the herbarium at Berlin from 1833 until 1860.

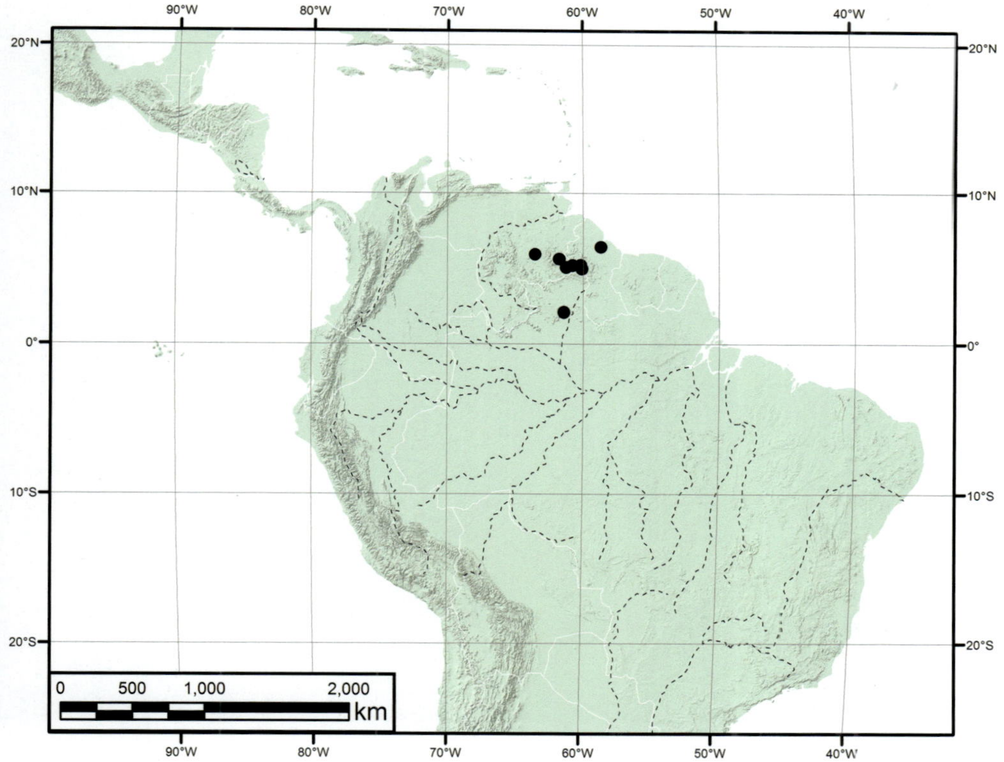

Map 29. Distribution of *Phragmipedium klotzschianum* in northern South America.

17. PHRAGMIPEDIUM CAUDATUM

Hipolito Ruiz (1754–1816), in his journal, noted that he and José Pavon (1750–1840) saw plants of this spectacular species in Peru around Tarma, at Cuchero and Muña, while at least one herbarium specimen was collected near Pillao. Ruiz prepared a description of the species, under the name *Cypripedium grandiflorum* during his stay at Muña between August and September 1786 but, unfortunately, this was never formally published. According to his diary this plant was illustrated by J. Brunete at the same time (Pupulin & Dressler, 2012; Pupulin, 2012). His illustration is now in the archives of the Jardín Botaníco de Madrid (Frontispiece).

John Lindley eventually described this species, as *Cypripedium caudatum*, in 1840 based upon a flower from a Peruvian collection made by Ruiz and Pavon, and sent by Andrew Matthews to William Hooker at Kew. Pupulin and Dressler (2012) consider that the attribution by Matthews was mistaken and that the flower was collected by Juan José Tafalla Navascués (1756–1811) and Juan Agustin Manzanilla (fl. 1793–1816). The specimen that Lindley examined was in poor condition and his description is very brief, distinguishing it by its oblong-lanceolate, acuminate sepals, elongate, lanceolate, long-acuminate petals that were pubescent on their outer surface, and by its lip with a hairy

Fig. 133 (left). *Phragmipedium caudatum* being examined by Dr Henry Oakeley in forest, Monopampa, near Muna, Huanuco Province, Peru. 2300 m asl. (HO); **Fig. 134** (right). Flowers from cultivated plant. (PC)

Fig. 135. *Phragmipedium caudatum*, growing epiphytically in montane forest, Monopampa, near Muna, Huanuco Province, Peru. 2300 metres asl. (HO).

Fig. 136. *Phragmipedium caudatum*, growing epiphytically in montane forest, Monopampa, near Muna, Huanuco Province, Peru. 2300 metres asl. (HO).

Plate 30. *Phragmipedium caudatum*. Drawn by Carol Woodin from a plant cultivated by Lynn & Bob Wellenstein, Candor, New York.

orifice. Mounted on the type sheet in the Lindley herbarium at Kew are three other flowers from a collection made between 1873 and 1876 by Walter Davis in Peru for the nursery of Messrs Veitch & Sons of King's Road, Chelsea.

William Hooker (1844) subsequently illustrated a collection from the Andes east of Lima made by William Lobb, although the drawing was based upon a three-flowered herbarium specimen.

When reviewing the generic limits of the slipper orchids, Rolfe (1896) transferred Lindley's name to the genus *Phragmipedium*. For many years, the name *P. caudatum* was widely used for all the tropical American slipper orchids with long petals, a transversely rhombic, sparsely hairy staminode and a saccate lip.

H.G. Reichenbach (1852) soon realised that more than a single species was involved, but his account has caused such confusion that it continues to create nomenclatural instability. In his article, he included Josef von Warszewicz's description of *Cypripedium humboldti* which he had collected in oak woods on Chiriqui in Panama. It differs from *P. caudatum* in having an obovoid lip with a distinctive incurved margin to the orifice and chocolate flushing on the lip and petals. Braem *et al.* (2004) interpreted Warszewicz's description as invalid and the later combination, *Phragmipedium humboldtii* (Humboldt ex Rchb.f.) J.T. Atwood & Dressler, as illegitimate. As a result, they coined the name *P. popowii* Braem, Ohlund & Quéné for this taxon. Dressler & Pupulin (2011) and Pupulin & Dressler (2011, 2012) have established the validity of *Cypripedium humboldti*, a decision with which we agree. Thus, *P. popowii* is considered as an illegitimate name as it is based upon the same type specimen.

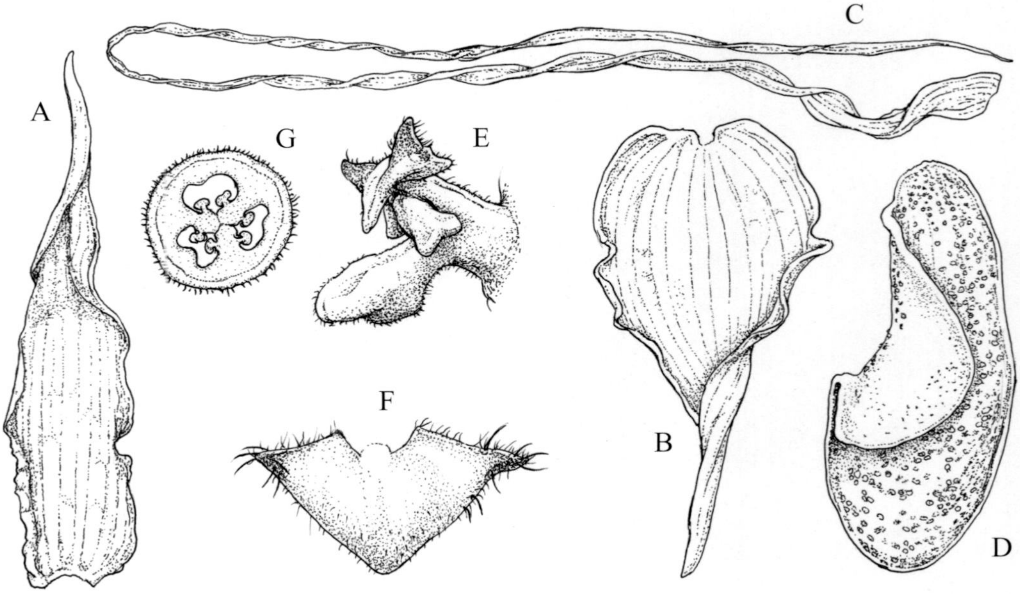

Fig. 137. *Phragmipedium caudatum*. **A**, dorsal sepal; **B**, synsepal; **C**, petal; **D**, lip, longitudinal section; **E**, column, side view; **F**, staminode; **G**, ovary, transverse section. All drawn by Carol Woodin from cult. Masur. A–D, × 1.5; E, F, × 6; G, × 9.

Currently three species of slipper orchids with long petals and a saccate lip are recognised: the Peruvian and Bolivian *P. caudatum*, *P. warszewiczianum* from the northern Andes of Venezuela, Ecuador and Colombia, and *P. humboldtii* from Central America. A full discussion of the nomenclatural problems is given under the account of *P. humboldtii* that follows.

Phragmipedium caudatum is an epiphyte that is closely allied to *P. warszewiczianum* and *P. humboldtii.* The former can be distinguished by its pink- or white-flushed lip, and the latter by its glossy chocolate brown lip with an inturned orifice. The distributions of these species do not seem to overlap. The Marañon gap seems to separate *P. caudatum* and *P. warszewiczianum* and the Darien gap separates *P. warszewiczianum* from *P. humboldtii.*

DESCRIPTION. Terrestrial, lithophyte, or epiphytic herbaceous perennial *herb*; rhizome up to 3 cm long, 6–8 mm in diam.; roots woolly, up to 5 mm in diam. *Leaves* lorate, obtuse, retuse-apiculate, oblique, often splitting along midvein at apex, 15–60 × 2.8–5.5 cm, thick and coriaceous, glabrous, light green, midvein prominent; margins pale green. *Inflorescence* 2–4- (5-) flowered, racemose, 30–60 cm high, with flowers blooming simultaneously, or nearly so; peduncle velutinous, light green; floral bracts ovate-elliptic, obtuse, finely retuse-apiculate at apex, oblique, 4–10 × 2.8–5 cm, much shorter than the ovary, the adaxial surface with sparse multicellular hairs, some having a round, glandular tip, the margins pale yellow-green, fused basally to 2 cm; sterile lower bracts absent. *Flowers* large, green, yellow and white, veined and suffused with green; outer surfaces mostly finely puberulous to glabrous, the inner surfaces mostly glabrous, the margins undulate, yellow and white; ovary and pedicel 100–201 mm long, tomentose-velutinous, yellow-green, hooked apically. *Dorsal sepal* ovate-lanceolate, acuminate, minutely hooded at apex, 90–170 × 20–30 mm, yellow and white with 7–11 green somewhat reticulate nerves, puberulent, the margins undulate, yellow and white; synsepal ovate, acuminate, hooded at the apex, 100–120 × 30–50 mm, nerves 13–17, yellow-green, puberulent, the margins undulate, white. *Petals* linear-elongate, bilobed, oblique at apex, 440–800 × 10–15 mm, white with green and reddish brown nerves, hanging parallel to lip, twisting, inner surface glabrous to puberulent basally, both surfaces becoming red-tomentose apically, the basal hairs long, white to reddish brown, arachnoid, the margins white, somewhat undulate, ciliate with red multi-cellular hairs. *Lip* urceolate, 50–70 × 20–30 mm, yellow-green, the opening obovate, the apical lip rim glabrous to puberulent, the apical lip lining to 10 mm deep, the lateral rim yellow-green with a row of dark reddish brown spots to the inside of a distinct rim of woolly pubescence 2–4 mm wide, spurs absent, the lateral lining to 14 mm wide, the claw face white, variously spotted with dark reddish brown, ribbed, tubercles absent; claw interior pilose. *Column* 4–5 mm long, yellow-green, often with red splotch above; pubescence dark red and white; staminode broadly trilobed-triangular, 5–6 × 9–17 mm, yellow-green with dark red lateral lobes, lobes acute to obtuse, often falcate; apical lobe acute to sparse, truncate, multicellular, reddish brown long hairs, marginal on apices of lateral lobes; stigma 4–5 mm wide. *Capsule* velutinous. Chromosome number: 2n=28 (Karasawa, 1980). **Plate 30. Figs. 133–137.**

Phragmipedium caudatum (Lindl.) Rolfe in Orchid Rev. 4: 332 (1896). Type: Peru, Pillao, *Ruiz & Pavon s.n.* (holo. K!; iso. BM!, G!, M!, W!).
Cypripedium caudatum Lindl., Gen. Sp. Orch. Pl. 531 (1840).

Selenipedium caudatum (Lindl.) Rchb.f., Xenia Orch. 1: 3 (1854).
Paphiopedilum caudatum (Lindl.) Pfitzer in Engler, Pflanzenfam., Orch. 2: 84 (1888).
Phragmopedilum caudatum (Lindl.) Rolfe *emend.* Pfitzer in Bot. Jahrb. Syst. 25: 518 (1898).

DISTRIBUTION. Peru and Bolivia. **Map 30**.

HABITAT. Plants may be terrestrial, epiphytes, or lithophytes; 1000–2100 m.

PHENOLOGY. Flowers have been reported in February and April to September in central and southern Peru.

CONSERVATION STATUS. Endangered (EN): A2cd+3cd+4cd; B2ab(ii,iii,v). It is a local species in Peru and Bolivia with a large distribution area. Some subpopulations are large but the trend of the population and the number of mature individuals are decreasing. The population size has declined over the last three generations and is predicted to decline further in the future to many threats including deforestation, road-construction, collection for regional and international horticultural trade by local collectors and orchid growers and habitat degradation. The estimated area of occupancy of the species is less than 500 km^2 with an estimated continuing decline in the number of mature individuals and the quality of habitats in many locations.

ETYMOLOGY. The specific epithet refers to the long, caudate lateral petals.

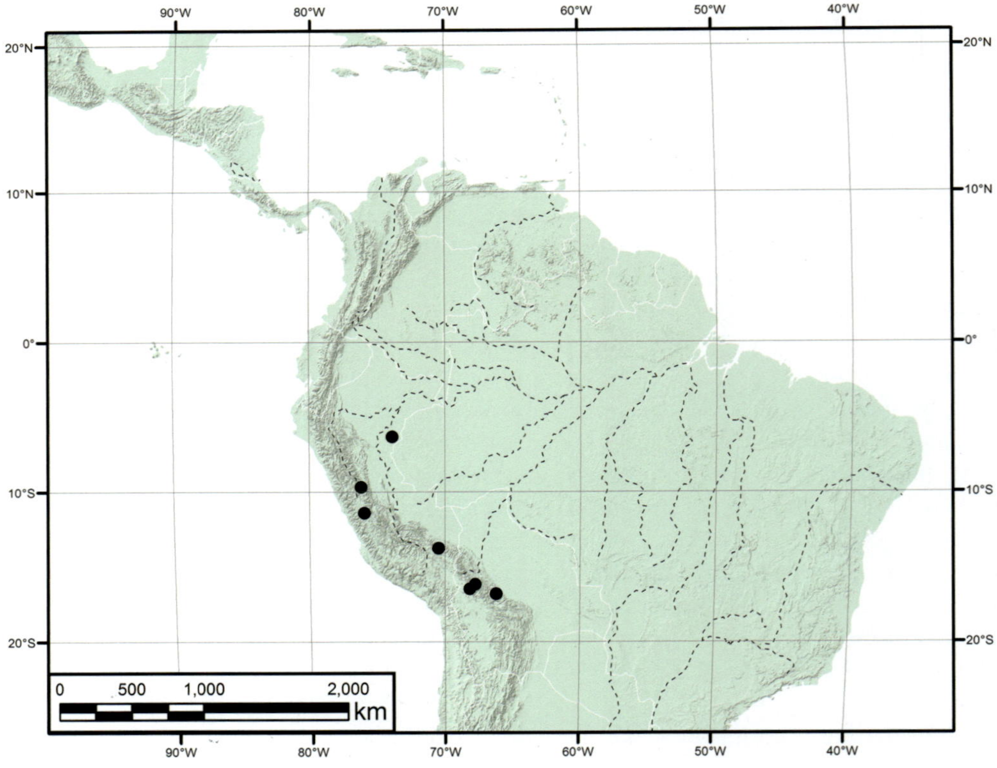

Map 30. Distribution of *Phragmipedium caudatum* in Peru and Bolivia.

18. PHRAGMIPEDIUM HUMBOLDTII

Phragmipedium humboldtii is allied to *P. caudatum* and *P. warszewiczianum* but can be readily distinguished by its flower colour, with the lip strongly flushed with chocolate brown, and distinctively ellipsoidal, rather than urceolate lip shape.

The nomenclature of this species is complicated. It has been variously treated as *P. caudatum* Lindl., *P. warszewiczii* (Rchb.f.) Christenson, *P. warszewiczianum* (Rchb.f.) Schltr., *P. popowii* Braem et al. and *P. exstaminodium* subsp. *warszewiczii* (Rchb.f.) Dressler. Josef von Warszewicz, an eminent plant collector in Central America, recognised as early as 1849 that this taxon was distinct from the Peruvian *P. caudatum* by reason of its lip shape and flower colouration. Later authors mostly agree. Atwood & Dressler (1999) published a new combination, *Phragmipedium humboldtii*, based upon *Cypripedium humboldtii* (as *humboldti*), a name with a short description given by Warszewicz and published in an article by H.G. Reichenbach (1852). Braem and Ohlund (2004) argued that *C. humboldtii* should be considered an invalid name, and thus be rejected. They, therefore, published a new name *P. popowii* for this taxon. Dressler (2005) argued that *P. popowii* was a later name for *P. exstaminodium*, the loss of a staminode being the only significant distinction between them. He, thus, proposed that the name *P. exstaminodium* be used for the species and recognised two varieties: *exstaminodium* for the taxon lacking a staminode, and *warszewiczii* for that with a staminode present.

Christenson (2006) agreed that *P. humboldtii* should be rejected but considered that Reichenbach's *Selenipedium warszewiczii* (Reichenbach, 1873) had been overlooked and thus proposed the new combination *Phragmipedium warszewiczii* (Rchb.f.) Christenson as the correct name. Atwood & Dressler (1998) lectotypified it, using a Warszewicz specimen collected on Chiriqui in Panama, and cited the name as *C. humboldtii* Warsz. ex Rchb.f. Such an interpretation of the authorship, however, invalidates the name under Article 34.1

Fig. 138. *Phragmipedium humboldtii* var. *humboldtii*. Column, front and side views. (FP)

Plate 31. *Phragmipedium humboldtii.* Drawn by Walter Hood Fitch for Robert Warner's *Select Orchidaceous Plants* t. 1 (1865) from a cultivated plant. (Kew collection)

of the International Code of Nomenclature (ICN), as argued by Braem and Ohlund (2004a), Christenson (2006) and Braem (2014). However, Dressler & Pupulin (2011) and Pupulin & Dressler (2011, 2012) rejected this attribution of the name since Reichenbach quite clearly cited the name as '*C. humboldti* Wzw.', thereby attributing the name to Warszewicz alone. They accepted it as validly published based upon Reichenbach's publication of Warszewicz's short diagnosis 'Blüthen stark gelb, rothvig, Lippe purpurenfleckig'. Thus, acceptance of the name attributed to Reichenbach becomes irrelevant under the ICN.

Archila (1999) described two species with obvious affinities to *P. humboldtii*, namely *P. monstruosum* based upon a specimen in which the flower and staminode closely resemble those of *P. humboldtii* but the lip is 15 cm long and *P. triandrum* which lacks a staminode which is replaced by a fertile stamen. These have been subsequently treated as subspecies of *P. humboldtii* by Mó Mó & Perez Garcia (2015, 2016). Based upon Archila's drawings and photographs of the types, both known from only a single specimen, we consider them to be no more than monstrous forms of *P. humboldtii* subsp. *humboldtii* and here reduce them to its synonymy. It could be argued that they occupy a similar relationship to the typical subspecies as subsp. *exstaminodium*, but that does appear to form a distinct population.

In Chiriqui, Panama, one population examined by McCook (1998) comprised mostly epiphytes on dead trees and tree stumps, while another population only three kilometres away was totally terrestrial (McCook, 1998). Dressler (2005) reported it growing on tall trees in wet montane forest, but also on isolated small trees in open pasture and on steep roadside cuttings, along with *P. longifolium*. Natural hybrids of *P. humboldtii* and *P. longifolium* were reported from Panama in 2002 (Dressler, 2005).

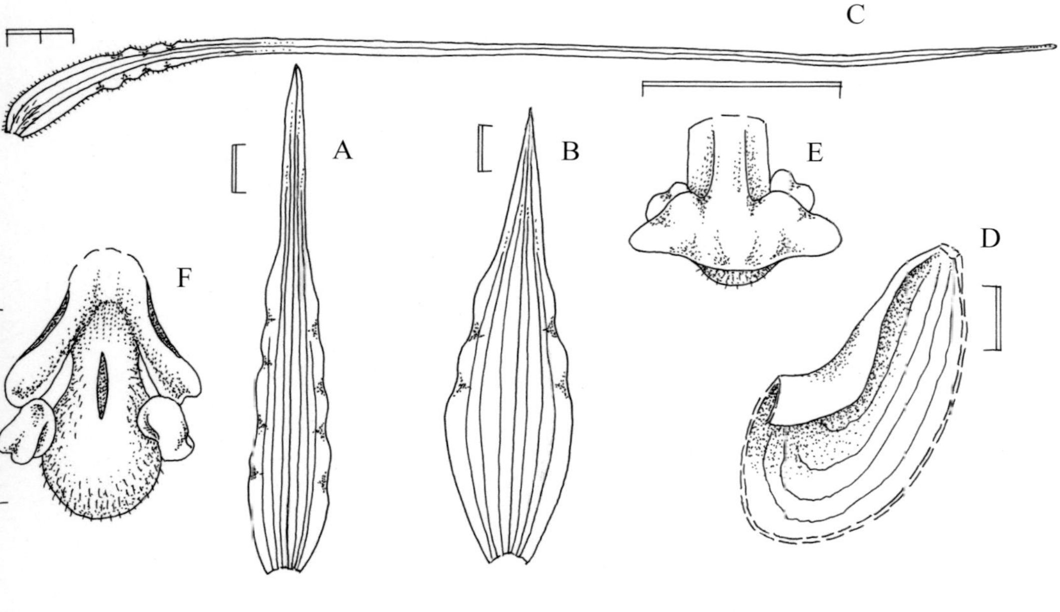

Fig. 139. ***Phragmipedium humboldtii*** var. ***humboldtii***. **A**, dorsal sepal; **B**, synsepal; **C**, petal; **D**, lip, longitudinal section; **E**, column and staminode, from above. ***P. humboldtii*** var. ***exstaminodium***. **F**, column, from beneath. A–E drawn by Judi Stone from *Powell 153*; F drawn after R. Lelen. Double bar = 10 mm.

Fig. 140. *Phragmipedium humboldtii var. humboldtii*. Cultivated plant. (FP).

DESCRIPTION. Terrestrial, lithophyte, or epiphytic herbaceous perennial *herb* growing in clumps of 3–4 fans; rhizome up to 3 cm long; roots woolly, up to 6 mm in diam., covered in red-brown hairs. *Leaves* thick and coriaceous, 5–6, lorate obtuse, retuse-apiculate, oblique and often splitting along mid-vein at apex, 15–60 × 2.8–5.5 cm, glabrous, light green, mid-vein prominent, the margins pale green. *Inflorescence* 30–60 cm long, 2–4 (5-) flowers, racemose, the flowers blooming simultaneously, or nearly so; peduncle velutinous, light green; floral bracts ovate-elliptic, obtuse, finely retuse-apiculate and oblique at the apex, 4–10 × 1.3–5 cm, much shorter than the ovary, the adaxial surface with sparse multicellular hairs, some having a round, glandular tip; margins pale yellow-green, fused basally to 2 cm; sterile lower bracts absent. *Flowers* deciduous, large, green, yellow and white suffused with reddish brown; outer surfaces mostly finely puberulous to glabrous; inner surfaces mostly glabrous; margins undulate, yellow and white; ovary and pedicel 10–21 cm long, tomentose-velutinous, yellow-green, hooked apically. *Dorsal sepal* ovate-lanceolate, long-acuminate, minutely hooded, 90–170 × 20–52 mm, yellow and white with 7–11 reddish brown somewhat reticulate nerves, puberulent, the margins undulate, yellow and white; synsepal ovate, acuminate, hooded, 90–120 × 20–52 mm, nerves 13–17, yellow-green, puberulent, the margins undulate, white. *Petals* linear-elongate, 130–800 × 7–15 mm, white with green and reddish brown nerves, hanging parallel to lip, twisting, inner surface glabrous to puberulent basally, both surfaces becoming tomentose apically; basal hairs long, white to reddish brown, arachnoid, the margins white, somewhat undulate, ciliate with red multicellular hairs, the apex bilobed, oblique, red-hairy. *Lip* urn-shaped, 32–76 × 20–30 mm, yellow-green suffused with reddish brown, the lip opening obovate, the apical rim glabrous to puberulent, the apical lining to 1 cm deep, the lateral rim red-brown with a row of dark reddish brown spots to the inside of a distinct rim of woolly pubescence 2–4 mm wide, the incurved sides up to 14 mm wide, the claw face white, variously spotted with dark reddish brown, ribbed, tubercles absent, its interior pilose. *Column* 10–14 mm long, yellow-green, marked with maroon-red above, with dark red and white pubescence; staminode widely trilobed-triangular, 5–6 × 9–17 mm, yellow with dark maroon-red side lobes, the lobes acute to obtuse, often falcate; apical lobe acute to truncate; hairs sparse, long, marginal on apices of lateral lobes, multi-cellular, reddish brown. *Capsule* velutinous. 2n=28 (Karasawa, 1980). **Plate 31. Figs. 138–143, 145.**

Phragmipedium humboldtii (Warsz.) J.T. Atwood & Dressler in Selbyana 19, 2: 246 (1998). Type: Panama, Chiriqui, *Warszewicz* 41 (lecto. W; holo. B†).

Cypripedium humboldtii Warsz. in Rchb.f. in Bot. Zeitung 10, 40: 691 (1852), as *C. humboldti*.

C. caudatum var. *roseum* Hort ex Delchevalerie in Rev. Hort. 39: 133 (1867). Type; hort. Guibert, Passy (holo. not preserved); Lindenia 2: t. 96 (neo., selected by Pupulin & Dressler, 2012)

Selenipedium warszewiczii Rchb.f., Xenia Orchid. 2: 189 (1873).

S. caudatum var. *giganteum* Carrière in Rev. Hort. 56: 367 (1884). Type: cult. Rougier-Chavière (holo. not preserved).

S. caudatum var. *roseum* L. Linden in Ill. Hort. 33: t. 506 (1886). Type: Ill. Hort. 33: t. 506 (1886), (lecto., designated by Pupulin & Dressler, 2012)

S. caudatum var. *albertianum* L. Linden in Lindenia 4: 64 (1888). Type: cult. Hye (probably not preserved; lecto., Lindenia 4: 64 (1888), selected by Pupulin & Dressler, 2012)

Cypripedium caudatum var. *albertianum* L. Linden in Lindenia 4: 64 (1888).

C. caudatum Lindl. var. *warszewiczii* (Rchb.f.) Veitch, Man. Orch. Pl. 4: 61 (1889).

Selenipedium caudatum (Lindl.) Rchb.f. var. *warszewiczii* (Rchb. f.) Pucci, 60–61 (1891).

Paphiopedilum caudatum (Lindl.) Pfitzer var. *warszewiczii* (Rchb.f.) Stein, Orchideenbuch, 461 (1892).

Phragmopedilum caudatum (Lindl.) Rolfe Pfitzer var. *warszewiczii* Pfitzer in Engler, Pflanzenr. 1, H.12: 53 (1903).

Phragmipedium warszewiczianum sensu Garay in Orchid Dig. 43: 140 (1979), *non* (Rchb.f.) Schltr.

P. monstruosum Archila in Revista Guatemalensis 2, 3: 5 (1999), **syn. nov.** Type: Guatemala, Alta Verapaz, Kaquipeck, *Archila s.n.* (holo. AGUAT).

P. triandrum Archila in Revista Guatemalensis 2, 3 (1999), **syn. nov.** Type: Guatemala, Alta Verapaz, *Archila s.n.* (holo. BIGUA).

P. popowii Braem, Ohlund & Quéné in Richardiana 4, 4: 185 (2004), nom. superfl. Based upon same type as *P. humboldtii* Warsz.

P. exstaminodium Castano, Hagsater & E. Aguirre subsp. *warszewiczii* Dressler in Orchid Digest 69, 2: 89 (2005). Type: Panama, Bocas del Toro, Cerro Colordado, 7 m W of Chame, along trail to Guaymi, *Croat* 69170 (holo. MO).

Figs. 141 & 142. *Phragmipedium humboldtii* var. *humboldtii*. Cultivated plant. (FP).

P. warscewiczii (Rchb.f.) Christenson in J. Orchideenfr. 13, 2: 142 (2006).
P. caudatum var. *albertianum* (L. Linden) Gruss & Wolff, Orchid Atlas: 337 (2007).
P. humboldtii (Warsc.) Atwood & Dressler subsp. *monstruosum* (Archila) Mó Mó & Perez Garcia in Die Orchidee 66, 3: 222 (2015) *emend.* Lankesteriana 16, 1: 39 (2016), **syn. nov.**
P. humboldtii (Warsc.) Atwood & Dressler subsp. *triandrum* (Archila) Mó Mó & Perez Garcia in Die Orchidee 66, 3: 222 (2015) *emend.* Lankesteriana 16, 1: 39 (2016), **syn. nov.**

DISTRIBUTION. Panama, Costa Rica, Nicaragua and Guatemala. **Map 31**.

HABITAT. Epiphyte with bromeliads and other orchids in moss and lichens on large trees, such as *Coccoloba*, in montane forest; 1500–1600 m

POLLINATION AND PHENOLOGY. In Chiriqui, Panama, a syrphid fly of the genus *Metasyrphus* (Grimaldi, pers. com.) was collected exiting a basal lip opening with an intact pollinium on its abdomen. Otherwise, pollination is unknown in this species. Flowers have been reported in April to July in Panama and Guatemala.

CONSERVATION STATUS. Endangered (EN): B2ab(ii,iii,v). It is a local species with a large but very fragmented distribution. A well-known population occurs in the Fortuna forest reserve in Panama where it is protected from collection. However, the population is decreasing and the number of mature individuals and the abundance of the species are low due to continuing collection for horticulture, agriculture intensification, forest fires and deforestation. The estimated area of occupancy of the species is less than 500 km^2 with an estimated continuing decline in the number of mature individuals and the quality of habitats in all the location.

VERNACULAR NAMES. 'Chinella' in Panama and 'Barbuda' in Guatemala.

ETYMOLOGY. Named for Alexander von Humboldt (1769–1859), who was a Prussian geographer, naturalist and explorer. Between 1799 and 1804, he travelled extensively in Latin America, exploring and describing it for the first time from a modern scientific point of view.

Fig. 143. *Phragmipedium humboldtii.* (FP)

18a. var. exstaminodium (Castaño, Hagsater & Aguirre) J.T. Atwood & Dressler ex J.M.H. Shaw in *Orchid Rev.* Suppl., 119(1296): 84 (2011) [*Quart. Suppl. Int. Reg. Orchid Hybrids*].

Phragmipedilum exstaminodium Castaño, Hagsater & Aguirre in Orquidea (Mex.) 9: 193 (1984). Type: Mexico, Chiapas, Tziscao, *Leleu s.n.* (holo. AMO!).

DESCRIPTION. Similar in habit to the typical variety. *Flowers* similar to P. *humboldtii* except for the *column* in which the pollinia are in contact with sides of stigmatic surface, and the staminode is much reduced or absent. Chromosome number unknown. **Fig. 144.**

DISTRIBUTION. Var. *exstaminodium* has been found in Chiapas, Mexico, and also in adjacent Guatemala and possibly Honduras. **Map 31.**

HABITAT. Epiphytic, clinging to the bark of trees 1.54 m above the ground in wet oak-conifer forest, a habitat that is rapidly being clear-felled; between 600 and 1700 m elevation.

PHENOLOGY. Castaño *et al.* (1984) give the flowering period as May to July.

CONSERVATION STATUS. Critically Endangered (CR): A2bc+3bc+4bc; B2ab(ii,iii,v); C1+2a(i, ii); D. It is a very rare, local subspecies with a restricted distribution. The trend of the population is decreasing, the number of mature individuals is very low and it is disappearing due to deforestation because the forests in southern Mexico and Guatemala are being cleared due to agriculture intensification, cattle raising, forest fires, overcollection and the current land use practices that have not allowed the vegetation to regenerate. The estimated area of occupancy of the species is less than 500 km^2 with an

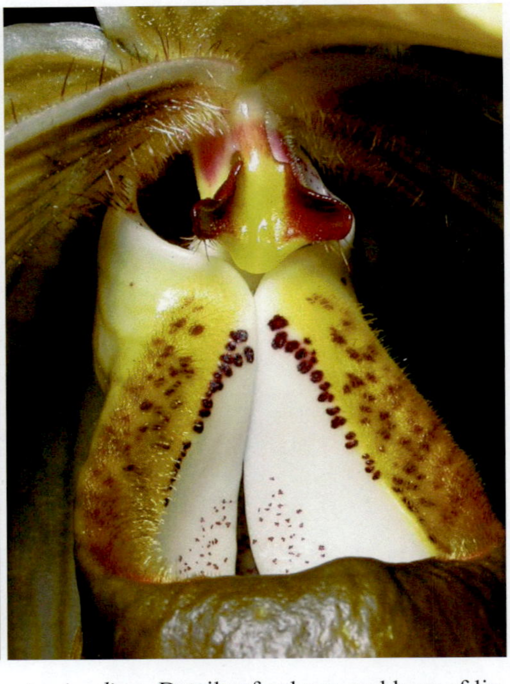

Fig. 144 (left). *Phragmipedium humboldtii* var. *exstaminodium*. Details of column and base of lip, from front. (FP) **Fig. 145** (right). *P. humboldtii* var. *humboldtii*. Details of column, staminode and base of lip, from front. (FP)

estimated continuing decline in the number of mature individuals and the quality of habitats in the single location. The population has declined by 95% over the last three generations and is projected to become extinct if conservation measures are not applied.

ETYMOLOGY. The species name refers to the absence from the column apex of a staminode, which is found in all other cypripedioid orchids.

VERNACULAR NAME. 'Tanal de Bigotes' (Castaño *et al.*, 1984; Dressler, 2005)

Compared to the other long-petalled species, little is known about var. *exstaminodium* and only a few plants exist in cultivation. The original Mexican population was in a wet montane forest that has been largely felled, and the taxon may well be lost in the wild in that locality. It is listed as an endangered species in Mexico, and may already be extinct there, only five plants remained when one of us (PC) visited the habitat in 1990. It also occurs in adjacent Guatemala and possibly Honduras, but it is likely that its habitats there are also under great pressure from logging and a rapidly increasing population.

The staminode, much reduced in this taxon, has been perceived as a critical structure in effecting pollination in cypripedioid orchids (Van der Pijl and Dodson, 1966). Although Castaño *et al.* (1984) suggested that it is not autogamous, the type specimen bears two pollinia that are in contact with the sides of the stigmatic lobes. The physical contact between stigmatic surface and pollinia suggests strongly that self-pollination is obligate in this species. If it is autogamous, the loss of its staminode may have followed.

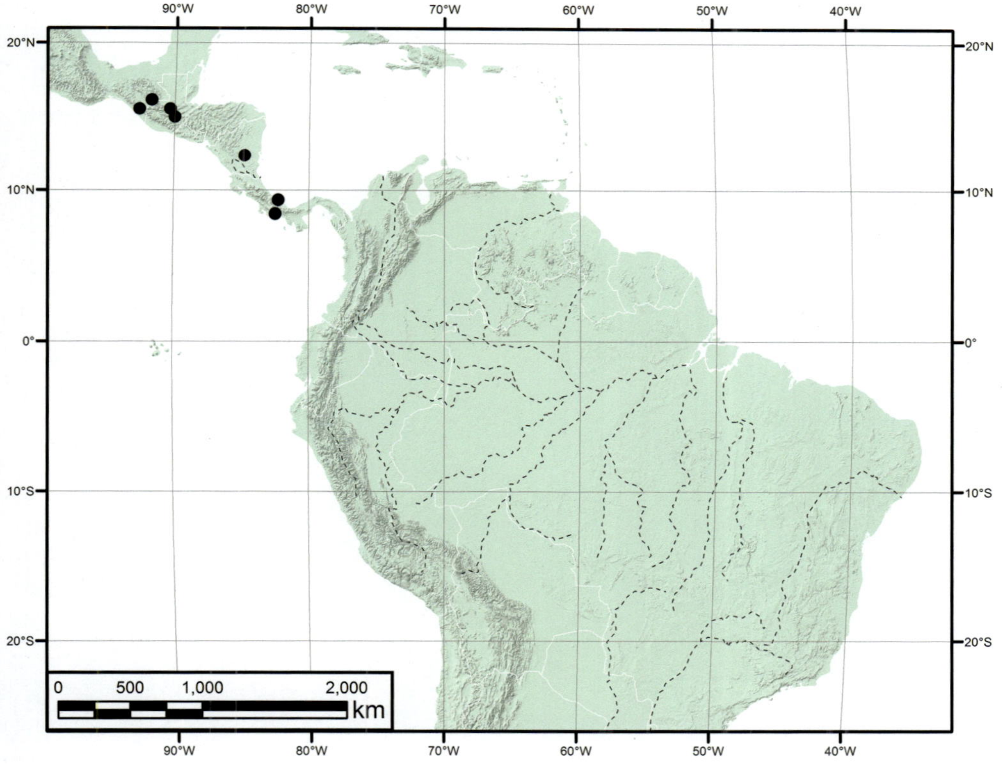

Map 31. Distribution of *Phragmipedium humboldtii*, including var. *exstaminodium*, in Central America.

19. PHRAGMIPEDIUM LINDENII

In 1846, John Lindley established *Uropedium* and described the monotypic *Uropedium lindenii* based upon Jean Linden's collection from Ocaña in Colombia. Robert Dressler and Norris Williams transferred it to the present genus in 1975. The genus *Uropedium*, based upon this species, predates *Phragmipedium*, but used to be rejected under the *International Code of Botanical Nomenclature* because it was based upon a teratological specimen. However, when the Code changed, *Phragmipedium* was conserved over *Uropedium*.

This is the most distinctive of all slipper orchids in that its flowers lack a pouch-like lip which is replaced by a petal-like organ. The loss of the slipper-shaped lip, so characteristic of the subfamily Cypripedioideae, has been interpreted as a mutation linked to self-pollination.

It is an obligately autogamous species, in which self-pollination is effected by a third functional stamen, which grows until it contacts the stigmatic surface. Virtually every flower in nature and in the glasshouse sets fruit.

DESCRIPTION. Terrestrial, herbaceous perennial *herb* up to 40 cm tall; rhizomes up to 1 cm long, 0.5–0.8 cm in diam.; roots striate, 0.3–0.5 cm in diam. *Leaves* lorate, obtuse, emarginated and oblique at apex, 25–40 × 2.5–4.3 cm, coriaceous; margins revolute, light green. *Inflorescence* 2–4-flowered, racemose; peduncle green, pilose; floral bracts ovate-elliptic, obtuse and apiculate at apex, 4–6.4 × 1.6–4 cm, much shorter than ovary, the adaxial surface sparsely pubescent with multi-cellular, gland-tipped hairs, the margins light green, fused basally; sterile lower bracts usually absent; if present, like leaves. *Flowers*

Fig. 146. *Phragmipedium lindenii.* Plant growing on old lava flow, lower slopes of Tunguragua, Ecuador. (PC). **Fig. 147** (inset). Column and staminode (JL).

Fig. 148. *Phragmipedium lindenii*. Plant growing on old lava flow, lower slopes of Tunguragua, Ecuador. (PC).

Plate 32. *Phragmipedium lindenii.* Drawn by Carol Woodin from a plant cultivated by Lynn and Bob Wellenstein, Candor, New York.

Fig. 149. The habitat of *Phragmipedium lindenii* on old lava flows above river, lower slopes of Tunguragua, Ecuador. (PC).

yellow-green becoming maroon apically, the outer surfaces glabrous to puberulent, the inner surfaces mostly glabrous, the margins yellow to white, somewhat undulate; pedicel and ovary 120–180 mm long, densely pilose. *Dorsal sepal* ovate-lanceolate, acuminate and minutely hooded at apex, 80–120 × 14–25 mm, yellow-green with 9–11 parallel to somewhat reticulate maroon nerves, glabrous; margins undulate; synsepal ovate-lanceolate, minutely hooded at apex, 80–100 × 15–35 mm, yellow-green with 14–17 parallel maroon nerves, the margins undulate, white to yellow. *Petals* lanceolate, acute, 180–410 × 5–13 mm, yellow-green, the inner surface glabrous basally, both surfaces becoming pilose-tomentose apically, the basal hairs reddish brown, the margins basally ciliate. *Lip* petaloid, ovate-lanceolate, not saccate, 230–410 × 10–15 mm, like lateral petals. *Column* up to 5 mm long, green with a red blotch above; staminode widely trilobed-triangular, 5–8 × 9–12 mm, yellow-green with maroon lateral lobes, the apical lobe broadly acute, with sparse, marginal, multi-cellular, dark red hairs on the side lobes; median stamen to 5 mm long, contacting stigmatic surface; stigma 5–7 mm wide. *Capsule* puberulent-pilose, curving apically. 2n=28 (Karasawa, 1980). **Plate 32. Figs. 146–150.**

Phragmipedium lindenii (Lindl.) Dressler & N.H. Williams in Taxon 24: 691 (1975). Type: Venezuela, savannah overlooking Lake Maracaibo, *Linden s.n.* (holotype K!).
Uropedium lindenii Lindl., Orch. Linden. 28 (1846).
Cypripedium lindenii (Lindl.) Van Houtte in Fl. Serres 18: 155 (1870).
C. caudatum Lindl. var. *lindenii* (Lindl.) Veitch, Man. Orch. Pl. 4: 60 (1889).
Selenipedium caudatum (Lindl.) Rchb.f. var. *lindenii* (Lindl.) Pucci, Les Cypripedium, 56 (1891).

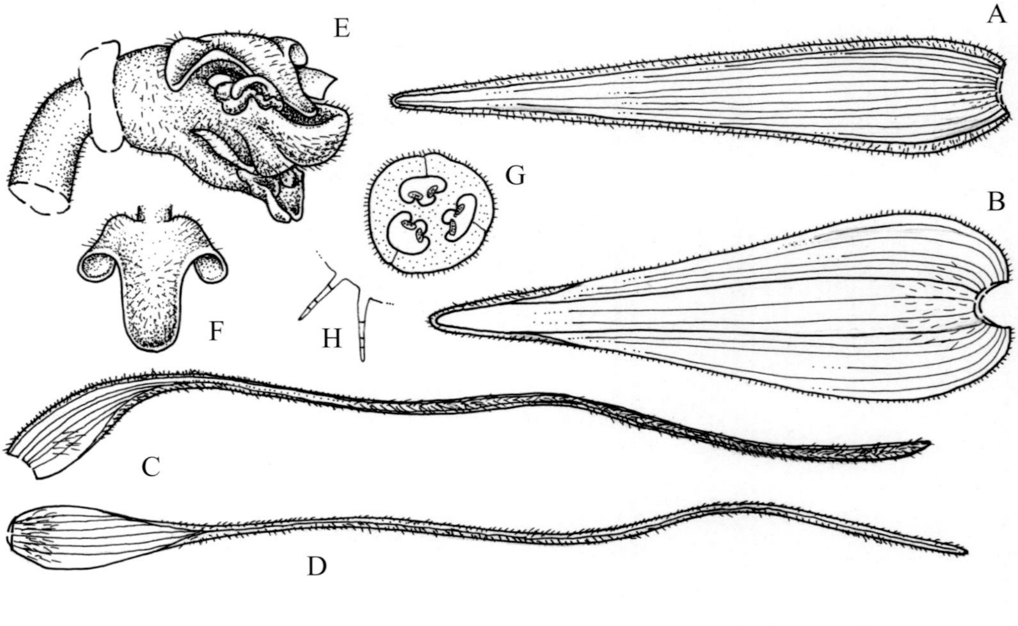

Fig. 150. ***Phragmipedium lindenii.*** **A**, dorsal sepal; **B**, synsepal; **C**, petal; **D**, lip; **E**, column, side view; **F**, staminode; **G**, ovary, transverse section; **H**, hairs on ovary. All drawn by Judi Stone from cult. Lynn and Bob Wellenstein, Candor, New York. A-D, × 1; E, F, × 3; G, × 4.5; H, × 48

S. caudatum (Lindl.) Rchb.f. var. *uropedium* Rolfe, Lindenia 7: 21 (1892).
Paphiopedilum caudatum (Lindl.) Pfitzer var. *lindenii* (Lindl.) Stein, Orchideenbuch, 460 (1892).
Phragmopedilum caudatum (Lindl.) Rolfe *emend.* Pfitzer var. *lindenii* (Lindl.) Pfitzer in Engler, Pflanzenr. 1, H.12: 52 (1903).

DISTRIBUTION. W Venezuela, Colombia and Ecuador, eastern flanks of the eastern Andean cordillera. **Map 32**.

HABITAT. Terrestrial on road embankments and in lava fields; 1400–2100 m in elevation.

PHENOLOGY. Flowering has been reported from March to November.

CONSERVATION STATUS. Least Concern (LC). It is rated as Endangered in Venezuela but it is relatively widespread and locally abundant in Colombia and Ecuador, where it is a primary coloniser of areas such as old lava flows. The trend of the population is stable, the estimated extent of occurrence and the estimated area of occupancy of the species are greater than 20000 km^2 and 2000 km^2, respectively. In addition, no major threats are affecting the species which could drive the species to extinction in the near future.

ETYMOLOGY. Jean Linden (1817–1898), born in Luxemburg, collected orchids commercially in South America and established a renowned orchid nursery, eventually called L'Horticulture Internationale, with branches in Brussels, Ghent and Paris.

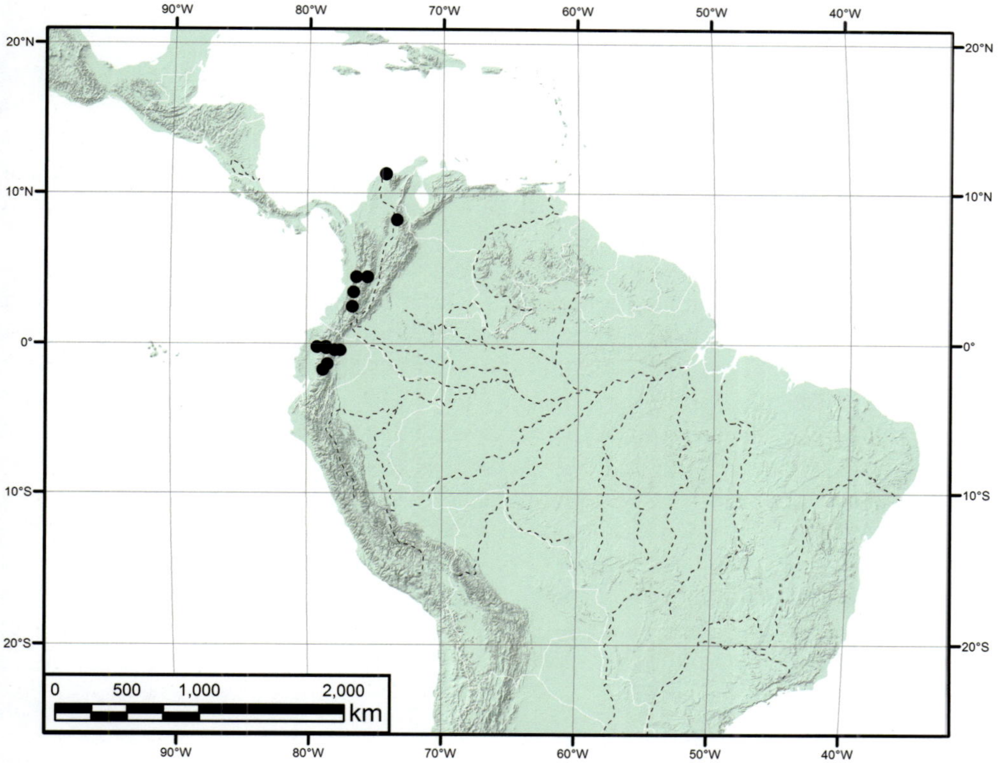

Map 32. Distribution of *Phragmipedium lindenii* in the northern Andes.

20. PHRAGMIPEDIUM WARSZEWICZIANUM

H.G. Reichenbach (1852) originally described it as *Cypripedium warszewiczianum* based upon a coloured illustration, published as *Cypripedium caudatum* by John Lindley (1844) in his and Paxton's *Flower Garden*. This shows a plant with a pale rose-pink lip, characteristic of the species that is now widely known as *P. wallisii*. Schlechter (1922) transferred it to *Phragmipedium* but created confusion by citing a Panamanian collection (*Powell* 153) in his discussion, thereby suggesting that the distinctive taxon found in Central America was *P. warszewiczianum*. Unfortunately, Reichenbach (1874) caused further confusion when he described a Warszewicz collection from Chiriqui in Panama as *Cypripedium warszewiczii*. On the basis that Reichenbach's epithets were merely orthographic variants referring to the same species, both epithets '*warszewiczii*' and '*warszewiczianum*' have subsequently been applied to the Panamanian species. Braem *et al.* (2004) clarified this confusion, showing that Reichenbach clearly considered the taxa from Central America and from Ecuador to be distinct, applying his *C. warszewiczianum* to the latter. (For further discussion of the Central American species readers are referred to the treatment of *P. humboldtii*.)

Phragmipedium warszewiczianum occurs in Ecuador, Colombia and western Venezuela. It differs from the Peruvian *P. caudatum* in its flower colour and lip in which the lower margin projects forwards as a narrow keel and has a hairy band above on each side, but with the hairs being small and inconspicuous. In *P. caudatum* the lip margins are not prominently forwards projecting and the hairs on the band on each side are large and conspicuous. Both differ from *P. humboldtii* which has a rich chocolate brown lip with incurved margins to the orifice that are not thickened.

H.G. Reichenbach (1874) described *Selenipedium wallisii* in the second volume of his *Xenia Orchidacea* based upon a collection from Ecuador made by Gustav Wallis, a collector for the nursery of Messrs James Veitch & Sons of Chelsea. He gave a detailed description of it, distinguishing it by its lack of bristles on the margins of the tepals. Leslie Garay (1979) transferred it to *Phragmipedium*, as *P. wallisii*. It is undoubtedly conspecific with *P. warszewiczianum*.

Dressler (2005) reduced it to subspecific rank within *P. lindenii*, a species that has a petaloid rather than a slipper-shaped lip. Dressler & Pupulin (2011) clearly outlined the nomenclatural problems of this taxon and, supported Dressler's decision to give it subspecific rank, citing the similarity of the artificial hybrids resulting from hybridising *P. lindenii* and *P. wallisii* with *P. longifolium*. We prefer, for the time being, to distinguish *P. warszewiczianum* from *P. lindenii* while recognising that the latter is probably a stabilised mutation of the former. However, this hypothesis is based on casual observation and has not been tested. Pupulin & Dressler indicated that the confusing name *P. warszewiczianum* should not be used until a proposal to conserve the well-known name *P. wallisii* over it is submitted and tested. The *International Code of Botanical Nomenclature* allows for well-established names of horticulturally and agriculturally important species to be conserved. As far as we are aware, they have not presented such a proposal and, in the web-based World Checklist of Flowering Plants, Govaerts (2014) has listed *P. warszewiczianum* as the accepted name for this species.

DESCRIPTION. Epiphytic or terrestrial herbaceous perennial *herb* up to 60 cm tall; rhizomes up to 2 cm long; roots 4–10 mm in diam. *Leaves* thick and coriaceous, lorate, obtuse,

Plate 33. *Phragmipedium warszewiczianum*. Drawn by Carol Woodin from a plant cultivated by Lynn & Bob Wellenstein, Candor, New York.

Fig. 151. *Phragmipedium warszewiczianum*. Flower from cultivated plant. (PC).

apiculate, oblique, often splitting at the apex, 30–50 × 3–4.5 cm, glabrous, the margins pale green. *Inflorescence* 3–4-flowered, racemose, flowers blooming simultaneously; peduncle velutinous; floral bracts ovate, obtuse, oblique, 4.5–9 × 2–4.2 cm, much shorter than ovary, the adaxial surface glabrous to finely puberulous with simple hairs lacking glandular tips, the margins pale green, fused basally; sterile lower bracts generally absent, when present, leaf-like. *Flowers* white suffused with yellow and pink, the outer surfaces mostly tomentose to velutinous, the inner surfaces mostly glabrous, the margins white, undulate; ovary and pedicel 140–210 mm long, pilose-velutinous, yellow-green. *Dorsal sepal* ovate-lanceolate, acute, minutely hooded at apex, 80–130 × 15–20 mm, yellow-green, with 10 or 11, parallel nerves, the margins white, undulate; synsepal ovate-lanceolate, minutely hooded at apex, 75–115 × 17–35 mm, yellow-green with 11–15 darker yellow parallel nerves, the margins white, undulate. *Petals* linear-caudate, minutely bifid at tomentose apex, 300–550 × 7–15 mm, pendent parallel to lip, the inner surface glabrous, the basal hairs white, matted, the margins undulate, white, lacking long pubescence. *Lip* obovate, 40–54 × 15–18 mm, white suffused with green and pink; mouth yellow, obcordate, the apical rim wide and flaring, glabrous, the apical lining to 1 cm deep, the lateral rim tomentose, hairs short pale and

Figs. 152 & 153. *Phragmipedium warszewiczianum.* Flowers from two cultivated plants. (PC, FP).

short, pubescent area to 1 mm wide, spurs absent, the lateral lining to 10 mm wide, the claw face white, ribbed, edged in reddish brown spots, the claw interior puberulent. *Column* up to 4 mm long, yellow-green, with a white indumentum; staminode widely triangular, lateral lobes somewhat falcate, reddish brown, with marginal, multicellular, reddish brown hairs on apex of lateral lobes; stigma up to 8 mm wide. *Capsule* pilose, curving apically. 2n=28 (Karasawa, 1980). **Plate 33. Figs. 151–154.**

Phragmipedium warszewiczianum (Rchb.f.) Schltr. in Repert. Sp. Nov. Regni Veg. Beih. 17: 9 (1922); Garay in Orchid Digest 43, 4: 140 (1979).

Cypripedium warszewiczianum Rchb.f. in Bot. Zeit. 10, 40: 689 (1852). Type: cult. Mrs Lawrence, illn in Lindley & Paxton's Flower Garden 1: t. 9 (1850).
Selenipedium wallisii Rchb.f., Xenia Orch. 2: 189 (1873).
Cypripedium wallisii (Rchb.f.) B.S. Williams, Orchid Grower's Manual ed. 6: 261 (1885).
C. caudatum Lindl. var. *wallisii* (Rchb.f.) Kent in Veitch, Man. Orch. Pl. 4: 61 (1889).
Selenipedium caudatum (Lindl.) Rchb.f. var. *wallisii* (Rchb.f.) Pucci, Les Cypripedium 60 (1891).
Paphiopedilum caudatum (Lindl.) Pfitzer var. *wallisii* (Rchb.f.) Stein, Orchideenbuch 460 (1892).
Phragmopedilum caudatum (Lindl.) Rolfe var. *wallisii* (Rchb.f.) Pfitzer in Engler, Pflanzenr. 4, 50, H.12: 53 (1903).

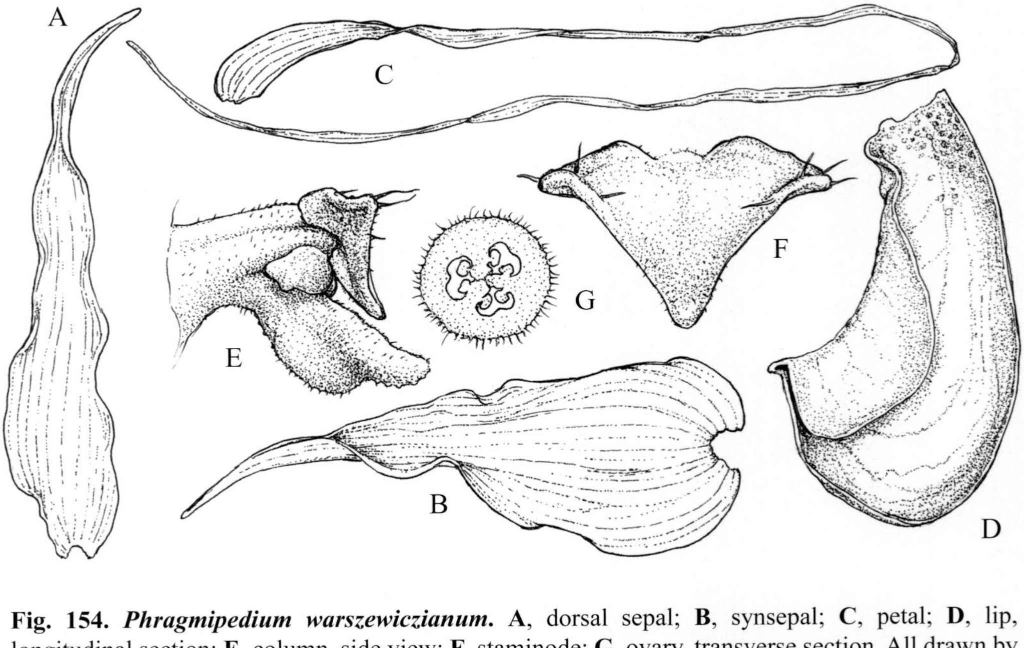

Fig. 154. ***Phragmipedium warszewiczianum.*** **A**, dorsal sepal; **B**, synsepal; **C**, petal; **D**, lip, longitudinal section; **E**, column, side view; **F**, staminode; **G**, ovary, transverse section. All drawn by Carol Woodin from cult. Hegedus. A–D, × 1.5; E,F, × 6; G, × 9.

Phragmipedium wallisii (Rchb.f.) Garay, Fl. Ecuador 9: 24 (1979). Type: Ecuador, Loja, Zamora, 800–1300 m, *Wallis s.n.* (holo. W); Ecuador, Zamora, *Lehmann* 6268 (neo. W!; isoneo. AMES, G, K, L, NY, US).

P. lindenii (Lindl.) Dressler & N.H. Williams subsp. *wallisii* (Rchb.f.) Dressler in Orchid Digest 69, 2: 89 (2005).

DISTRIBUTION Colombia, Ecuador and W Venezuela. **Map 33**.

HABITAT. Terrestrial or epiphyte in montane forest, often on wet banks and cuttings; 800–1700 m.

PHENOLOGY. November and July. The range of flowering time is unknown, as are pollinators.

CONSERVATION STATUS. Vulnerable (VU): A3cd+4cd; B2ab(ii,iii,v). It is a local species with a large fragmented range. The trend of the population is decreasing, the number of mature individuals and the abundance of the species are decreasing due to agriculture intensification, forest fires, over-collection and deforestation. The estimated area of occupancy of the species is less than 500 km^2 with an estimated continuing decline in the number of mature individuals and the quality of habitats in all the locations. The population has declined in recent years and is projected to decline further in the future.

ETYMOLOGY. Named for Josef von Warszewicz (1812–1866), a renowned plant collector in the tropical Americas, originally for the Belgian nursery of van Houtte and later as an independent collector.

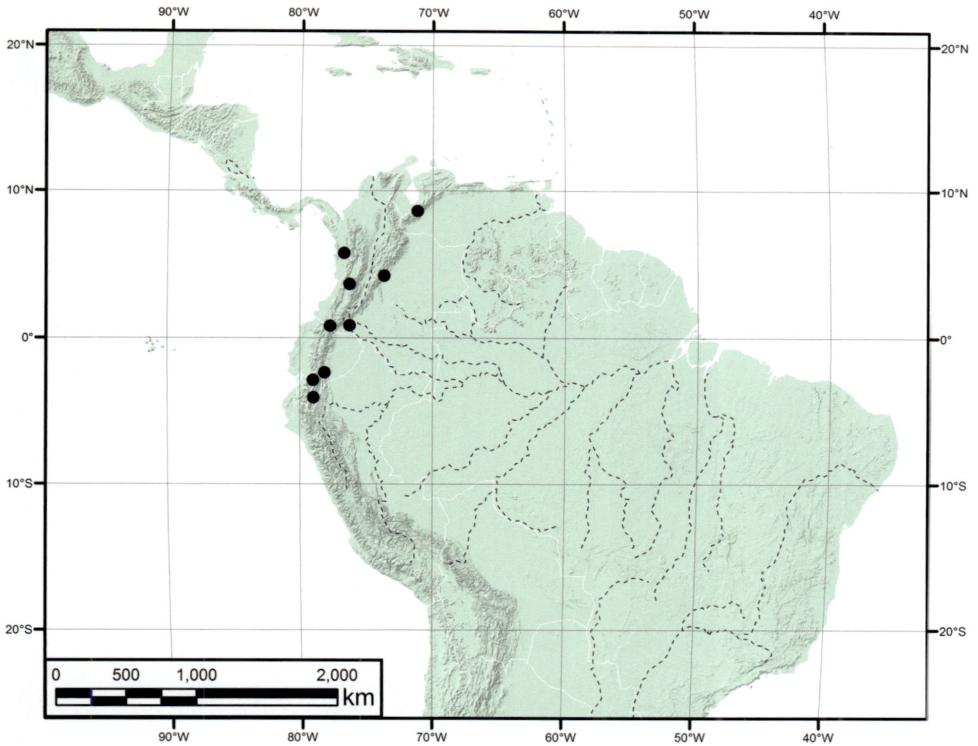

Map 33. Distribution of *Phragmipedium warszewiczianum* in the northern Andes.

21. PHRAGMIPEDIUM GUIANENSE

Sambin and Braem (2015b) described *P. guianense* in the journal *Richardiana* based upon a collection made by Jean-Pierre Bikaeff in 2012 and subsequently flowered in cultivation in Cayenne. This species has been known for some time but has been misidentified by Chiron & Bellone (2005) and Szlachetko *et al.* (2011) as *P. caudatum*, a species native to Peru. According to the authors, it is similar to that and to the Central American *P. humboldtii* but differs in having smaller flowers with shorter sepals, much shorter petals 10–70 cm long (vs 80 cm or more long), and a lip with a flared orifice like *P. caudatum* but coloured like that of *P. humboldtii*.

Its flowers have much in common with *P. klotzschianum*, particularly in lip shape and sepal and petal length. It relationship to that species needs further investigation.

Phragmipedium guianense Sambin & Braem in Richardiana 15: 4 (2015). Type: French Guyana, near Saül, *Jean-Pierre Bikaeff in C. Saül* 003 AS 02 (holo. CAY).

DESCRIPTION. Tufted epiphytic *herb* up to 24 cm tall; roots brownish red, hairy; rhizome very short; stems. *Leaves* somewhat coriaceous, conduplicate, 3–5 oblong-ligulate, up to 24 cm long, 1.9–2.3 cm wide, clear green, with entire margins. *Inflorescence* almost as long as or slightly longer than the leaves, up to 24.5 cm long, 1–2 flowered for 10 to 11 days and rarely opening at the same time; peduncle cylindrical, slightly laterally compressed, 63 mm long, 4 mm in diameter, pubescent, with a clear green sterile bract, 45 mm long, 16 mm wide, asymmetric at the tip; floral bracts 20 mm long, 8 mm broad, with an asymmetric

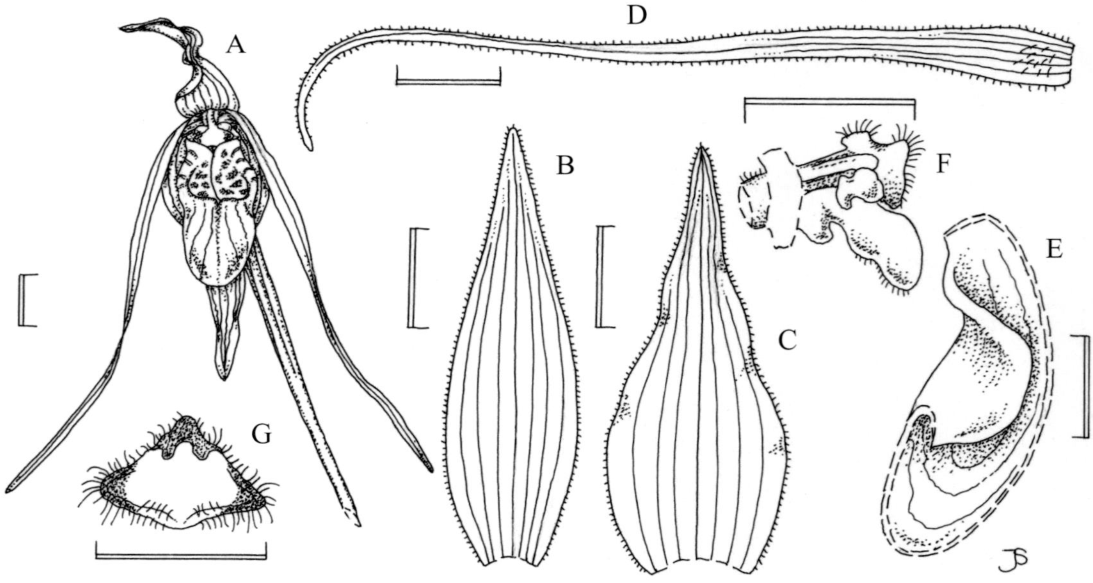

Fig. 155. *Phragmipedium guianense*. **A**, flower; **B**, dorsal sepal; **C**, synsepal; **D**, petal; **E**, lip, longitudinal section; **F**, column, side view; **G**, staminode. All drawn by Judi Stone after drawings of type by A. Sambin. Double bar = 10 mm.

tip. *Flowers* 10.5 cm across, the sepals cream-coloured to yellow-green with 8–9 purple veins and a purple or brown margin, the petals purple-brown with a greenish base, the lip with a purple border, green base veined with purple, white to yellow with large purple spots on the inside; pedicel and ovary cylindrical, 100–105 mm long, 4 mm in diam., green with purple margins and a brownish base. *Dorsal sepal* oblong-lanceolate, 45 × 9 mm, pubescent on outer surface, margins undulate; synsepal concave, ovate-elliptic, acute, 42 × 16 mm, pubescent on outer surface. *Petals* narrowly lanceolate to linear, 72–103 × 5 mm. *Lip* deeply saccate, 28 × 12 mm, the side lobes strongly incurved. *Column* 6–6.5 mm long; staminode transversely rhombic, 3 × 7 mm, ciliate, yellow with purple tips; stigma rhombic, 3 × 3.2 mm. **Fig. 155.**

DISTRIBUTION. Central French Guyana only. **Map 34.**

HABITAT. Known as an epiphyte on large trees in evergreen rain forest on deep rich clay soil over basic rock; 80–500 m elevation.

FLOWERING. August until the end of November.

CONSERVATION STATUS. Critically Endangered (CR): A3cd+4cd; B1ab(ii,iii,v)+2ab(ii,iii,v); C1+2a(i,ii). This species is a very local and rare endemic, with a population comprising a few small subpopulations near Saul. Populations are threatened with extinction by over-collection for horticulture.

ETYMOLOGY. Named for the country of origin.

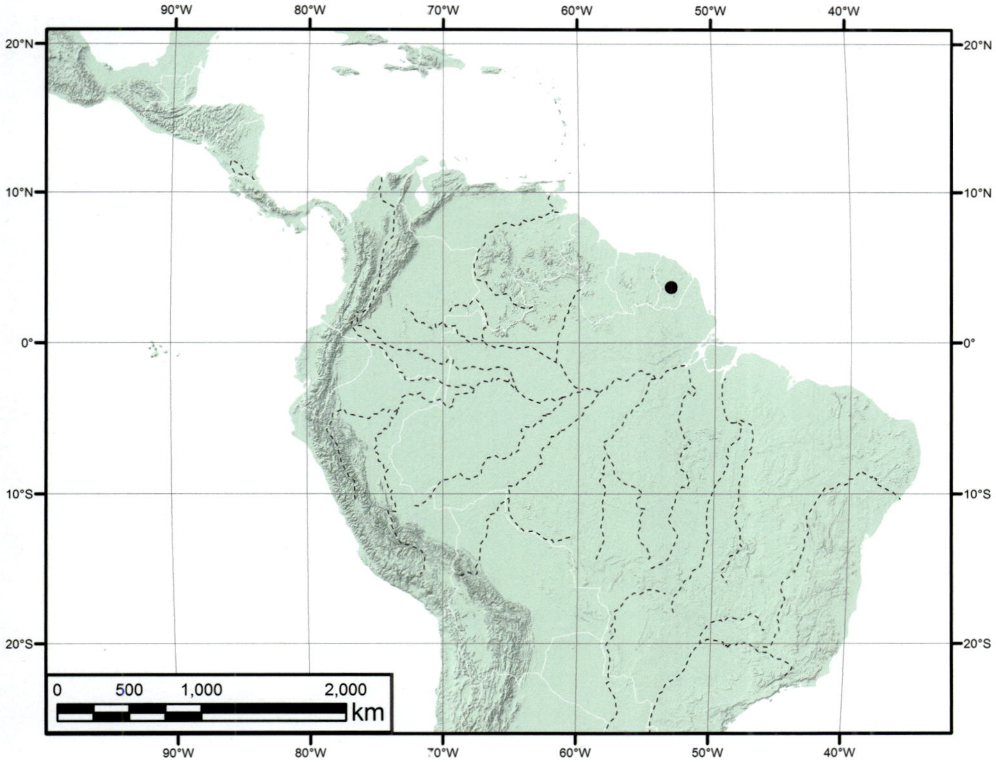

Map 34. Distribution of *Phragmipedium guianense* in French Guyana.

NATURAL HYBRIDS AND DUBIOUS SPECIES

A number of natural hybrids of *Phragmipedium* have been described or postulated over the years and are discussed below. Some were initially described as species but their intermediate morphology between probable sympatric species suggests otherwise.

PHRAGMIPEDIUM × BRASILIENSE

When described, the authors stated that two plants of this taxon had been purchased by John Fitch from Margaret Ilgenfritz, who had originally imported them from Brazil. The plants were then sent by Ron Ciesinki of Taylor Orchids to Jerry Fischer of Orchids Limited who crossed them and found that the progeny were uniform in appearance. Thus, the authors described this taxon as a new species. Because of their resemblance to the Brazilian species *P. vittatum*, Quené & Gruss (2003) compared it with that species and its hybrid with *P. longifolium* (*P.* Arthur) and noted its differences from both. In a lecture at the World Orchid Conference in Singapore, Gruss suggested a parentage of *P. sargentianum* and *P. christiansenianum*. However, given its floral morphology, inflorescence length and branching, it is surprising that neither compared the plants with *P.* Memoria Adele Bishop (*P. vittatum* × *sargentianum*) which it closely resembles. Given the unknown provenance of the material now at Orchids Limited and its similarity to the artificial hybrid, we think that this is best treated as a hybrid, possibly a natural one.

Phragmipedium × brasiliense Quené & Gruss in Orchid Digest 67, 4: 242 (2003), **stat. nov.** Type: ex Brazil, cult. Jerry Fischer, Orchids Limited, *Quené 11403* (holo. IRVC).
Postulated parentage: *P. vittatum* (Vellozo) Rolfe × *P. sargentianum* (Rolfe) Rolfe

PHRAGMIPEDIUM CHRISTIANSENIANUM

The type specimen was cultivated and flowered by Jurgen Roeth who received it from Hans Christiansen, the eminent Danish orchid grower, after whom it was named. Christiansen received it in a batch of *P. longifolium* said to have originated from Colombia. It was originally thought to be a hybrid but, according to Christiansen, breeds true. A number of other plants in cultivation, whose provenance is not clearly known, have been described as species over the years and continue to be problematic for the taxonomist. This taxon is certainly similar to *P. longifolium* in its flower size and morphology but also has hints of *P. pearcei* or a similar species in its lip side lobes, staminode and the crinkled margins of the floral segments. Further information is needed before an accurate assessment of this taxon can be made. For the time being, we prefer to consider it amongst the dubious species and natural hybrids where it may or may not belong.

Phragmipedium christiansenianum Gruss & Roeth in Die Orchidee 52, 1: 76 (2001). Type: ex Colombia, fl. 16 August 1999, cult. J. Roeth HAL083737 (holo. HAL).

DESCRIPTION. A terrestrial *herb*. *Leaves* 6, linear, acute, up to 30 × 1.2 cm. *Inflorescence* up to 40 cm long, bearing several flowers in succession; fertile bracts conduplicate, lanceolate, acute, up to 4.2 × 1.8 cm. *Flowers* greenish yellow with red margins to the petals, an ochre-brown lip with cream side lobes, marked with green and brownish spots, and a greenish staminode with a basal margin of dark purple hairs; pedicel and ovary 52 mm long. *Dorsal sepal* ovate, acute, 52 × 19 mm, with a twisted apex and undulate margins in the upper two-thirds; synsepal elliptic-ovate, acute, 42 × 28 mm, with undulate side margins. *Petals* spreading at 45° below the horizontal, linear-tapering, acute, 84 × 9 mm, spirally twisted and with undulate margins in the basal half. *Lip* deeply saccate, 42 × 18 mm, slightly broader across the basal half; side lobes sharply incurved, adnate below staminode. *Column* 0.5 cm long; staminode transversely elliptic, 5–6 × 8–10 mm.

PHRAGMIPEDIUM × COLOMBIANUM

The published photographs of this hybrid are somewhat intermediate between its alleged parents. However, it differs from typical *P. schlimii* only in having a dark pink lip. On balance, it may be no more than a well-coloured form of *P. schlimii*.

Phragmipedium × colombianum Gruss in Die Orchidee 62, 1: 30 (2011). Type: Colombia, without exact locality, cult. Franz Glanz, Gruss 2010–09–25 (holo. HAL).
Postulated parentage: *P.* [*schlimii* var.] *manzurii* W.E. Higgins & P. Viveros × *P. schlimii* (Linden & Rchb.f.) Rolfe.

PHRAGMIPEDIUM × MERINOI

This natural hybrid was described by Gruss (2010) based upon a plant collected by Gilberto Merino in Pastaza province of Ecuador and flowered by the Ecuagenera nursery of Gualaceo, Ecuador. It is intermediate in floral features between its putative parental species, namely *P. boissierianum* and *P. pearcei*. The description is accompanied by a description and excellent photograph which includes a detailed floral dissection. It needs to be compared with *P.* × *richteri*.

Phragmipedium × merinoi O. Gruss in Die Orchidee 61, 3: 176 (2010). Type: Ecuador, Pastaza Prov., without exact loc., cult. Ecuagenera, *Merino 2010–03–25* (holo. QCNE).
Postulated parentage: *P. boissierianum* (Rchb.f.) Rolfe var. *reticulatum* (Rchb.f.) Rolfe × *P. pearcei* (Rchb.f.) Rauh & Senghas

PHRAGMIPEDIUM × PFITZERIANUM

Jurgen Röth (2012) originally identified this as the natural hybrid *P.* × *grande*, based upon an Ecuadorian plant that flowered in cultivation in Germany for Heinrich Buchwald. Röth gave its parentage as *P. longifolium*, a common Ecuadorian species, and *P. humboldtii*, a species that does not occur in Ecuador but is restricted in its distribution to Central America. In that case the epithet '*grande*' cannot be used for this plant as it is a later homonym of the artificial hybrid *P.* Grande which was raised and flowered by the Veitch nursery in 1894 from *P. humboldtii* (as *P. caudatum*) crossed with *P. longifolium*.

Gruss (2013) realised Röth's error and published a new name, *P.* × *pfitzerianum*, for this hybrid, citing its parentage as *P. warszewiczianum*, the common long-petalled species found in Ecuador, crossed with *P. longifolium*.

Phragmipedium × pfitzerianum O.Gruss in Orchideen J. 20, 1: 22 (2013).
Based upon *P.* × *grande* Roeth in Die Orchidee 63, 5: 403 (2012), *nom. inval.*
Postulated parentage: *P. warszewiczianum* (Rchb.f.) Schltr. × *P. longifolium* (Rchb.f. & Warsz.) Rolfe

PHRAGMIPEDIUM × RICHTERI

Jurgen Röth and Olaf Gruss (1994) described *P. richteri* based upon a plant cultivated by Franz Glanz and said to have come from Peru. They distinguished it by its 'sedge-like leaves up to 70 cm in length; a reverse-elliptic staminode with a short horn at the lower margin; intensely black-brown hairs at the basal and lateral margin and short hairs elsewhere; and an occasionally branched inflorescence'. The authors admitted that they were sceptical that *P. richteri* was a good species when they described it. McCook (1998) and Christenson (2003) considered it to be a natural hybrid of *P. pearcei* × *boissierianum*.

This taxon had been documented by Bennett and Christenson (1993) based upon a plant collected by J. Schunke in August 1985 in the district of La Banda de Shilcayo in Peru at an elevation of 600 m on wet rocks in montane rain forest. However, Gruss (2003) stated that growers have asserted that it breeds true, considered an unlikely occurrence for a primary hybrid. Nevertheless, hybrids can stabilise and at least *P. boissierianum*, one of the putative parental species, is found in the region where Schunke collected orchids. It is certainly intermediate in size and morphologically in both its habit and leaf and floral morphology between the postulated parents. The plant and flowers are similar to some of the larger specimens of *P. pearcei* found in Ecuador. Petal undulation which is pronounced in *P. richteri* is evident to a lesser extent in *P. pearcei*. Given the stable nature of the former we think it best to treat it as a hybrid but its true status still needs to be confirmed.

Phragmipedium × richteri Roeth & O. Gruss in Die Orchidee 45, 3: back cover (1994).
Type: ex Peru, cult. F. Glanz (holo. HAL 072657).
Postulated parentage: *P. pearcei* (Rchb.f.) Rauh & Senghas × *P. boissierianum* (Rchb.f.) Rolfe.

Figs. 156 & 157. *Phragmipedium × richteri*. Plant cultivated by D. Martin, Sunbury, England.

DESCRIPTION. *Plant* 10-leaved. *Leaves* linear, acute, up to 65 × 2.2 cm. Inflorescence erect, sometimes branched, bearing foliaceous sterile bracts. *Flowers* borne in succession, sepals and petals yellowish green with darker veins, petals with a purple margin, lip yellow-green with darker green veins and a pale rim; ovary pubescent. *Dorsal sepal* ovate, subacute, 38–44 × 11–13 mm, with undulate margins and a somewhat obliquely recurved tip, synsepal concave, elliptic, obtuse, 32–35 × 18–20 mm. *Petals* arcuate-pendent linear-tapering to an acute tip, spirally twisted, 66–81 × 7–10 mm, margins strongly undulate-crispate. *Lip* deeply urceolate, 38–40 × 15–20 mm, the incurved side lobes adjacent in front of staminode, the rim of the orifice sharp and out-curved. *Column* short; *staminode* transversely elliptic, shortly apiculate at tip, 5–6 × 9–9.5 mm, green with the basal margin densely shortly dark purple-black pubescent. **Figs. 156 & 157.**

DISTRIBUTION. N Peru (San Martin Region).

HABITAT. Probably growing on rocks by and in fast-flowing streams in lower montane forest; c. 600 m (Bennett & Christenson, 1993).

PHRAGMIPEDIUM × ROETHIANUM

The taxon is intermediate in its habit, leaves and flowers between its putative parents, *P. hirtzii* and *P. longifolium*, species which are sympatric in north-western Ecuador. It is similar to *P. longifolium* in its long broad leaves, in the colouring of its flower and its staminode which has a prominent basal fringe of short dark purple hairs. It has the lip shape and long spirally twisted petals of *P. hirtzii*.

P. × roethianum O. Gruss & Kalina in Orchidee (Hamburg) 49: 245 (243–245, fig.]. Type: Ecuador, cult. T. Kalina, Northbrook, Illinois, *Kalina & Gruss* HAL 079921 (holo. HAL). Postulated parentage: *P. hirtzii* Dodson × *P. longifolium* (Rchb.f. & Warsz.) Pfitz.

DESCRIPTION. Leaves up to 40 × 2 cm; inflorescence up to 60 cm long; flowers 5–7, up to 13 cm in diam.; dorsal sepal up to 4.5 × 1.3 cm; synsepal up to 4.2 × 2.5 cm; petals spirally twisted, up to 7.8 × 0.7 cm; lip 3.4 × 1.8 cm; staminode transversely elliptic-subrhombic, 5 × 5 mm, with short dark purple hairs along the basal margins.

PHRAGMIPEDIUM × TETZLAFFIANUM

Olaf Gruss was sent this orchid from Canada in 1999 by Allan Tetzlaff who flowered it in his collection, having had it in cultivation for over ten years under the name *Phragmipedium klotzschianum*. Gruss (2000) subsequently described it as a new species of the section *Himantopetalum*, naming it for Norman Leo Tetzlaff, Allan's father. It is a vigorous plant and grows easily. And when it flowers it bears beautiful large flowers on an erect inflorescence which can have 1–4 flowers open simultaneously (up to 8 in older larger plants). According to Gruss this species is closest to *Phragmipedium richteri* Roeth & Gruss from Peru. Discussion on the internet remarks on its close resemblance to the artificial hybrid *Phragmipedium* Simon Marcotte (*P. ecuadorense* [*pearcei*]× Nitidissimum) and to the unregistered grex *P. richteri* × *klotzschianum*. Braem *et al.* (2014) list it as an artificial hybrid and we are inclined to agree.

DESCRIPTION. A *terrestrial herb*; rhizome short, bearing shoots 4–10 mm apart. *Stem* short, completely enclosed by six sheaths, imbricate at the base, laid out as a fan. *Leaves* coriaceous, lanceolate, acute, up to 50 cm long and 1.7 cm wide, erect then arched, with a smooth, slightly cartilaginous margin, central veining sunken, acutely ribbed on the lower part, dark green above, light green below, smooth. *Inflorescence* up to 35 cm tall, branched in the mature plants, erect, with 3–10 flowers, 1–4 open simultaneously, in very large plants up to 8 flowers open simultaneously, in sequence from the bottom upwards; peduncle round in cross section reddish brown, covered with fine whitish hairs; lowermost sterile bract 7.1 × 1.5 cm, the successive ones smaller, thin, lanceolate, rounded, vaginate, green with an indistinct nerve, reddish brown at the base; floral bract 4.1 × 1.8 cm, shorter than the ovary, thin, ovate, rounded, vaginate at base, green with indistinct veining, lightly red-brown at the base, with a membranous margin, central veining ribbed, brownish in the fourth basal. *Flowers* up to 24 cm wide, 10.5–12 cm high; sepals light green, with a white margin and green or brownish veins, externally green at the base, with green and brownish veins; petals yellow-greenish in the lower third, with green veining, a central darker vein and red margins, later brownish-red; lip brown-olive, lightly suffused with green near the opening, with darker veins, side lobes creamy yellow with fine green and brown spots, larger irregular light-brown spots inside, inside lip cream yellow with fine brown spots; staminode yellowish-brown, upper border with thick violet-brown hairs, lateral and lower borders with short violet-brown hairs; ovary 5.2 cm long, fusiform, slightly curved, dark red-brown, finely hairy. *Dorsal sepal* thin-textured, ovate, rounded, erect or curved forward, 5.4–6 × 1.5–1.6 cm, margins strongly folded, touching at the centre; synsepalum ovate,

Plate 34. *Phragmipedium* × *tetzlaffianum*. Drawn by Carol Woodin from a plant cultivated by Lynn & Bob Wellenstein, Candor, New York.

rounded at the front, concave, 5.25–6 × 2.5 cm, slightly wider and longer than the lip, with undulate and folded margins. *Petals* are up to 12 × 0.8 cm, linear-lanceolate with a rounded tip, separate 30–45° oblique, coiled up to 6 times, margins undulate at the base, in the basal third few small dark hairs on the upper margin, more substantial red hairs on the inferior margin, in the third external short red hairs on both sides. *Lip* saccate, 3.54 × 1.5–2 cm, smooth; side lobes strongly folded at the base, the edges touching; side margins slightly hollow, front border cartilaginous, with fine whitish smooth hairs. *Column* c. 6 mm long; staminode transversally oval, lumpy, 9–10 × 5–6 mm; stigma transversally oval, 7 × 4.5 mm, from ivory white to pale yellow, short whitish hairs on the lower margin. **Plate 34. Fig 157.**

Phragmipedium × tetzlaffianum Gruss in Caesiana 15: 33–40 (2001). Type: provenance unknown, cult. A. Tetzlaff *s.n.* (holo. HAL).

Putative parentage: ?*P. pearcei* (Rolfe) Rauh & Senghas × *P.* Nitidissimum.

DISTRIBUTION: Without exact provenance. Plants were said to have been imported from Venezuela but may not have been collected there

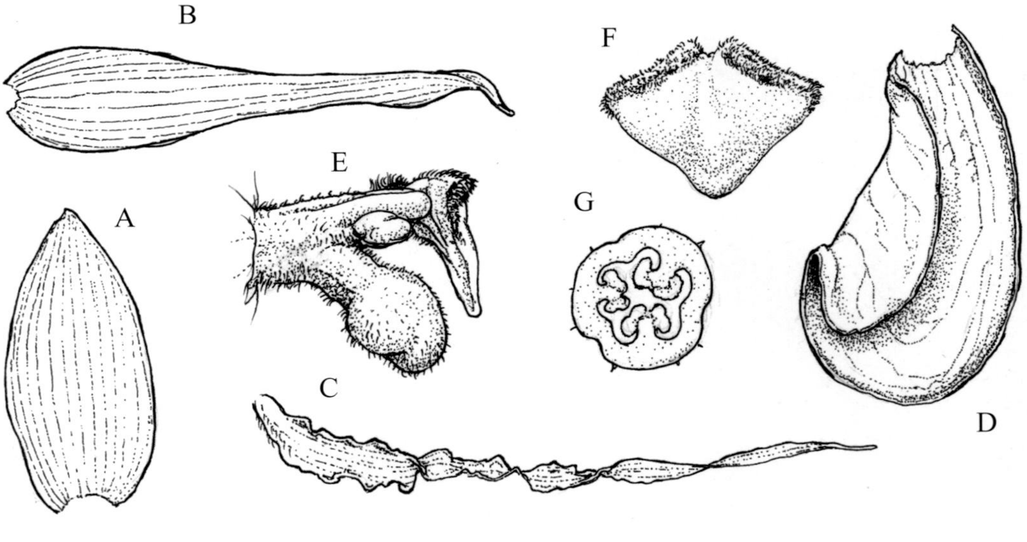

Fig. 158. ***Phragmipedium* × *tetzlaffianum*.** **A**, dorsal sepal; **B**, synsepal; **C**, petal; **D**, lip, longitudinal section; **E**, column, side view; **F**, staminode; **G**, ovary, transverse section. Drawn by Paul Harwood from plant cultivated by Bob & Lynn Wellenstein, New York. (Kew Spirit Collection no. 57271). A–D, × 1; E, F, × 3; G, × 4.5.

ARTIFICIAL HYBRIDISATION

Although a *Phragmipedium* hybrid appeared not long after *Paphiopedilum* Harrisianum, the first artificial hybrid slipper orchid which flowered in 1869, it took a very distant second place to the latter genus thereafter. *Phragmipedium* Dominianum (*P. caricifolium* × *caudatum*), the first artificial *Phragmipedium*, was raised and flowered by John Dominy in 1870 in the Veitch nursery in Chelsea. The well-known *P.* Sedenii (*P. longifolium* × *schlimii*), a hybrid still commonly found in collections, followed in 1873. Three years later, Veitch produced *P.* Stenophyllum (*P. caricinum* × *schlimii*) and this was closely followed in 1877 by *P.* Albopurpureum (*P.* Dominianum × *schlimii*), the first complex hybrid. Two popular hybrids followed: *P.* Grande (*P. caudatum* × *longifolium*) appeared in 1881, and the similar *P.* Giganteum (*P. caudatum* × Grande) in 1894. Both survive in collections to the present day.

By the turn of the century only some 30 *Phragmipedium* hybrids had been flowered (Table 4), contrasting markedly with the hundreds of *Paphiopedilum* hybrids already available. This reflected the fewer species of *Phragmipedium*, their more testing cultural requirements and the length of time it took for their hybrids to flower. The Veitch nursery dominated orchid hybridisation at the time, half of the flowered hybrids emerging from their establishment, thanks to their capable growers and hybridisers John Dominy and John Seden. Seven further hybrids were made from 1893 onwards by Frederick Sander's nursery at St Albans, whilst a few were made by amateur growers such as Bowring, Cookson and Measures. By 1900, eight species had been used in breeding.

Phragmipedium breeding more or less marked time until the late 1980s when the discovery of the startling scarlet-flowered *P. besseae* breathed fresh life into the art. The nursery at the heart of this revival was the Eric Young Orchid Foundation in Jersey which was given a plant of *P. besseae* at the 12th World Orchid Conference in Tokyo in 1987. Their breeder Alan Moon successfully grew and flowered it on his return and used it to breed the first of an amazingly successful line of award-winning scarlet, orange and pink hybrids.

Plate 35. *Phragmipedium* Sedenii. Drawn by John Day on 25 August 1873 at the Veitch nursery in Chelsea. (Kew collection).

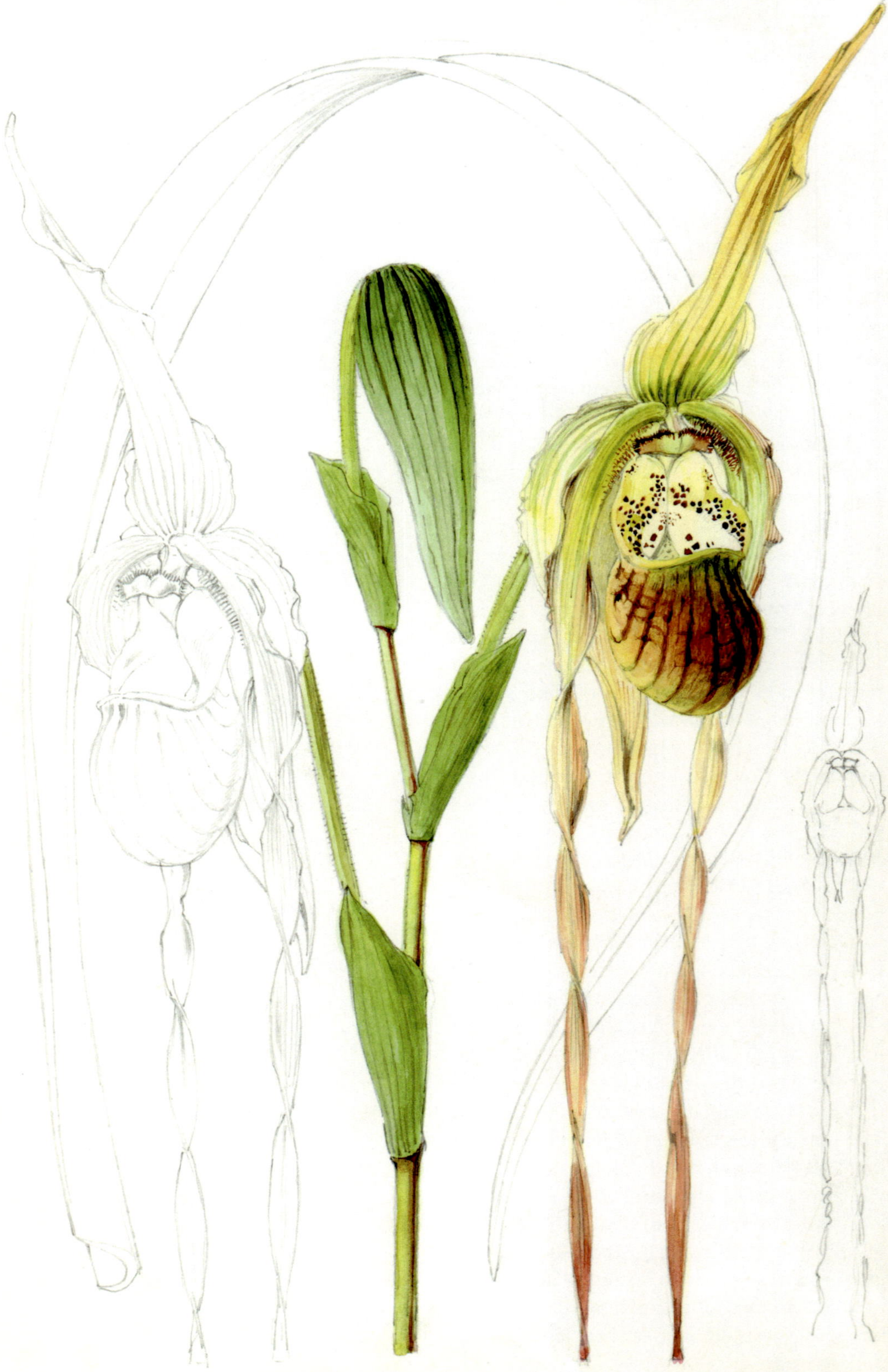

Plate 36. *Phragmipedium* Dominianum. Drawn by W.G. Smith at the nursery of B.S. Williams, 14 July 1870. (Kew collection).

Table 4. *Phragmipedium* hybrids before 1900.

Year	Hybrid	Female parent	Male parent	Breeder
1870	Dominianum	*caricinum*	*caudatum*	Veitch
1873	Sedenii	*longifolium*	*schlimii*	Veitch
1876	Stenophyllum	*caricinum*	*schlimii*	Veitch
1877	Albopurpureum	Dominianum	*schlimii*	Veitch
1881	Conchiferum	*caricinum*	*longifolium*	Bowring
	Grande	*caudatum*	*longifolium*	Veitch
1882	Schroderae	*caudatum*	Sedenii	Veitch
	Cardinale	Sedenii	*schlimii*	Veitch
1883	Calurum (= Ainsworthii)	*longifolium*	Sedenii	Veitch
1887	Saundersianum	*caudatum*	*schlimii*	Bull
1888	Lemoinierianum	Calurum	Sedenii	Lemoinier
	Nitidissimum	*warszewiczii*	Conchiferum	Cookson
1891	Cleola	*boissierianum*	*schlimii*	Veitch
1892	Brysa	*boissierianum*	Sedenii	Veitch
	Elsteadianum	Conchiferum	Grande	Ingram
	Hardyanum	Calurum	*caudatum*	Hardy
	Perseus	*lindleyanum*	Sedenii	Veitch
1893	Eva	Cardinale	*longifolium*	Measures
	Hanischianum	*boissierianum*	Cardinale	Sander
	Pulchellum	Grande	Sedenii	Vanner
	Raymond Faroult	*caricinum*	Sedenii	Faroult
1894	Giganteum	*caudatum*	Grande	Veitch
	Stella	*schlimii*	*vittatum*	Sander
1895	Finetianum	Cardinale	*caudatum*	Finet
	L'Unique	*lindleyanum*	*schlimii*	Ingram
1896	Uranus	Grande	*lindleyanum*	Sander
	Urgandiae	*lindleyanum*	*longifolium*	Mantin
1899	Geralda	*caudatum*	*lindleyanum*	Rockefeller
1900	Cardinal-Schlim	Cardinale	*schlimii*	Measures
	Mantinii	Conchiferum	*longifolium*	Mantin

Table 5. Introduction of species into hybridisation programmes since 1980.

Year	Hybrid	Female parent	Male parent	Breeder
1980	Carol Kanzer	*pearcei*	*schlimii*	Seagulls L.O.
1985	Bethava	*caudatum*	*pearcei*	R.J.Rands
1991	Ecua-Bess	*ecuadorense*	*besseae*	Joe Kunitsch
	Eric Young	*besseae*	*longifolium*	Eric Young Orchid Foundation
	Mary Bess	*besseae*	*caricinum*	F. Glanz
	Hanne Popow	*besseae*	*schlimii*	H. Doll
1992	Andean Fire	*besseae*	*lindleyanum*	L. Schodje
	Memoria Ed Murphy	*kaieteurum* (= *lindleyanum* var. *kaieteurum*)	*longifolium*	D. Olivas
1993	Rosalie Dixler	*besseae*	*kaieteurum*	J. Dixler
	Gary Baker	*czerwiakowianum* (= *boissierianum*)	*kaieteurum*	Baker & Chantry
	Rosalie Dixler	*besseae*	*kaieteurum*	J. Dixler
	Dennis Kleinbach	*czerwiakowianum*	*longifolium*	Baker & Chantry
	Gary Baker	*czerwiakowianum*	*kaieteurum*	Baker & Chantry
1994	Wössen	*richteri*	*schlimii*	F. Glanz
1995	Franz Glanz	*richteri*	*besseae*	O. Gruss
1997	Mountain Maid	*besseae*	*hirtzii*	Fox Valley
1998	Barbara LeAnn	*besseae*	*fischeri*	J. Fischer
2000	Purple Wave	Memoria Dick Clements	*fischeri*	J. Fischer
2002	Fantasy Dream Chaser	*hirtzii*	*sargentianum*	F. & M. Kaufmann
2004	Jersey	*dalessandroi*	*besseae*	Eric Young Orchid Foundation
	Nicholle Tower	*dalessandroi*	*longifolium*	Eric Young Orchid Foundation
	Petit Arquette	*dalessandroi*	*pearcei*	Eric Young Orchid Foundation
	Seymour Tower	*dalessandroi*	*schlimii*	Eric Young Orchid Foundation
	Grosnez	*sargentianum*	*dalessandroi*	Eric Young Orchid Foundation

Year	Hybrid	Female parent	Male parent	Breeder
	Mont Cambrai	*caricinum*	*dalessandroi*	Eric Young Orchid Foundation
	Paternoster	*wallisii*	*dalessandroi*	Eric Young Orchid Foundation
	Point des Pas	*caudatum*	*dalessandroi*	Eric Young Orchid Foundation
	Mardi Gras	*brasiliense*	*fischeri*	Orchids Ltd.
2007	Peruflora's Eumelia Arias	*kovachii*	*schlimii*	Peruflora
	Fritz Schomburg	*kovachii*	*besseae*	Piping Rock Orchids
	Hayley Decker	*kovachii*	Saint Ouen	Piping Rock Orchids
2008	Petite Quenelle	*besseae*	*andreettae*	Eric Young Orchid Foundation
	Woodstream	*richteri*	*exstaminodium* (= *humboldtii* var. *exstamino-dium*)	W. Goldner

Phragmipedium breeding more or less marked time until the late 1980s. There are a number of possible reasons why the early *Phragmipedium* hybridising lost impetus. The relatively few species available to work with allied with the limited colour range could partly explain this. However, the difference in chromosome numbers between the species (2n=18-30) critically led to second and third generation crosses having low fertility or sterility problems, thus stalling the production of new hybrids.

The discovery of the startling scarlet-flowered *P. besseae* in 1981 breathed fresh life into the art (Table 5). The nursery at the heart of this revival was the Eric Young Orchid Foundation (EYOF) in Jersey which was given a plant of *P. besseae* at the 12th World Orchid Conference in Tokyo 1987. Alan Moon, the curator at EYOF, successfully grew and flowered it on his return. At around the same time, EYOF also received seed of *P. dalessandroi* from a site in Ecuador from Dr Calaway Dodson. These two closely allied species were used to breed an amazingly successful line of award-winning scarlet, orange and pink hybrids.

At the time, *P. dalessandroi* was considered a variety of *P. besseae*. Thus the original hybrids made were all given the same grex name regardless of whether *P. besseae* or *P. dalessandroi* was used. This gave rise to distinctly morphologically distinct hybrids bearing the same grex name. In 2004 the Royal Horticultural Society (RHS) International Orchid Registrar began to accept *P. dalessandroi* as distinct for orchid registration purposes. Therefore, many of the early hybrids with *P. dalessandroi* in their background gained new grex names.

This move proved highly controversial at the time but, in our opinion, made sense as the hybrids made at the Foundation were manifestly different depending upon which species had been used. From the outset the Foundation pursued two separate lines of breeding, and maintained meticulous records of the specific parents used.

Around the same time as these new *P. besseae* hybrids were being produced, the late Professor Don Wimber, on a sabbatical year at Kew, was working with Eric Young Orchid Foundation (EYOF) on ground-breaking research to create tetraploid plants. In other genera it had been shown that the creation of tetraploids, using colchicine, had overcome sterility problems. Consequently, many of the diploid crosses made by EYOF were treated with colchicine at the protocorm stage in order to convert some of them to tetraploids. Although in other orchid genera these tetraploids have enabled multiple generations to be produced, in phragmipediums it has not proved to be quite so successful. However, a consequence of this conversion is that tetraploid plants produce flowers of increased size, deeper colour, heavier substance and fuller appearance; all desirable traits when breeding for award. This said, the tetraploid plants are slower to grow than their diploid or triploid counterparts.

Fig. 159. *Phragmipedium dalessandroi*. The lower flower is diploid (2n) and the upper one tetraploid (4n). All following photos are by EYOF.

Fig. 160 (left). Diploid *Phragmipedium* Eric Young. **Fig. 161** (right). Tetraploid *Phragmipedium* Eric Young. (EYOF)

The first *P. besseae* hybrid registered by the EYOF was *P.* Eric Young, a cross between *P. besseae* and *P. longifolium*, and named in honour of Eric Young, the Foundation's founder. The first flowering plants were all diploid. This was followed in close succession by tetraploid plants, of which the cultivar 'Mont Millais' received an Award of Merit and the George Moore Medal from the Royal Horticultural Society in 1992. By modern standards the flower might seem decidedly average in quality, but it was the best cross in this new line of *P. besseae* breeding in its day.

The hybrids *P.* Eric Young and *P.* Nicholle Tower have proved to be good parents and highly influential in the EYOF breeding programme. The next step in these hybrid lines was to back cross *P.* Eric Young onto *P. besseae*, and *P.* Nicholle Tower onto *P. dalessandroi*. In both cases the aim was to intensify the depth of colour, but also to create diverging types of *P. besseae*-influenced hybrids..

Fig. 162. *Phragmipedium* Eric Young 'Trinity' AM/RHS, a diploid clone in 1991.

Fig. 163 (left). *Phragmipedium* Nicholle Tower 'Saint Brelade' (*P. dalessandroi* with *P. longifolium*). This plant is heavily influenced by its *P. dalessandroi* parent having down-swept petals and a branched inflorescence. **Fig. 164** (right above). *P.* Don Wimber (tetraploid). **Fig. 165** (right below). *P.* Ouaisne (tetraploid).

Arguably the best flower in terms of shape has been the hybrid *P.* Don Wimber (*P.* Eric Young × *P. besseae*), named in honour of Don Wimber shortly before his death. The cross exhibits the classic shape of *P. besseae* reflecting the double dose of the species in its background. The tetraploid forms have flowers approaching 12 cm across with great depth of colour. However, the plant typically has an unbranched, single-flowered inflorescence.

By using *P. dalessandroi* crossed with *P.* Eric Young to make the cross *P.* Ouaisne, EYOF produced a hybrid with a branching inflorescence capable of holding eight or more flowers simultaneously. The flower shape also clearly shows the influence of *P. dalessandroi*. The hybrid *P.* Mont Sohier (*P.* Nicholle Tower × *P. dalessandroi*) contains no *P. besseae* in its background, but the double dose of *P. dalessandroi* is evident in the very down-swept petals.

These hybrids are still some of the best orange-flowered *Phragmipedium* produced to date, and highly desirable plants. It is unlikely they will be surpassed, as subsequent generations have yet to yield anything approaching their quality.

The early hybridisers had access to *P. schlimii*, the only pink-flowered species known at the time. Early combinations with it produced the first 'pink' hybrids, and it was also used to make *P.* Hanne Popow, the first *P. besseae* hybrid that was registered in 1991. Many cultivars in this grex may exhibit an uneven colouration across the petals, and sometimes the upper half can be darker than the lower. This hybrid has probably influenced modern-day crosses more so than pure *P. schlimii*. The Foundation created tetraploid forms of this cross and the fine variety *P.* Hanne Popow 'Isle of Jersey' awarded an FCC by the American Orchid Society (AOS) back in 1996 is illustrated here. This plant has been used extensively in our breeding programme.

One of the best pale pinks is *P.* Petit Port (*P.* Hanne Popow × *P. schlimii*). The double dose of *P. schlimii* is evident here, and although this species can sometimes result in offspring with reflexed or twisted petals, the conversion of the hybrid to a tetraploid has overcome this negative characteristic.

The next step was to take tetraploid forms of the two primary hybrids, *P.* Eric Young and *P.* Hanne Popow, and combine them to produce *P.* Grouville. This grex gives a wide palette of pink tones, and a flower size approaching 11 cm in diameter.

Fig. 166 (left). *Phragmipedium* Hanne Popow tetraploid 'flavum type'. **Fig. 167** (right). *P.* Petit Port 'Jersey' AM/RHS.

Continuing on the hybrids derived from *P.* Hanne Popow; EYOF crossed a diploid *P.* Hanne Popow with a diploid *P. longifolium* and converted this to make a tetraploid *P.* Longueville. Like *P.* Grouville, this grex comes in a wide range of pink, and is reminiscent of the early tetraploid *P.* Eric Young plants in shape.

So far we have discussed hybrids in various combinations of *P. besseae*, *P. schlimii*, and *P. longifolium*. Another key species within our *Phragmipedium* hybridising programme has been *P. sargentianum*. It is fair to say that both *P. longifolium* and *P. sargentianum* are vigorous plants, and when used in hybridising impart this vigour to their progeny. The hybrid between them, *P.* Sorcerer's Apprentice, is capable of producing enormous plants with equally large flowers, especially in its tetraploid form.

When *P.* Sorcerer's Apprentice is combined with *P.* Memoria Dick Clements it gives *P.* Portelet, yielding deep maroon-red flowers on a branched inflorescence. When crossed with *P.* Jersey, however, it produces *P.* Greve de Lecq which has a very different flower shape and a much more vibrant colour. Both crosses have tall flower spikes, as one would expect, but on a more compact plant than *P.* Sorcerer's Apprentice.

Perhaps the most significant primary hybrid made using *P. sargentianum* was the groundbreaking red *P.* Memoria Dick Clements. In its day it was the darkest red seen, and subsequently garnered many awards around the world. Both diploid and converted

Fig. 168 (above left). *Phragmipedium* Grouville 'Trinity' AM/RHS (Hanne Popow × Eric Young). **Fig. 169** (above right). *P.* Grouville. **Fig. 170** (below left). *P.* Grouville dark-coloured clone. **Fig. 171** (below right). *P.* Grouville pale tetraploid clone.

Fig. 172. *Phragmipedium* Jason Fischer flowers in a box.

tetraploid forms were raised at EYOF. Though the diploids were very good, the tetraploids exceeded all expectations with much larger, better-shaped flowers and an even deeper red in colour than its diploid counterpart.

With a view to intensifying the red colouration, the next step was to cross *P. besseae* and *P. dalessandroi* onto *P.* Memoria Dick Clements. The latter cross was made to try to impart a branching habit to the next generation, but sadly this proved to be at the expense of flower shape. However, using the pollen from a tetraploid *P. besseae* originating from Mr. Terry Root of the Orchid Zone USA, onto one of the EYOF's finest tetraploid clones of *P.* Memoria Dick Clements, yielded *P.* Jason Fischer, which is, in our opinion, the finest hybrid seen to date. The intensity of colour, allied to near perfect shape, good size and flatness of flower set this tetraploid grex head and shoulders above its contemporaries. It has been one of the most highly awarded *Phragmipedium* grexes, and has not been bettered in this colour range. Sadly, Jason Fischer seems to be a reluctant parent, even in its tetraploid form, and crosses with it rarely hold seed. Those that do exhibit poor germination, and have never yielded flowers equal in quality to their *P.* Jason Fischer parent.

Another fruitful line of breeding with *P. sargentianum* has been to combine it with *P.* Hanne Popow to make *P.* Beauport. Whilst these have given rise to large flowers in a range of colours from peach to deep pink, their shape is not the best and one particular clone has earned the name 'Peach Ugly' on the Foundation's list.

We saw *P.* Beauport as a parent with potential to give a broad colour range. Initially, we chose to cross *P. dalessandroi* onto the dark *P.* Beauport clone pictured above to make *P.* La Hougette. This grex has given highly floriferous plants, with the characteristic down-swept petals of *P. dalessandroi*. Interestingly, the flowers have a deep pink background overlaid by deep red hairs, giving a vibrant purple tone; a colour absent in the genus prior to the discovery of *P. kovachii*.

In an attempt to create paler flowers, we took a tetraploid pale form of *P.* Longueville with a tetraploid peach form of *P.* Beauport, and were rewarded with the delicately coloured *P.* Anne Port. This stunning hybrid has flowers in excess of 12 cm across, is lightly branched, and is capable of holding several flowers at once. This third generation cross contains *P. schlimii*, *P. besseae*, *P. longifolium*, and *P. sargentianum* in its background, and is one of the more complex hybrids EYOF has made to date.

Fig. 173 (top left). *Phragmipedium* Hanne Popow tetraploid 'flavum type'. **Fig. 174** (middle left). *P.* La Hougette. **Fig. 175.** (bottom left). *P.* La Hougette. **Fig. 176.** (right). *P.* La Hougette 'Gorey'

Fig. 177. *Phragmipedium* Anne Port.

PALE-COLOURED HYBRIDS

In recent years the EYOF has pursued the goal of creating yellow and pastel shades, through to near whites. It is an area which has proved to be difficult; particularly as going back to the yellow *P. besseae* immediately presents the problems of poor vigour and relatively small flower size. This is also compounded by the lack of a tetraploid yellow *P. besseae*. Numerous attempts to convert the protocorms have failed to yield plants that have subsequently produced leaves, and continued to grow. Those that perhaps had been converted appear unable to produce chlorophyll, and eventually die. Those that continued to grow have been diploid. However, unperturbed, we have made some promising hybrids, and are now starting to see some highly satisfying results. Of course many of these hybrids are not new grexes, as most have already been made and registered with the more familiar orange-red *P. besseae*.

The most obvious starting point was to remake *P.* Eric Young with the yellow *P. besseae*, seen here in diploid version. We also made *P.* Hanne Popow using the same *P. besseae*. Taking diploid forms of these pale hybrids, we then crossed them together to recreate *P.* Grouville, but of course this time with both parents made with the yellow *P. besseae*. The result was a line of extremely pale plants ranging from ivory to near white with a pale pink pouch, from light apricot, through to yellow.

Fig. 178 (left). *Phragmipedium* Eric Young (diploid). **Fig. 179** (top right). *P.* Eric Young (diploid). **Fig. 180** (bottom right). *P.* Eric Young tetraploid 'flavum type'.

Another successful remake was that of *P.* Don Wimber. This used a tetraploid *P.* Eric Young made with the yellow *P. besseae*, crossed back with a diploid yellow *P. besseae*. It has given a wonderful pale yellow flower with the merest hint of pink and is a good size (13 cm in diameter).

Two further lines were developed using the pale peach form of *P.* Longueville as one parent. The first, crossed with the yellow *P. besseae*, gave *P.* Augres. The second, with *P.* Hanne Popow crossed with the yellow *P. besseae*, gave *P.* Havre des Pas. Both hybrids exhibit a good degree of branching once the plant has reached maturity, and have been successful on the show bench. However, as with the remade *P.* Don Wimber, these two delicately coloured hybrids are triploid, and therefore will not breed further.

A more recent addition to this range is the hybrid *P.* Les Platons. Here, once again, the pale orange tetraploid *P.* Beauport was combined with a tetraploid yellow *P. besseae* derived *P.* Hanne Popow. It has produced 'two-tone' flowers in subtle shades ranging from shell pink to apricot, the base colour overlaid by darker hairs producing a striking effect. As usual, the *P.* Beauport parent has given a round pouch with full petals giving a well-balanced flower.

Fig. 181 (top left). *Phragmipedium* Don Wimber 'Saint Aubin'. AM/RHS. **Fig. 182** (middle left). *P.* Don Wimber (tetraploid). **Fig. 183** (bottom left). *P.* Don Wimber (tetraploid). **Fig. 184** (right). *P.* Don Wimber 'Saint Martin'.

Fig. 185. *Phragmipedium* Don Wimber group display in Eric Young Foundation's display house.

Fig. 186 (top right). *Phragmipedium* Don Wimber 'flavum type'. **Fig. 187** (bottom). *P.* Don Wimber on bench at Eric Young Foundation.

LONG-PETALLED HYBRIDS

The hybrid *Phragmipedium* Grande (*P. caudatum* × *P. longifolium*) hails from the late 1800s but still takes pride of place in many modern collections. The plants are vigorous and produce extremely large flowers. Diploid forms are easily capable of having petals over 60 cm in length. When converted to tetraploid, the petals become shorter but wider, and there is a general increase in other floral segments. Because the flower has greater substance, it also proves to be more robust than its diploid counterpart.

Although using *P. caudatum* in its own right for hybridising would seem the obvious route, experience has shown that *P.* Grande will yield superior hybrids. *P. caudatum* can prove difficult as a parent, especially when used with species or hybrids from section *Micropetalum*.

Fig. 188 (left). *Phragmipedium* Grande 'Victoria Village' AM/RHS. **Fig. 189** (right). *P.* Sorcerer's Apprentice.

EYOF has predominantly used the tetraploid *P.* Grande in its breeding programme. A fine example of this breeding is *P.* Les Dirouilles, a cross between *P.* Sorcerer's Apprentice and *P.* Grande. In our opinion, this plant is superior to *P.* Sorcerer's Apprentice, the inflorescences being even taller and the stems thicker, almost 2 cm in diameter at the base. Furthermore it produces larger flowers with longer petals, with a rich mahogany colour.

To add colour to the long-petalled forms, we initially used species or hybrids containing *P. schlimii* or *P. besseae*. Taking two hybrids originally from the 1800s, we crossed *P.* Grande onto the pink *P.* Calurum and made the unusually coloured *P.* Archirondel. The petals take their pink colouration from the *P. schlimii* influence, and the exterior olive-green on the pouch from *P. longifolium*. In this instance the petals have retained a reasonable length. Usually long-petalled crosses when combined with those from section Micropetalum result in flowers with considerably reduced petal length, and this is especially the case when using tetraploid parents. This is demonstrated perfectly in the hybrid *P.* Les Landes, a cross between *P.* Hanne Popow and *P.* Grande. To date, this is still one of the better long-petalled hybrids with strong pink colouration.

Fig. 190. (left). *Phragmipedium* Les Dirouilles. **Figs. 191 & 192** (right). *P.* Les Landes.

Continuing the concept of branched inflorescences we used *P. dalessandroi* to get the magnificent *P.* Quetivel Mill (formerly known as *P.* China Dragon). This gave an abundance of deep red flowers with a bright yellow inner pouch. Bearing in mind the wonderful red colour of *P.* Memoria Dick Clements, it was logical to cross this with *P.* Grande. The resultant hybrid (*P.* Fliquet) did not give the depth of red we expected, but did give longer petals and, with less twisting, presenting a highly symmetrical flower.

Finally, we have commenced breeding toward paler forms of long-petalled hybrids. One example is the creation of *P.* Bouley Bay using *P.* Grande crossed onto *P.* Eric Young (made with the yellow *P. besseae*). This has resulted in some highly desirable paler forms but is very much a work in progress. The same is true for the tetraploid hybrid *P.* Mourier Point. However, this pale yellow cross is a little different. Rather than using the yellow *P. besseae* for its colour, we have used an unstable *P.* Nicholle Tower plant that occasionally produces single yellow flowers. This was then crossed onto a *P.* Grande that was made from a very pale *P. caudatum sanderae*. It has given the kind of result we require, but obviously via a different route.

Fig. 193 (left). *Phragmipedium* Quetivel Mill 'Jersey'. **Fig. 194** (right). *P.* Memoria Dick Clements 'Trinity'.

Fig. 195 (top left). *Phragmipedium* Fliquet. **Fig. 196** (top right). *P.* Fliquet 'St Helier'.
Fig. 197. (bottom left). *P.* Nicholle Tower 'St Brelade'. **Fig. 198** (bottom right). *P.* Bouley Bay.

THE RECENTLY DISCOVERED PHRAGMIPEDIUM SPECIES

Much has been written of *P. kovachii* and its highly controversial discovery and subsequent naming. Suffice to say the size and colour of this truly stunning species are beyond most people's wildest dreams. Obviously, this new species is highly desirable in its own right but also for its potential to make a whole new range of hybrids.

The first legal *P. kovachii* plants in Europe were flasks of seedlings released around the time of the 18th World Orchid Conference in Dijon, France in 2005. These were quickly snapped up by eager growers, but sadly many proved difficult to grow from flask. The EYOF acquired flasks at this time and have patiently grown the plants to flowering size. The first flowerings were quickly utilised in our hybridising programme, but hybrid plants raised in Peru and latterly in the USA had a head start, and, consequently, the hybrid grexes registered have reflected this. Our patience in growing our original importations of *P. kovachii* proved worthwhile when we presented *P. kovachii* 'Trinity' FCC/RHS, our finest cultivar to date. Of all the plants we have flowered this one is distinctive in its ability to retain the depth of colour and resist the reflexing and waviness of petals so commonly seen on other plants. When put up for award it measured 16.5 cm in diameter. But this was on a freshly opened, not fully mature, flower; in a previous flowering it had measured a staggering 18.5 cm.

To date, *P. kovachii* has yet to prove itself as an outstanding parent. The hybrids seen so far have not lived up to expectations. The wonderful purple colouration does not appear

Fig.199. *Phragmipedium kovachii* 'Trinity' FCC/RHS.

to transfer and, although some unusual pink and mauve pink colours have been produced, nothing close to purple has appeared. Furthermore, and not surprisingly, flower size is diminished, though subsequent backcrosses may rectify this situation.

The flowers of the first generation hybrids are not of aesthetically pleasing proportions; either the petals are disproportionately large, the pouch and/or synsepal is too small. However, without sounding too negative, it is early days in these lines of breeding especially when you compare the early *P. besseae* hybrids to their modern counterparts. At EYOF we have treated flasks with colchicine in order to create potential tetraploids. We are close to flowering what we believe to be tetraploid plants. These will hopefully form the basis of a whole generation of tetraploid *P. kovachii* hybrids.

A hybrid we had high hopes for was *P.* Les Varines, a cross with *P. sargentianum*. One would have expected a *P. kovachii* equivalent of *P.* Memoria Dick Clements; dark, well proportioned, purple flowers. Unfortunately, neither the shape nor colour of *P. kovachii* comes through, although the plant produces a branched inflorescence over one metre tall and the flower has interesting markings in the pouch.

Fig. 200 (left). *Phragmipedium* Les Varines 'Keppel Tower'. **Figs. 201 & 202** (right). *P.* Les Varines.

A more successful example of an early hybrid from EYOF which has recently received two Awards of Merit from the RHS is the grex of *P.* La Vingtaine; a cross with *P.* Memoria Dick Clements. This has given a consistently good shape on a flower 15 cm in diameter. The colour is mainly a deep raspberry pink, but some do exhibit more orange-pink tones from the *P. besseae* in its background.

Although attention has almost exclusively focused on *P. kovachii* and its hybrids in the last few years, there has still been scope to pursue other avenues. *Phragmipedium fischeri*, *P. andreettae* and, more recently, *P. manzurii* (*P. schlimii* var. *manzurii*) all have potential to produce beautiful, more delicate blooms. Our first hybrid with *P. andreettae* has produced *P.* Petite Queillette; interestingly a pink bloom given that we use the yellow *P. besseae* as the other parent.

There are exciting times ahead as new combinations are made using all of these newly discovered species. We suspect that *Phragmipedium* hybridising will be polarised into those who believe 'bigger is better' whilst others believe 'the best things come in small packages'. Either way we can look forward to an exciting future.

Fig. 203. (left). *Phragmipedium* La Vingtaine. **Fig. 204** (above right). *P.* Petite Queillette (tetraploid). **Fig. 205** (below right). *P.* Petite Queillette.

CULTIVATION

The small flowers and leggy plants of *Selenipedium* make them poor subjects for cultivation. Their thin pleated leaves dry out rapidly when they are dug up and the plants do not seem to have many reserves in their stems and roots to enable them to recover. A few plants have been grown (e.g. Dunsterville, pers. comm.; Zelenko, pers. comm.) but only wild-collected mature plants and they do not do well in cultivation. However, experience with cypripediums (Frosch & Cribb, 2012), which were also considered tricky to grow, suggest that they should be relatively easy to grow given appropriate composts, fertilisation and watering regimes and temperate climate.. The following account focuses on *Phragmipedium* culture, while *Mexipedium* thrives under similar conditions.

Phragmipedium species grow in a variety of habitats. A few species, such as *P. caudatum* and *P. exstaminodium,* are epiphytic but the majority of species are terrestrial or lithophytic. The habitats of the terrestrial and lithophytic species can vary from the splash zones of waterfalls and rapids to stream and river banks. *Phragmipedium caricinum*, *P. hirtzii*, *P. klotzschianum*, *P. lindleyanum* var. *kaieteurum*, *P. pearcei* and *P. sargentianum* all grow on boulders in or by waterfalls. *P. longifolium* and *P. boissierianum* grow happily on roadside banks. *P. besseae* and *P. kovachii* grow above streams amongst vegetation on wet rocky cliffs. Detailed accounts of the cultivation of phragmipediums have been provided by Le Doux (2003).

Figs. 206 & 207. Two views of *Phragmipedium* hybrids growing at the Eric Young Orchid Foundation in Jersey.

Fig. 208. View of *Phragmipedium* hybrids growing at the Eric Young Orchid Foundation in Jersey.

Most have proven to be very amenable in cultivation, despite initial fears about species such as *P. besseae* and *Mexipedium xerophyticum* when they were first introduced. Trial and error, combined with a better knowledge of their native habitats, has quickly shown them to be relatively easy to grow, providing certain parameters are fulfilled.

Composts

Most species can be happily grown in plastic pots with additional drainage holes cut around the base. They benefit from an open, free-draining compost but with adequate moisture-retentive components, most do not tolerate excessive drying out. A variety of media have been successfully used to grow the species, including bark, rock wool, sphagnum moss and coconut husk. The addition of perlite to retain moisture and charcoal to freshen the compost is recommended. Le Doux (2003) recommends a compost of 9 parts fine redwood bark; 3 parts perlite or sponge rock, 3 parts fluffed peat or Pro-Mix HP (a sphagnum peat-based mix). Bone meal is added as a fertiliser. For 12.5 cm (5 inch) pots or larger she adds medium grade bark to make up a quarter to a third of the mix. Rockwool is not recommended for *Mexipedium xerophyticum* for which she recommends adding dolomitic lime and more charcoal to the bark, perlite and peat mix given above.

Water

Phragmipediums are intolerant of poor quality water. Tap water with low dissolved solids is tolerated but rain-water or water produced by reverse osmosis or deionisation is much better for them. The pots should be flushed through every time they are watered and should never be allowed to stand in the run-off. Most species should be kept moist at all times, the exceptions being the epiphytic species *Phragmipedium caudatum* and *P. humboldtii* and *Mexipedium xerophyticum* which should be allowed to dry out a bit between waterings.

Fig. 209. Propagation by air-layering a plant at the Eric Young Orchid Foundation in Jersey.'

Humidity and air circulation

Most species prefer to grow in a humid environment with the relative humidity above 50%. Growing plants on a gravel tray in which the gravel is kept moist at all times, but not the pot bases, is ideal. Humidity levels can be dropped a bit in winter when watering is reduced.

Light

The epiphytic species of *Phragmipedium* and *Mexipedium xerophyticum* need good light levels similar to that required by cattleyas and vandas. However, most species thrive under medium light levels, such as those suitable for phalaenopsis.

Temperature

Phragmipediums grow well in an intermediate greenhouse with day-time temperatures in the mid to high 20°Cs and night-time temperatures between 15 and 18°C. *Mexipedium xerophyticum* grows well at temperatures towards the upper end of this spectrum.

Fertiliser

Phragmipediums are greedy feeders and can tolerate higher levels of fertiliser than paphiopedilums. Le Doux (2003) recommends half to quarter strength fertiliser containing micro-nutrients, with a thorough flushing of the compost with water every four or five feedings.

Pests

Although relatively pest free, phragmipediums can suffer from spider mite infestation and should be treated with a suitable spray at the first sign of infestation. Slugs and snails can also be damaging, especially to the flowers and regular checking to reduce these pests is necessary.

BIBLIOGRAPHY

Adanson, M. (1763). *Familles des Plantes*. Vincent, Paris.

Aiton, W. (1789). *Hortus Kewensis*. G. Nichol, London.

Albert, V. (1994). Cladistic relationships of the slipper orchids (Cypripedioideae: Orchidaceae) from congruent morphological and molecular data. *Lindleyana* 9: 115–132.

Albert, V. & M.W. Chase (1992). *Mexipedium*: A new genus of Slipper orchid (Cypripedioideae: Orchidaceae). *Lindleyana* 7: 172–176.

Albert, V.A. & B. Pettersson (1994). Expansion of the genus *Paphiopedilum* Pfitzer to include all conduplicate-leaved slipper orchids (Cypripedioideae: Orchidaceae). *Lindleyana* 9(2): 133–140.

Anon. (1922). Self-fertilization of *Cypripedium schlimii*. *Orchid Review* 30: 108.

Anon. (2009a). *Mexipedium* (*Phragmipedium*) *xerophyticum*. Antec Laboratory and Ladyslipper Farm. http://www.ladyslipper.com/mexipedium.htm.

Anon. (2009b). *Phrag.* (*Mexipedium*) *xerophyticum*. Orchids Limited. http://www.orchidweb.com/orchidofweek.aspx?id=397.

Archila, F. (1999). Dos nuevos entidades del genero *Phragmipedium* Rolfe para Guatemala: *Phragmipedium monstruosum* Archila y *Phragmipedium triandrum* Archila. *Revista Guatemalensis* 2, 3: 3–8.

Atwood, J.T. (1984). The relationships of the slipper orchids (Subfamily Cypripedioideae, Orchidaceae). *Selbyana* 7: 129–247.

Atwood, J.T. & R.L. Dressler. (1998). Clarifications and new combinations in the *Phragmipedium caudatum* complex from Central America. *Selbyana* 19: 245–248.

Atwood, J.T. & R.L. Dressler (1999). Clarification and new combinations in the *Phragmipedium caudatum* complex from Central America. *Selbyana* 19, 2: 246.

Bänziger, H., S. Pumikong & K. Srimuang (2012). The missing link: bee pollination in wild lady slipper orchids *Paphiopedilum thaianum* and *P. niveum* (Orchidaceae) in Thailand. *Bull. Soc. Ent. Suisse* 85: 1–26.

Barbosa Rodrigues, J. (1877). *Selenipedium. Genera et Species Orchidarum Novarum* 1: 201.

Barbosa Rodrigues, J. (1937). *Selenipedium vanillicarpum. Rodriguesia* 2, 8: 39.

Barretto, G., P.J. Cribb & S. Gale (2009). *The Wild Orchids of Hong Kong*. Natural History Publications (Borneo), Kota Kinabalu.

Batista, J.A.N., L.B. Bianchetti & K.F. Pellizaro (2005). Orchidaceae da Reserva Ecológica do Guará, DF, Brasil. *Acta Botanica Brasilica* 19(2): 221–232.

Baumann, H., S. Kunkele & R. Lorenz (1989). Die nomenklatorischen Typen der von Linnaeus veröffentlichen Namen europäischer Orchideen. *Mit. Arbeitskreis Heimische Orchid. Baden-Württemberg* 21: 452.

Bennett, D. & E. Christenson (1993, 1995). *Icones Orchidacearum Peruvianum*. A. Pastorelli de Bennett, Lima.

Bentham, G. & J.D. Hooker (1883). Cypripedieae. *Genera Plantarum* 3: 487–488.

Braem, G. (2004). *Phragmipedium kovachii*, Schluckebieria — nouvelle section du genre *Phragmipedium* – et reflexions sur les practiques taxonomiques. *Richardiana* 4, 3: 89–102.

Braem, G. (2011). A re-evaluation of the infrageneric taxonomy of the genus *Phragmipedium*. *Richardiana* 12, 1: 16–24.
Braem, G. (2014). The enigmatic (?) case of the most beautiful *Phragmipedium popowii*. *Orchid Digest* 78, 2: 101–112.
Braem, G. (2015). Validation of *Phragmipedium* subgenus *Schluckebieria* (Orchidaceae: Cypripedioideae). *Richardiana* 15, 1: 289–290.
Braem, G. (2016). The *Phragmipedium longifolium* complex. *Orchid Digest* 80, 1: 36–43.
Braem, G. & S. Ohlund (2004). A language trap – *Phragmipedium caudatum, Phragmipedium warscewiczianum, Phragmipedium humboldtii. Australian Orchid Review* 69, 5: 19–21.
Braem, G., S. Ohlund & R.-J. Quéné (2005). Le vrai *Phragmipedium warscewiczianum* pourrait-il se presenter, s'il vous plait? Un nouveau cas d'identification erronée – Le complexe *Phragmipedium caudatum*, écalaircissements, et description d'une nouvelle espèce. *Richardiana* 4, 4: 171–196.
Braem, G.J., Tesón, E. & Faust, J.-P. (2016). *Phragmipedium anchicayense*. A new slipper orchid from Colombia. *Australian Orchid Review* 81, 5: 9–13.
Braem, G.J., E. Tesón & D.M. Macias (2014). *Phragmipedium anguloi*, a new species from Colombia belonging to *Phragmipedium* subgenus *Micropetalum* (Orchidaceae). *Richardiana* 14: 289–292.
Brieger, F.G. (1973). 2. Tribus Phragmipedieae. In F.G. Brieger, R. Maatsch & K.H. Senghas, (eds), R. Schlechter, *Die Orchideen* ed. 3, 9–12: 168–171.
Brown, N.E. (1885). *Selenipedium kaieteurum* N.E. Br., n. sp. *Gardeners' Chronicle* n.s. 24: 262.
Brown, R. (1813). *Cypripedium*. In Aiton, *Hortus Kewensis* ed. 2, 5: 220–222.
Cameron, K.M. & M.W. Chase (2000). Nuclear 18s rDNA sequences of Orchidaceae confirm the subfamilial status and circumscription of Vanilloideae. In K.L. Wilson & D.A. Morrison (eds), *Monocots: Systematics and Evolution*. CSIRO, Melbourne, Australia.
Campacci, M.A. and R. Takase. (2000). *Phragmipedium chapadense* Campacci & Takase. *J. Hokkaido Orchid Soc.* 29 (Suppl.): 1–3.
Carrière, E.A. (1884). *Selenipedium caudataum* var. *giganteum*. *Revue Horticole* 56: 367.
Castaño, R., E. Hagsater & E. Aguirre (1984). *Orquidea* (Mexico City) 9(2): 193.
Cevallos-Ferriz, S.R.S. & E.A. González-Torres (2005). Geological setting and phytodiversity in Mexico. In: F.J. Vega, T.G. Nyborg, M. del C. Perrilliat, M. Montellano-Ballesteros, S.R. Cevallos-Ferriz y S.A. Quiroz-Barroso (eds). *Studies on Mexican Paleontology* pp. 1–18. Springer, Amsterdam.
Chase, M.W., K.M. Cameron, H.G. Hills & D. Jarrell (1994). DNA sequences and phylogenetics of the Orchidaceae and other monocots. A. Pridgeon (ed.), *Proc. 14th World Orchid Conference* pp. 61–73.
Chase, M.W., K.M. Cameron, R.L. Barrett & J.V. Freudenstein (2003). DNA data and Orchidaceae systematics: a new phylogenetic classification. In K.W. Dixon, S.P. Kell, R.L. Barrett & P.J. Cribb (eds), *Orchid Conservation* pp. 69–89. Natural History Publications (Borneo), Kota Kinabalu.
Chen, S.C. & Y.Z. Xi (1987). Chinese cypripediums, with a discussion on the classification of the genus. In *Proc.12th World Orchid Conference, Tokyo* pp. 141–146.

Chiron, G. & R. Bellone. (2005). *Les orchidées de Guyane française*. Tropicalia & AFCEV, France. p. 218

Chiron, G.R., A. Sambin & G.J. Braem (2015). *Apedium*, un nouveau genre dans la sous-famille Cypripedioideae (Orchidaceae). *Richardiana* 15: 243–254.

Christenson, E.A. (1997). Split slippers: botany. *American Orchid Society Bulletin* 66, 12: 1278–1281.

Christenson E.A. (2003). *Phragmipedium richteri*, a natural hybrid sporadically seen in cultivation. *Orchids* 72: 210–213.

Christenson, E.A. (2006). The former *Phragmipedium caudatum* from Panama. *Journal für den Orchideenfreund* 13, 2: 141–143.

Cogniaux, E. (1893). Tribe Cypripedilinae Pfitz. In Martius, F., *Flora Brasiliensis* 3, 4: 10–16.

Cox, A. (1995). *The utility of 5S r-DNA in phylogenetic reconstructions: development of the polymerase chain reaction in plant systematics*. Ph.D. thesis, University of Reading.

Cox, A.V. (1997). Cytological characterization of *Mexipedium xerophyticum*. *Lindleyana* 12: 162–165.

Cox, A.V., A.M. Pridgeon, V. Albert & M.W Chase (1997). Phylogenetics of the slipper orchids (Cypripedioideae: Orchidaceae); nuclear rDNA sequences. *Plant Systematics and Evolution* 208: 197–223.

Cox, A.V., G.J. Abdelnour, M.D. Bennett & I.J. Leitch (1998). Genome size and karyotype evolution in the slipper orchids (Cypripedioideae: Orchidaceae). *American Journal of Botany* 85: 681–687.

Cribb, P.J. (1987). *The genus* Paphiopedilum. Timber Press, Portland, Oregon.

Cribb, P.J. (1997). The Genus *Cypripedium*. Timber Press, Portland, Oregon.

Cribb, P.J. (1998). The genus *Paphiopedilum*. Natural History Publications (Borneo), Kota Kinabalu & Royal Botanic Gardens, Kew.

Cribb, P.J. (2009). The genus *Selenipedium*. *Curtis's Botanical Magazine* 26, 1–2: 5–20.

Cribb, P.J. & A. Schuiteman. (2015). A new species of *Selenipedium* (Orchidaceae: Cypripedioideae) from Ecuador. *Lankesteriana* 15, 3: 179–182.

Damian, M.N., N. Williams & M. Whitten (2004). *Phragmipedium kovachii* — molecular systematic of a New World orchid. *Orchids* 2: 132–137.

Dodson, C.H. (1966). Studies in orchid pollination, *Cypripedium*, *Phragmipedium* and allied genera. *American Orchid Society Bulletin* 35: 125–128.

Dodson, C.H. (1999). *Selenipedium* Rchb.f. In P.M. Jørgensen & S. Leon-Yánez, eds., *Catalogue of the Vascular Plants of Ecuador Cypripediaceae*. Missouri Botanic Garden Press.

Dodson, C.H. (2004). *Native Ecuadorian Orchids. Volume V: Rodriguezia-Zygosepalum*. Dodson Publ., Sarasota, Florida.

Dodson, C.H. & O. Gruss (1996). *Phragmipedium dalessandroi* Dodson & Gruss spec. nov. — Eine neue Art der Gattung *Phragmipedium* aus Ekuador. *Die Orchidee* 47(4): 216–220.

Dodson, C.H. & C.A. Luer (2005). Orchidaceae. In G. Harling & L. Andersson (eds.), *Flora of Ecuador* 225, 2. Botanical Institute, Göteborg University.

Dressler, R.L. (1981). *The Orchids: Natural History and Classification*. Harvard University Press.

Dressler, R.L. (1983). Classification of the Orchidaceae and their probable origin. *Telopea* 2(4): 413–424.

Dressler, R.L. (1993a). *Phylogeny and Classification of the Orchid Family*. Cambridge University Press.

Dressler, R.L. (1993b). *Field Guide to the Orchids of Costa Rica and Panama. Comstock Publishing Associates, Ithaca, New York.*

Dressler, R.L. (2005). Le Chinela, alive and well in western Panama. *Orchid Digest* 69, 2: 86–91.

Dressler, R.L. & F. Pupulin (2011). *Phragmipedium* section *Phragmipedium*. The long-petaled tropical American lady's slipper orchids. *Orchids* (*Lindleyana*) 80: 626–630.

Dressler, R.L. & N.H. Williams. (1975). (408). Proposal for the conservation of the generic name 1393b *Phragmipedium* Rolfe (1896) (Orchidaceae), against *Uropedium* Lindley (1846). *Taxon* 24: 691 (1975).

Duncan, R.E. (1959). List of chromosome numbers in orchids. In C. Withner (ed.), *The Orchids*: 529–587. *Chronica Botanica* 32, Ronald Press, New York.

Dunsterville, G.C.K. (1970). *Phragmipedium klotzscheanum* (Rchb.f. ex Schomb.) Rolfe. *Orchid Review* 78: 292–294. p. 186

Dunsterville, G.C.K. & E. (1971). *Selenipedium steyermarki* Foldats. *Orchid Digest* 35: 51–52.

Escobar, R. (1998). *Native Colombian Orchids*. Volume 6 Suppl. *Leucohyle – Zootrophiion*. Dodson Publ., Sarasota, Florida.

Fessel, H.S., E. Lückel & J. Röth. (2000). *Phragmipedium × grande* – ein neuer Naturbastard aus Ecuador. *Die Orchidee* 51, 2: 164–165.

Fighetti C., S. Eloe & P. Bechtel. (2004). Judging *Phragmipedium kovachii*. *Orchids* 73, 1: 40–43.

Filho, J.A.S. & M. Tabarelli (2006). Bromeliads species of the Atlantic Forest of northeast Brazil: losses of critical populations of endemic species. *Oryx* 40: 218–223.

Foldats, E. (1961). Contribucion a la Orquideoflora de Venezuela II. *Boletin de la Sociedad Venezolana de Ciencias Naturales* 22, 100: 253–276.

Forzza, R.C., A.V. Christianin, M.G.L. Wanderley & S. Buzato (2003). *Encholirium* (Pitcairnioideae–Bromeliaceae): conhecimento atual e sugestões para conservação. *Vidalia* 1: 7–20.

Fowlie, J.A. (1970a). Orchid collecting in the Colombian Andes. Some lessons learned in the dry barrancas of Ocaña. *Orchid Digest* 34(1): 6–8, 12.

Fowlie, J.A. (1970b). *Phragmipedium longifolium* (Rchb.f.) Rolfe. *Orchid Digest* 34: 249–250.

Fowlie, J.A. (1973). With Ghillany in Brazil: Part V. In search of *Phragmipedium vittatum* in the peat bogs of the Planalto. *Orchid Digest* 37: 111.

Fowlie, J.A. (1986). In Brazil; Part XXXIII. *Phragmipedium sargentianum* atop granite chapadas in the peat bogs in Pernambuco. *Orchid Digest* 50: 104–109

Frosch, W. & P.J. Cribb (2012). *Hardy Cypripediums: Species, Hybrids and Culture.* Royal Botanic Gardens, Kew.

Gallardo-Cruz, J.A., E.A. Pérez-García y J.A. Meave (2009). β-Diversity and vegetation structure as influenced by slope aspect and altitude in a seasonally dry tropical landscape. *Landscape Ecology* 24: 473–482.

Garay, L.A. (1960). On the origin of the Orchidaceae. *Botanical Museum Leaflets* 19: 57–95.

Garay, L.A. (1972). On the origin of the Orchidaceae. II. *Journal of the Arnold Arboretum* 53: 202–215.

Garay, L.A. (1978). *Phragmipedium. Opera Botanica, B. Flora of Ecuador* 9 (225:1): 24.

Garay, L.A. (1979). The genus *Phragmipedium. Orchid Digest* 43, 4: 133–148.

Ghillany, A. (1972). In search of the elusive *Phragmipedium sargentianum* in Brazil. *Orchid Digest* 36: 111–115.

Gorinsky, P.D. (1973). Habitat notes from a collection in Guyana of *Phragmipedium klotzschianum. Orchid Digest* 37, 2: 57–62.

Govaerts, R. (2014). *World Checklist of Orchidaceae*. Facilitated by the Royal Botanic Gardens, Kew. Available at: http://apps.kew.org/wcsp/ (accessed 3 June 2014).

Gruss, O. (1996). Die Gattung *Phragmipedium* (Teil 5). Sektion Micropetalum (Hallier) Garay. *Die Orchidee* 47(1): 13–17.

Gruss, O. (2000). *Phragmipedium chapadense* Campacci & Takase. *Die Orchidee* 51, 5: 625–627.

Gruss, O. (2003). A checklist of the genus *Phragmipedium. Orchid Digest* 67, 4: 213–241.

Gruss, O. (2010). *Phragmipedium × merinoi* – eine neue Naturhybride aus Ecuaodor. *Die Orchidee* 61, 3: 176–177.

Gruss, O. (2011). *Phragmipedium × colombianum* – eine neue Naturhybride aus Kolombien. *Die Orchidee* 62, 1: 30 (2011)

Gruss. O. (2012). *Phragmipedium hirtzii* var. *luteoviridis*, eine neue Farbform aus Ecuador. *Die Orchidee* 63, 5: 382.

Gruss, O. (2013). *Phragmipedium fischeri* Braem & Mohr, Teil 1. Artbeschreibung. *Orchideenjournal* 20, 1: 4–10.

Gruss, O. (2014). *Lateinamerikanishe Frauenshuhe: Phragmipedium, Mexipedium, Selenipedium, Cypripedium*. Orchideenzauber, Ruhmannsfelden.

Hágsater, E. (1996). Mexico (Regional Account). In E. Hágsater (ed.), IUCN/SSC Orchid Specialist Group. *Orchids — Status Survey and Conservation Action Plan*. IUCN.

Hágsater, E. & M.A. Soto Arenas (1998). Orchid conservation in Mexico. *Selbyana* 19: 15–19.

Hágsater, E., M.A. Soto Arenas, G.A. Salazar Chávez, R. Jiménez Machorro, M.A. López Rosas & R.L. Dressler (2005). *Las Orquídeas de México*. Instituto Chinoín, México, 304 pp.

Hallier, H. (1896). Ueber *Paphiopedilum amabile* und die Hochgebirgsflora des bergs K'Lamm, nebst einer Uebersicht die Gattung *Paphiopedilum. Annales Jardin Botanique Buitenzorg* 14: 18–52. p. 86

Hegedus, L.S. and F.R. Stermitz. (1986). Further facts on *Phragmipedium besseae. American Orchid Society Bulletin* 55, 4: 367–369.

Hooker, W. (1844). *Cypripedium caudatum* Lindl. *Hooker's Icones Plantarum* 7: 658–659.

International Orchid Register. Available at http://apps.rhs.org.uk/horticulturaldatabase/orchidregister/orchidregister.asp.

IUCN. (2001). *IUCN Red List Categories and Criteria*: Version 3.1. IUCN Species Survival Commission, IUCN, Gland, Switzerland.

IUCN. (2003). *Guidelines for Application of IUCN Red List Criteria at Regional Levels:* Version 3.0. IUCN Species Survival Commission. IUCN, Gland and Cambridge.

IUCN. (2010). *Guidelines for Using the IUCN Red List Categories and Criteria*: Version 8.0. IUCN Standards and Petitions Subcommittee, IUCN, Gland and Cambridge.

Janssen, T. & K. Bremer (2004). The age of major monocot groups inferred from 800 + rbcL sequences. *Botanical Journal of the Linnean Society* 146: 385–398.

Jarvis, C.E., F.R. Barrie, D.M. Allan & J.L. Reveal (1993). *A List of Linnaean Generic Names and their Types*. Koeltz, Koenigstein.

Karasawa, K. (1979). Karyomorphological studies in *Paphiopedilum*, Orchidaceae. *Bulletin of the Hiroshima Botanic Garden* 2: 1–149.

Karasawa, K. (1980). Karyomorphological studies in *Phragmipedum*, Orchidaceae. *Bulletin of the Hiroshima Botanic Garden* 3: 1–35.

Karasawa, K. (1988). Karyomorphological observations on *Phragmipedium besseae* Dodson & Kuhn Orchidaceae. *Bulletin of the Hiroshima Botanic Garden*. 10: 47–50. p. 17

Karasawa, K. (2005). A karyomorphological study on *Selenipedium aequinoctiale* Garay, Orchidaceae. *Bulletin of the Hiroshima Botanic Garden*. 22–23: 1–4. p. 17

Karasawa, K. & Tanaka, R. (1976). 2n = 18 in *Phragmipedium boissierianum*. *CIS* 20: 13–14.

Karasawa, K., M. Aoyama & G. Ishida (2005). A karyomorphological study on *Selenipedium aequinoctiale* Garay, Orchidaceae. *Bulletin of the Hiroshima Botanic Garden* 22–23: 1–4.

Klink, C.A. & R.B. Machado (2005). Conservation of the Brazilian Cerrado. *Conservation Biology* 19(3): 707–713.

Koopowitz, H. (1995). *Phragmipedium xerophyticum* and its culture. *Orchid Digest* 59: 108–110.

Koopowitz, H. (2003). *Phragmipedium kovachii* in the wild. *Orchid Dig.* 67, 4: 248–255.

Koopowitz, H. (2008). *The Tropical Slipper Orchids. Paphiopedilum and Phragmipedium Species and Hybrids.* Timber Press, Portland, Oregon.

Linden, J. and H.G. Reichenbach. (1854). *Orchidaceae Schlimianae. Bonplandia* 2: 277–284.

Koopowitz, H., A. Thornhill and M. Anderson (1993). Species distribution profiles of the Neotropical orchids *Masdevallia* and *Dracula* (Pleurothallidinae, Orchidaceae); implications for conservation. *Biodiversity and Conservation* 2: 681–690.

Kraenzlin, F. (1901). *Cypripedium* L. *Orchidacearum Genera et Species* 1: 11–86.

Le Doux, M. (1996). The diminutive *Phragmipedium xerophyticum*. *Orchid Digest* 60: 122–128.

Le Doux, M. (2003). Orchids 101 – *Phragmipedium* culture. *Orchid Digest* 67, 4: 244–246.

Leitch, I.J., I. Kahandawala, J. Suda, L. Hanson, M.J. Ingrouille, M.W. Chase & M.F. Fay (2009). Genome size diversity in orchids: consequences and evolution. *Annals of Botany* 104: 469–481.

Linden, L. (1886). *Selenipedium caudatum* var. *roseum*. *Illustration Horticole* 33: t. 506

Linden, L. (1888). *Selenipedium caudatum* var. *albertianum*. *Lindenia* 4: 64.

Lindley, J. (1840). *Cypripedium*. In *Genera and Species of Orchidaceous Plants* pp. 525–532. Ridgways, London.

Lindley, J. (1846). *Cypripedium irapeanum*. Edwards's Botanical Register 32: t. 58.

Lindley, J. (1850). The long-tailed lady's slipper (*Cypripedium caudatum*). *Lindley & Paxton's Flower Garden* 1: t. 9.

Lindley, J. (1853). *The Vegetable Kingdom*. Bradbury and Evans, London.

Linnaeus, C. (1753). *Species Plantarum*. L. Salvius, Stockholm.

Manza, S. (2006). *Phragmipedium vittatum* (Vellozo) Rolfe. Available at: http://www.slipperorchids.info/phragdatasheets/vittatum/index.html.

McCook, L.M. (1989). The genus *Phragmipedium* Part 1. *American Orchid Society Bulletin* 58(10): 1095–1100.

McCook, L. (1989a). Systematics of *Phragmipedium* (Cypripedioideae; Orchidaceae). Ph.D. thesis, Cornell University.

McCook, L.M. (1989b). The genus *Phragmipedium* Part 1. *American Orchid Society Bulletin* 58, 10: 1095–1100.

Mc Cook, L.M. (1990). The genus *Phragmipedium* Part 2. *American Orchid Society Bulletin* 59, 2: 153–158.

McCook, L. (1989). Systematics of *Phragmipedium* (Cypripedioideae: Orchidaceae). Ph.D. thesis, Cornell University.

McCook, L. (1998). An annotated checklist of the genus *Phragmipedium*. *Orchid Digest* Spec. Publ. 2–12.

McGough, H.N., D.L. Roberts, C. Brodie & J. Kowalczyk (2006). *CITES and Slipper Orchids – An introduction to slipper orchids covered by the Convention on International Trade in Endangered Species*. Royal Botanic Gardens, Kew.

Menezes, L. (2004). On "*Orchids from the Brazilian Central Plateau*". Sao Paolo. Available at: http://www.delfinadearaujo.com/on/on23/pages/lou02en.htm.

Menezes, L.C. (1989). *Phragmipedium vittatum*: An orchid in the process of extinction. *Orchid Digest* 53(2): 77–78.

Menezes, L.C. (2004). *Orquideas Planalto Central Brasileiro*. Ed. Ibama, Brasilia.

Menezes, L.C. (2014). *Orquideas Planalto Central Brasileiro, exp. Inst. Brasileiros do Melo Ambiente e dos Recursos Naturais Renovaveis, Brasilia.*

Menezes, L. & G. Braem. (1993). *Phragmipedium vittatum* (Vell.) Rolfe, die Bedeutung von Büschbranden für deren Ökologie. *Schlechteriana* 4: 119.

Miller, D. (2008). *The Organ Mountain, its history and its orchids*. Ed. Scart, Novo Friburgo.

Miller, D. & I. & R. Warren (2002). Two rare orchids on one mountain. Brazilian Orchids. *Equatorial Plants Newsletter* 19, 2: 9–12.

Miller, D., R. Warren, I.M. Miller & H. Seehawer (2008). *The Organ Mountain Range: Its History and its Orchids*. Ed. Scart, Nova Friburgo.

Moat, J. (2007). *Conservation Assessment Tools Extension for ArcView 3x*. Version 1.2. GIS Suite, Royal Botanic Gardens, Kew.

Mó Mó, E.A. & E.A. Perez Garcia (2015). Die Cypripedioideae von MesoAmerika. Teil 2b Phragmipedium. *Die Orchidee* 66, 3: 216–225.

Mó Mó, E.A. & E.A. Perez Garcia (2016). Clarification of two taxon names in *Phragmipedium* (Orchidaceae, Cypripedioideae). *Lankesteriana* 16, 1: 39.

Moon, A. & P.J. Cribb. (1998). Thoughts on the identity of the red Phragmipediums. *Orchid Review* 105: 227–230.

Oakeley, H. (2004). *Phragmipedium kovachii. Orchid Society Great Britain* J. 53: 24–25.

Pabst, G.F.J. & F. Dungs (1975), (1977). *Orchidaceae Brasilienses*. Brücke-Verlag Schmersow, Hildesheim.

Pasetti, M. (1995). *Le Cypripedioideae.* Associazione Lombarda Amatori Orchidee, Paris.

Pemberton, R. (2011). Pollination mechanisms in phragmipediums: hover-fly (Syrphidae) pollination and mechanical self-pollination (autogamy) in *Phragmipedium* species. *Orchids* 80: 364–367.

Pemberton, R. (2013). Pollination of slipper orchids (Cypripedioideae): a review. *Lankesteriana* 13, 1–2: 65–73.

Perez Garcia, E. (2010). El redescubrimiento de *Mexipedium xerophyticum* (Soto Arenas, Salazar & Hagstater) V.A. Albert & M.A. Chase. *Lankesteriana* 9, 3: 557–563.

Perez-Garcia, E.A. (2011). The rediscovery of *Phragmipedium xerophyticum*. *Slipper Orchid Alliance Newsletter* 12, 1: 10–14.

Pérez-García, E.A. & J.A. Meave (2004). Heterogeneity of xerophytic vegetation of limestone outcrops in a tropical deciduous forest region. *Plant Ecology* 175: 147–163.

Perez Garcia, E.A. & E.A. Mo Mo (2014). Die Cypripedioideae von Mesoamerika, Teil 1a. *Cypripedium*. *Die Orchidee* 65, 6: 476–482.

Perez Garcia, E.A. & E.A. Mo Mo (2015a). Die Cypripedioideae von Mesoamerika, Teil 1b. *Selenipedium*. *Die Orchidee* 66, 1: 54–67.

Perez Garcia, E.A. & E.A. Mo Mo (2015b). Die Cypripedioideae von Mesoamerika, Teil 2a. *Mexipedium*. *Die Orchidee* 66, 2: 128–133.

Perez Garcia, A. & E.A. Mo Mo (2015c). Die Cypripedioideae von Mesoamerika, Teil 2b. *Phragmipedium*. *Die Orchidee* 66, 3: 216–225.

Pérez-García, E.A., A.C. Sevilha, J.A. Meave & A. Scariot (2009). Floristic differentiation in continental Neotropical limestone outcrops of southern Mexico and central Brazil: a Beta diversity approach. *Boletín de la Sociedad Botánica de México* 84: 45–58.

Pittman, C. (2012). *The Scent and Scandal: Greed, Betrayal, and the World's Most Beautiful Orchid*. Univ. of Florida Press.

Pfitzer, E. (1889). Diandrae Cypripedilinae. In A. Engler & K. Prantl (eds), *Die Natürlichen Pflanzenfamilien* II, 6: 82.

Pfitzer, E. (1894). Beitrage zur Systematik der Orchidaceen. In A. Engler ed., *Botanische Jahrbücher* 19: 1–42.

Pfitzer, E. (1903). *Cypripedium*. In A. Engler, *Das Pflanzenreich* IV, 50 (Heft 12) Orchidaceae Pleonandrae: 28–42.

Pridgeon, A.M., P.J. Cribb, M.W. Chase & F. Rasmussen (1999). *Genera Orchidacearum* 1. Oxford Univ. Press.

Pupulin, F. (2012). The Orchidaceae of Ruiz & Pavon's "Flora Peruviana et Chilensis". A taxonomic study. 1. *Anales del Jardín Botaníco de Madrid* 69, 1: 21–79 & 2, 143–186.

Pupulin, F. & R.L. Dressler (2011). A nomenclatural note on *Cypripedium humboldtii* (Orchidaceae: Cypripedioideae). *Harvard Papers in Botany* 16, 1: 167–179.

Pupulin, F. & R.L. Dressler (2012). On the meaning of some old names in *Phragmipedium* (Orchidaceae: Cypripedioideae) with a few typifications. *Die Orchidee* 63, 5: 383–393.

Quené, R.-J.W. & O. Gruss (2003). *Phragmipedium brasiliense*, a new species of the section Lorifolia from Brazil. *Orchid Digest* 67, 4: 242–243.

Rafinesque, C.S. (1819). *Criosanthes*. *J. Phys. Chim. Hist. Nat.* 89: 102.

Rafinesque, C.S. (1838). *Sacodon*. In *Flora Telluriana* 4: 46. Philadelphia.

Rasmussen, F. (1985). Orchids. In R.M.L. Dahlgren et al. (eds), *The Families of the Monocotyledons — Structure, Evolution and Taxonomy. Springer,* Berlin, pp.: 247–274.

Rauh, W. & K.H. Senghas. (1975). *Phragmipedium caricinum* und *Ph. pearcei. Die Orchidee* 26: 56–62.

Reddy, V. (2008). *Mexipedium xerophyticum. Orchid Forum, The Newsletter of the Orchid Forum of Sacramento* April: 3–4.

Reichenbach, H.G. (1852). Neue Orchiden der Expedition des Herrn J. de Warscewicz. *Botanische Zeitung (Berlin)* 10, 40: 691.

Reichenbach, H.G. (1854–1858). *Xenia Orchidacea* 1. F.A. Brockhaus, Leipzig.

Reichenbach, H.G. (1865). Ueber einige Garten-Orchideen. *Hamburger Garten- und Blumenzeitung* 21: 293–301.

Reichenbach, H.G. (1871). *Selenipedium roezlii* Rchb.f. *Gartenflora* 20: 164.

Reichenbach, H.G. (1873). Tafel 189. *Selenipedium wallisii* Rchb. fil. *Xenia Orchidacea* 2: 189.

Reichenbach, H.G. (1874). *Selenipedium reticulatum. Xenia Orchidacea* 2, 10: 223.

Reichenbach, H.G. (1878). New Garden Plants. *Gardeners' Chronicle* n.s. 9: 202.

Reichenbach, H.G. (1882). New Garden Plants. *Gardeners' Chronicle* n.s. 18: 521–522.

Reichenbach, H.G. (1886). *Cypripedium lindleyanum. Gardeners' Chronicle* n.s. 25: 680.

Robbins, S. (1989). *Phragmipedium besseae. The Kew Magazine* (*Curtis's Botanical Magazine*) 6, 3: t. 135.

Rodriguez Salas, M., J. Benavides Ranilla & J.R. Epinoza (2007). Genetic relationships of *Phragmipedium* species (Orchdiaceae) using amplified fragment length polymorphism (AFLP) fragments. *Lankesteriana* 7, 3: 493–496.

Rolfe, R.A. (1896). The *Cypripedium* group. *Orchid Rev.* 4: 327–334, 363–367.

Röth, J. (2012). *Phragmipedium* × *grande*, ein neuer Naturbastard aus Ecuador. *Die Orchidee* 63, 5: 403–404.

Röth, J. and O. Gruss. (1994). *Phragmipedium* x *richteri* Roeth & O. Gruss. *Die Orchidee* 45, 3: back cover.

Saddi, E.M., R. Conrado Lopes & R.H. Potsch Andreata (2005). Floristics and Conservation of Orchidaceae at Rio das Pedras Reserve. *Selbyana* 26(1): 318–325.

Salazar Chávez, G. & E. Hágsater (1997). *Diversidad y Conservación de orquídeas de la región de Chimalapa, Oaxaca, México*. México. Informe final del Proyecto G024 presentado a la CONABIO.

Sambin, A. & G. Braem (2015a). *Selenipedium chironianum*, une intéressante addition à la flore d'orchidées de Guyane française. *Richardiana* 15: 183–189.

Sambin, A. & G. Braem (2015b). Une excitante nouvelle espèce de Guyane: *Phragmipedium guinanense* (Orchidaceae: Cypripedioideae). *Richardiana* 15: 2–6.

Sandoval, E., T. Terrazas, G. Salazar, A. Vallejo & B. Estrada. (2003). Anatomía vegetativa de *Mexipedium xerophyticum* (Soto, Salazar & Hágsater) V.A. Albert & M.W. Chase y géneros relacionados (Orchidaceae, Cypripedioideae) *Lankesteriana* 7: 54–56.

Schlechter, R. (1922). Beiträge zur Orchideenkunde von Zentralamerika 1. Orchidaceae Powellianae Panamenses. *Repertorium Specierum Novarum Regni Vegetabilium* Beihefte 17: 9.

Schlechter, R. (1926). Das System der Orchidaceen. *Notizblatt der Botanisches Garten Berlin-Dahlem* 9: 563–591.

Shefferson, R.P., D.L. Taylor, M. Weiss, S. Garnica, M.K. McCormick, S. Adams, H.M. Gray, J.W. McFarland, T. Kull, K. Tali, T. Yukawa, T. Kawahara, K. Miyoshi & Y.-I Lee (2007). The evolutionary history of mycorrhizal specificity among lady's slipper orchids. *Evolution* 61: 1380–1390.

Silva, J.F., M.R. Farinas, J.M. Felfili & C.A. Klink (2006). Spatial heterogeneity, land use and conservation in the cerrado region of Brazil. *Journal of Biogeography* 33: 536–548.

Soto-Arenas, M.A. (1996). Mexico (Regional Account). In E. Hágsater & V. Dumont (eds), *Orchids — Status Survey and Conservation Action Plan*. IUCN/SSC Orchid Specialist Group, pp. 53–58. IUCN, Gland, Suiza y Cambridge, R.U.

Soto-Arenas, M.A. (2003). *Mexipedium xerophyticum* (Soto Arenas, Salazar & Hágsater) V.A. Albert & M.W. Chase. In E. Hágsater & M. Soto (eds), *Icones Orchidacearum Fascicles 5 & 6, Orchids of Mexico Parts 2 & 3*. Herbario AMO, México D.F.

Soto-Arenas, M.A. & G.A. Salazar (2004). Orquídeas. In: A.J. García-Mendoza, M.J. Ordónez y M. Briones-Salas (eds),. *Biodiversidad de Oaxaca* pp. 271–295. Instituto de Biología, Universidad Nacional Autónoma de México, Fondo Oaxaqueño para la Conservación de la Naturaleza y World Wildlife Fund. México, D.F.

Soto-Arenas, M.A. & A.R. Solano-Gómez (2007). Ficha técnica de Mexipedium xerophyticum. In M.A. Soto-Arenas, *Información actualizada sobre las especies de orquídeas del PROY-NOM-059-ECOL-2000*. Instituto Chinoin A.C., Herbario de la Asociación Mexicana de Orquideología A.C. Bases de datos SNIB-CONABIO. Proyecto No. W029. México. D.F.

Soto-Arenas, M.A., G. Salazar-Chávez & E. Hágsater (1990). *Phragmipedium xerophyticum*, una nueva especie del sur de México. *Orquídea* (México) 12: 1–10.

Soto Arenas, M.A., R. Solano Gómez & E. Hágsater (2007). Risk of extinction and patterns of diversity loss in Mexican orchids. *Lankesteriana* 7: 114–121.

Sprunger, S., P.J. Cribb & A. Toscano de Brito (1996). *João Barbosa Rodrigues: Iconographie des Orchidées du Brésil*. F. Reinhardt Verlag, Basel.

Stein, B. (1892). *Stein's Orchideenbuch*. Paul Parey Verlag, Berlin.

Steyermark, J. (1981). Erroneous citations of Venezuelan localities. *Taxon* 30, 4: 816–817.

Swartz, O. (1800). *Cypripedium* L. *Kungliga Svenska Vetenskapsakademiens Nya Handlinga* 21: 250–251.

Swartz, O. (1805). Genera et Species Orchidearum. *Neues Journal de Botanique* 1: 1–108.

Szlachetko, D.L. & M. Kolanowska. (2012). *Orchids of French Guiana*. Gantner, Ruggell.

Szlachetko, D.L. & M. Kolanowska. (2016). Notes on the genus *Selenipedium* (Orchidaceae, Cypripedioideae) with descriptions of new taxa. *Systematic Botany* 41, 1: 142–159.

Tanaka, R. & H. Kamemoto (1974). List of chromosome numbers in species of the Orchidaceae. In C. Withner (ed.), *The Orchids: Scientific Studies* pp. 411–483. John Wiley & Sons, New York.

Tiffney, B.H. (1985a). Perspectives on the origin of the floristic similarity between eastern Asia and eastern North America. *Journal of the Arnold Arboretum* 66: 73–94.

Tiffney, B.H. (1985b). The Eocene North Atlantic land bridge: Its importance in Tertiary and Modern phytogeography of the Northern Hemisphere. *Journal of the Arnold Arboretum* 66: 243–273.

Torres-Colín, L.M. (1989). *Estudio Florístico y Descripción de la Vegetación del Cerro Guiengola, en el Istmo de Tehuantepec, Oaxaca*. Tesis de Licenciatura (Biología). Universidad Nacional Autónoma de México. Los Reyes, Iztacala, Estado de México. 81 pp.

Vermeulen, P. (1966). The system of the Orchidales. *Acta Botanica Neerlandica* 15: 224–253.

Wendt, T. (1989). Las selvas de Uxpanapa, Veracruz-Oaxaca: Evidencia de refugios florísticos cenozoicos. *Anales del Instituto de Biología serie Botánica* 58: 29–54.

Wimber, D.E. (1994). The chromosome revolution for slipper orchids. *Proc. 14th World Orchid Conference* pp. 228–232.

Van der Pijl, L. & C.H. Dodson. (1966). *Orchid Flowers: their Pollination and Evolution*. Univ. of Miami Press.

Veitch, J. (1889). *Manual of Cultivated Orchids*. Veitch, Chelsea.

Zuiderwijk, R. (2009). *Mexipedium xerophyticum*. http://www.phragweb.info/mexipedium (consulted October 2009).

APPENDIX 1: SPECIMENS EXAMINED

Phragmipedium andreettae
SW COLOMBIA/ NW ECUADOR: sine prov., hort. Ecuagenera, November 2005, cult. *Portilla s.n.* (holotype: QCA!).

P. besseae
PERU: San Martin, road from Tarapoto to Yurimaguas, 1000 m, 13 July 1981, *Halton & Besse* 62 (holo. SEL!); cult. Powell 1987 (K, spirit coll.); *Whalley s.n.* (K, spirit coll.); cult. Kunisch (K, spirit coll.). ECUADOR: Morona Santiago, S of Gualaquiza, 1900 m, March 1986, *Dalström* 1488 (SEL); Azuay Prov., near Paute, March 1992, *Andreetta in Dalström* 1689 (SEL); Cuenca to Cola de San Pablo, Guarumales, 1300 m, 2 Dec. 1984, *Dodson et al.* 15501 (MO, QNCE) & 9 March 1985, *Dodson et al.* 15778 (RPSC, OCNE); Cuenca to Guarumales, between dam and Casa de Maquinas, 1500 m, 25 March 1991, *Luer et al.* 15257A (MO); also Cribb (sight record). Cult. Kew, 18 Oct. 1996 (K); cult. Doll, 24 Nov. 1987 (K0.

P. dalessandroi
ECUADOR: Zamora-Chinchipe, Rio Bombuscara, *D'Alessandro* 745 (RPSC); Rio Jamboe, *Dalessandro* 745A (holo. RPSC); Zamora-Gualaquiza, 900 m, 3 March 1986, *D'Alessandro* 518A (RPSC); cult. Eric Young Orchid Foundation (K, spirit coll.).

P. fischeri
ECUADOR: Carchi, Tulcan to Maldonado, beyond Madonda via Chical, 1000-1500 m, *J. Portilla s.n.* (RPSC). CULT.: SW COLOMBIA: cult. *Fischer s.n.* (K, Spirit coll.).

P. kovachii
PERU: without exact provenance, *Villena in Bennett* 7996 (holo. of *P. peruvianum*, SMF; photo. K); 2 km NE of Aguas Claras on road between Moyobamba and Nueva Cajamarca, 5 June 2002, *Kovach s.n.* (holo. SMF; photo. K) also Cribb (sight record).

P. schlimii
COLOMBIA: Ocaña, 1280m, *Schlim* 407 (lecto. W!, K! ; syn. K!, G!, P!) & *Schlim s.n.* (P!); Ocaña, Quebrada del Tigre, *sin. prov., s.n.* (W); Valle de Cauca, near El Querimal, *Espinal T.* 1833 (AMES); El Valle, road from Cali to Buenaventura, *Barkley & Willard* 35461 (AMES); km 54 on old road, *Dodson & Hills* 3125 (SEL); Putomayo, Cerro de Portachuelo, *Schultes & Smith* 3050 (AMES); Cauca, El Tambo, *von Sneidern* 131 (S); Rio Grande, W of Don Matias, 1500-1800 m, *Lehmann* 7259 (K); without exact provenance, *Purdie s.n.* (K); Antioquia, Angostura, 1400-1700 m, *Lehmann* 8047 (K) & 8048 (K); *von Sneidern* 132 (S); 1872, *Shuttleworth s.n.* (BML, W); La Elsa, *Espinal* T1901 (AMES); sin. loc., ex Herb. Rchb.f. (BM). Cult. Kew, March 1889 (K); Nov. 1889 (K); Feb. 1920 (K); cult. Kramer, Oct. 1889 (K).

P. boissierianum
PERU: Huanuco Dept., Leoncio Prado Prov., Divisonia Pass, E side, 2.5 km below summit, 1570 m, 21 July 1964, *Hutchinson, Wright & Straw RM* 6014 (K, P!); 2 km NE of Aguas

Claras on road between Moyobamba and Nueva Cajamarca, *Cribb* (sight record); Huanuco from Tingo Maria to Pucallpa, 1310 m, 4 June 1990, *Croat & Simpson* 81740 (MO, SEL); Rioja Prov., San Martin, road from Rioja to Pedro Ruiz, 1170 m, 23 March 1998, *Gray, Vasquez & Rojas* 15475 (MO, SEL); San Martin Prov., Tarapoto to Yurimaguas, 1000 m, 12-20 July 1981, *Halton & Besse* 109 (SEL) & 12 July 1981, *Besse, Halton & Luther* 787 (SEL, neotype and isoneotype of *P. cajamarcae*); Tingo Maria, cult. Marie Selby B.G., *Atwood* 615-2 (SEL); Pawao Portago, an hour below Pomobamba, Rio Apurimac, 1500 m, 7 July 1981, *Davis, Marshall, Tichmoor & Lessner* 1265 (SEL, Field); ex Peru, cult. 12 July 1966, *Mason* 1671 (K). ECUADOR: Loja, Rio Bombuscara and Timbara, 800-1200 m, *Lehmann* 8045 (AMES, K 12 sheets). MISC.: cult. Weathers, Dec. 1888 (BM); cult. Sander, 21 March 1921 (K).

P. longifolium

COSTA RICA: Alajuela, Near Rio Barranca, *Brenes* 21396 *(AMES,* F, NY); Cartago, Base of Cerro Doan, *Lent* 1600 (CR, F, NY); Turrialba, 1867, *Endres s.n.* (W, illustration); 1868, *Endres* 112 (W); ex Costa Rica, cult. Castelfranco, June 2001 (K). PANAMA: Chiriqui, Boquete, flowered in cult., *N. Williams s.n.* (SEL); Boquette, 24 Jan. 1988, *Christenson* 1688 (K, SEL); La Fortuna hydroelectric project, *Hammel 2095* (MO)*, Correa, Escobar & Mendoza* 2133 (MO)*, Folsom, Dressler & Dressler* 5593 (MO)*, Hammel* 2292 (MO), *Atwood* 781 (SEL, flowered in cult.); La Fortuna, cult. Marie Selby B.G. 10 Sept. 1886, *Christenson in Atwood* 781-9 (SEL); west of Caldera *River, Powell 308* (AMES); sine prov., *s. n.* (AMES 20461); Coclé, El Valle de Anton on road to La Mesa, *Atwood* 77173 (SEL, PLAS); El Valle de Anton, 24 Jan. 1988, *Christenson* 1689 (K, SEL); Cerro Jefe, Rio Indio, cult. 1 Jan. 1982, *Atwood 77177-5* (as var. *gracile* Atwood & Dressler SEL); Boquette, 1 Dec. 1984, *Christenson* 1625 (SEL); San Blas, Playon Chico, Rio Ukupsoni, 50 m, 29 Oct. 1991, *Herrera, Morris & Molita 1019* (MO, SEL). COLOMBIA: *Patin s.n.* (K); *Seeman s.n.* (K); 1876, cult. Ortgies, June 1889, *Wallis s.n.* (K); Bay of Choco, Cape Corrientes, Oct. 1847, *Seeman* 1111 (K); Rio Cauca, 300 m, 29 April 1878, *Lehmann* 6118 (K) & Rio Cauca, 0-300 m, *Lehmann* 8046 (K). ECUADOR: Fulfun, San Nicolas, Cascada del Colombia, junction of Rio Azul, and Rio Calima, 4 Aug. 181962, *Hugh-Jones* 162 (K); Cuyambe, *Lehmann* BT11 (K); Esmeraldas Prov., Lita to Esmeraldas rd, 12 May 1990, *Dodson, Gentry, Boyle & Rubio* 18279 (MO, QCNE); Alto Tambo, 400 m, 9 Sept. 1990, *Dodson, Rubio & Quelal* 702 (MO, QCNE); Lita, Alto Tambo, cult. 15 Aug. 1989, *Whitten* 1002 (QCNE) & 900 m, f. Cult. 18 June 1990, *Whitten* 90324 (QNCE); Pichincha Prov., Las Mercedes to Alluriquin rd, 1050 m, 19 Feb. 1995, *Ollgaard & Navarette* 863 (AAU, QCNE); cult. Bull, 1872 (K); cult. Dec. 1890, Burbidge (K); cult. Glasnevin B.G., April 1887 (K), 24 June 1887 (K), March 1888 (K); cult. Gower (K); cult. Kent, 1 June 1887 (K), 19 Sept. 1887 (K); cult. Kew, Feb. 1879 (K), Nov. 1886, May 1887 (K), July 1887 (K), Oct. 1887 (K), Nov. 1887 (K), April 1888 & Aug. 1889 (K), 20 Feb. 1917 (K); cult. Sander, Dec. 1890 (K); cult. Tautz, 16 Aug. 1889 (K); cult. Veitch, ex herb. Rchb.f. (BM), cult. Veitch, 1881 (BM), Sept. 1887 (K) & Feb. 1889 (K); cult. Herbert Veitch, Nov. 1886 (K); cult. 17 Nov. 1889, Weathers s.n. (BM).

P. caricinum

BOLIVIA: La Paz, S Yungas, Serrania de Nachani,70 km S of San Burja on rd to Palos Blancos, 23 July 1967, *Killeen* 2823 (SEL); Beni, Ballivian Prov., Beni, Caranavi to Yucumo road, 138 km from Caranavi, 660 m, 19 June 1989, *Smith & Buddensick* 13522 (K, MO); Santa Cruz Distr., Rio Espoja, cult. 22 Jan. 1875, *Fuchs SEL* 26-70-7 (SEL); Villa Tunari,

NE of Cochabamba, 26 Nov. 1978, *Fuchs* 5478 (SEL); E of Cochabamaba, *Fuchs SEL* 156-1976-2 (SEL); Santa Cruz Reg., Ichila, 26 Aug. 1986, *Taxiva* 11 (SEL); USBA, 5 March 1970, *Vasquez* 371 (SEL); Santa Cruz Dept., Ichila Prov., Amboro N.P., c. 15km SE up Rio Pitasama from Rio Sururu, 700 m, 26 Aug. 1983, *Solomon & Urcullo* 14076 (MO, SEL); Chapare, 3 May 1984, *Vasquez* 391 (SEL).

P. pearcei

ECUADOR: Morona-Santiago, between Limon and Macas, *Luer* 8197A (SEL); Napo, between Dureño and Santa Cecilia, Rio Aguarico, *Pinckley* 5633 (AMES, SEL); Napo-Pastaza, Rio Napo, Santa Rosa, *Harling* 3624 (S); Pastaza, between Topo and Mera, on rd from Baños to Puyo, *Hirtz* 1146 (SEL); Puyo, cult. Fuchs s.n. (SEL); Napo Prov., Canton Archidota, Hollin to Loreto rd, Rio Pucuno, 1000 m, 21 Aug. 1989, *Ceron & Factos* 7420 (MO, QCNE) & Rio Hollin, 1100 m, 14-22 Feb. 1989, *Hurtado & Shiguango* 1658 (MO, QCNE) & 23 Jan. 2000, *Whitten* 1598 (QCNE); Rio Tena, 500 m, 7 March 1987, *Dodson* 17055 (MO, QCNE); Rio Pucuno, 1150 m, 23 Sept. 1989, *Neill & Iguago* 9050 (MO, QCNE) & 1100 m, 19 Oct. 1989, *Neill & Palacios* 9093 (MO, QCNE). PERU: San Martin, rd from Tarapoto to Yurimaguas, *Schunke* 9825 (NY, SEL); Pongo de Caynarachi, Rio Cainarachi, *Klug* 2634 (AMES, BM, K, MO, NY, S, US); Tarapoto, *Spruce* 3955 (AMES, BM, G, K, OXF, W); Chazuta, Rio Huallaga, *Klug* 4099 (AMES, BM, CAS, K, MO, NY, S, UC, US); cult. Van Houtte, *Pearce s.n.* (K (photo.), W); cult. Glasnevin B.G. 12 Aug. 1887 (K); cult. Kew, 30 Aug. 1888 (K), & 25 Aug. 1925 (K); Tarapoto to Yurimagua rd., Caimarachi, 18 Dec. 1977, *Schuncke* 2828 (SEL); Tarapoto Distr., *Spruce* 3955 (K, P); Alto Beni, 22 June 1964, *Badcock* 212 (K).

P. hirtzii

ECUADOR: Esmeraldas, San Lorenzo Canton, 500m, 20 Feb 1991, *Palacios* 6924 (SEL); San Lorenzo Canton, 2 km E of El Placer, 500 m, 20 Feb. 1991, *Palacios* 6926 (MO, QCNE) & 350 m, *Neill, Toapanta, Shihuango, Pay & Casaluzan* 13887 (MO, QCNE); San Lorenzo Canton, Parroquia Mataje, 200 m, 21 Sept. 1992, *Aulestia, Tipaz, Burbano & Canticuz* 549 (MO, QCNE); Lita to Alto Tambo rd, 19 May 1987, 750-850 m, *Dodson, van der Werf & Palacios* 17134 (MO, QCNE, SEL); Lita, railroad from Ibarra to San Lorenzo, 600 m, 15 Aug. 1986, *Hirtz* 2919 (K, MO, QCNE); Carchi Prov., *Dalström* 1757 (SEL); Carchi Prov., El Pilon, 45 km W of Maldonao, 800 m, 3 Dec. 1979, *Madison & Besse* 7277 (SEL); Alto Tambo, 2 km S of Lita to San Lorenzo rd, 350-600 m, 13 Feb. 2003, *Clark* 7180 (QCNE); Imbabura, Sector Tabla Chupa, Canton Cotacachi, 900 m, 18 June 1992, *Tipaz & Gudiño* 1210 (QNCE); Tulcan Canton, Parroquia Tobar Donoso, 650-1000 m, 19-28 June 1992, *Tipaz, Zuleta & Guanga* 1293 (MO, QCNE), 900 m, 21 May 1992, *Quelal, Tipaz & Taicuz* 556 (MO, QCNE), 900 m, 22 Nov. 1993, *Aulestia, Aulestia & Guanga* 681 (MO, QCNE); Carchi, Reserva Awa, Sabalera, 900 m, 22 Nov. 1992, *C. Aulestia et al.* 549 (QNCE); Reserva Etnica Awa, Communidad, de Gualpi Medio, 900 m, 21 May 1992, C. *Quelal et al.* 556 (MO, QNCE); Reserva Awa, Sabalera, 650 m, 19 June 1992, *Tipaz et al.* 1293 (QCNE).

P. vittatum

BRAZIL: São Paulo, 18 March 1857, *Regnell III* 1206 (P!); Distr. Federal, Brasilia, Sector Industrial, 23 Jan. 1966, *Pabst, Pereiera & Apparicio 8754* (K); Chapada dos Veadeiros, NW of Veadeiros, 1000 m, 16 Feb. 1966, *Irwin, Grear, Souza & Reis dos Santos* 12950 (K). BOLIVIA: Santa Cruz, *Vasquez* 2845 (LPB).

P. klotzschianum

GUIANA: Roraima, Cabaceras del Rio Caroni, *Schomburgk* 33 (K, P), 718 (K, P, type), 727 (K, P) & 1065 (P); Aroie Creek, Cotinga R., *Im Thurn* 5 (K); Cotinga R., Oct. 1884-Jan. 12885, *Im Thurn* 31 (K); Krang Distr., Orinduik, Takaika Creek, 7 Nov. 1959, *Graham* 320 (K); Roraima, Cako R., 1863-4, *Appun* 1212 (K); Atabopo R., Autumn 1894, *Quelch & McConnell* 144 (K); Orinduik, 8 Jan. 1961, *Harris* K143 (K). VENEZUELA: Auyantepui, El Oso, *Foldats* 2635 (VEN) & *Schnee* 1669 (VEN); Rio Pacairo, NE of Sta Teresita de Cavanayen, *Steyermark* 60487 (MO, VEN); Rio Carun, *Cardona 2451* (VEN); Rio Venamo, *G & E Dunsterville 836* (VEN) & *Steyermark, G & E Dunsterville* 92429 (VEN); Uaitantepui, *Koyama & Agostini* 7512 (VEN); Gran Sabana, cumbres de Uaipap, *Phelps & Hitchcock* 427 (MO); Est. Bolivar State, 21 April 1988, *Leisner* 23559 (MO) & 1 May 1988, *Leisner* 24112 (MO); 29 Dec. 1978, *Xena* 304 (MO); Rio Cuqunomi, Dec. 1909, *Ule* 8569 (K). Cult. Moore, Dec. 1890 (K).

P. lindleyanum

VENEZUELA: Est. Bolivar, Roscio Distr., 450 m, 19 March 1959, *O. Renz* 9128 (RENZ) & 9395 (RENZ); 1 Aug. 1993, *Carnevali, Lorea, Warford, Whitlock & Fleming* 3220 (MO); 9 May 1986, *Liesner & Holst* 20728 (MO); 17 Aug. 1993, *Ramírez, Hokche, Raimundez, Gil, Rodriguez & Briceño* 4456 (MO); Rio Arabapo, *McConnell & Quelch* 144 (MO); Cerro Soropantepuy, *Steyermark* 60104 (MO, VEN); Cerro Guiaquinima, *Cardona* 444 (VEN); Cumbre del Auyantepui, Foldats 2626; 100 km S of El Dorado, *Foldats* 2667 (VEN);& *G & E Dunsterville* 417 (AMES, VEN) & *Antonio* 710 (VEN); cabaceras del Rio Morroco, *Rutkis s.n.* (MO). GUYANA: Kanuku Mts, 22 Jan. 1984, cult. E. Ross (K); Roraima, *Im Thurn* 58 (K); Roraima, *Schomburgk s.n.* (K), 727 (K) & 1097 (K); Roraima, cult. Veitch, 1881, *Burke s.n.* (K); *Hahn, Judziewicz, Tiwari & Gopaul* 4646 (MO); 1660 m, *Boom & Samuels* 9109 (MO); 550 m, *de Granville & Cremers* 1809 (MO); Demerara, Jan. 1882, *Jenman* 879 (K); bottom of Potaro Gorge near Kaieteur Falls, 200 m, 13 March 1962, *Cowan & Soderstrom* 2160 (K); Kaieteur Falls, *Jenman* 879 (K, type of *P. kaiteurianum*); cult. 1994, *J. Kunitch s.n.* (K). FRENCH GUYANA: Mts. De la Trinité, Bassin de la Mena, 200 m, 16 March 1997, *de Granville* 13402 (P); Inselberg Talouaken, Massif de Tumuc-Humac, 620 m,10 Aug. 1993, *de Granville, Acevedo, Boyer & Hollenberg* 12189 (P); Massif de Emerillons, 400 m, 10 Sept. 1980, *de Granville* 3803 (P); Monts de la Trinité, 4 Aug. 1981, *de Granville* 4742 (P); Mts des Nourages, Bassin de l'Arataye, April 1989, *Larpin* 627 (P). SURINAM: *Glaziou* 879 (P). Cult. Weathers, Nov. 1889 (BM). Cult. Sander, 23 March 1887 (BM). Cult. Charlesworth, Dec. 1891, *Shuttleworth s.n.* (K); Glasnevin B.G. (K); cult. Kew 1894 (K) & 15 Feb. 1895 (K).

P. sargentianum

BRAZIL: cult. Kew 1894 (K) & 15 Feb. 1895 (K).

P. caudatum

PERU: *Ruiz & Pavon s.n.* (K, W); San Martin Prov., Juan Guerra, Tarapoto to Juanjui road, 300-400 m, 22 Dec. 1977, *Schunke* 9812 (MO, SEL); Muna, *Davis s.n.*, comm.. Veitch, July 1884 (K); Ollachoa to San Caboa, 17-24 July 1973, *Dillon* 1127 (SEL); Huanuco Prov., Cordillera de Carpish, 2800-2900 m, 15 June 1958, *Humbert 30953* (P). BOLIVIA: Chapare, CBBA, *Vasquez* 5C, (SEL); sine loc., *Pearce* s.n. (M); Sacramento, *Bang* 81 (BM, K) & La Paz, *Bang* 82 (BM, K, SEL).

P. humboldtii

MEXICO: Chiapas, Sierra Madre, Ocozocoautla, 1 May 2008, *Martinez-Ovando* 20 (AMO, OAX). GUATEMALA: between San Agustin Aguascatlan and Salama, *Lewis* 127 (AMES): Alta Verapaz, Xuchaneb, 12; Pansamala, 1200 m, June 1888, *von Tuerkheim* 1248 (K). PANAMA: June 1884, cult. T. Harcourt (BM); Bocas de Torro prov., Cerro Colorado, above Copper Mine, June 1988, *Luther, Besse, Halton & Kress* 1072 (SEL); Cerro Colorado, between Rio San Felix and mining camp 7 km W of Chamé, 1500 m, 8 July 1988, *Croat* 69170 (MO, SEL); Chiriqui, above Fortuna dam, 1100 m, June 1986, *Luther, Besse, Halton & Kress* 1104 (SEL); Chiriqui, La Fortuna, cult. 1 April 1982, *Dressler in Atwood N-7* (SEL); Chiriqui, Rio Caldera, Rio Caldera, *Lehmann* BT12 (K); Chiriqui, cult. Gower (K); Chiriqui, 4 May 1920, *Powell* 153 (K). COLOMBIA: Intendencia Meta, Dec. 1937, *O. Renz* 3266 (RENZ); cult. Sander 6 April 1892 (K). Cult. Drewett, July 1889 (K); cult. Glasnevin, 19 May 1887 (K) * 1890 (K); cult. Kew, May 1887 (K) & May 1889 (K); hort. Warner, 1864 (K).

P. humboldtii var. exstaminodium

MEXICO: Chiapas, Tziscao, *Leleu* 9.n. (AMQ); *McCook & Greenwood* (live collection); HONDURAS: comm. Veitch, July 1884, *Skinner s.n.* (K).

P. lindenii

COLOMBIA: Cajamarca, W Andes of Roldanillo, 1500-1600 m, *Lehmann* HK1300 (K); W Cordillera of Cali, 1600-2000 m, *Lehmann* 1873 (K); Cauca, Popayan, *Klaboch* s.n. (W); *Klaboch s.n.* (W); near Cali, *Lehmann* 217 (BM); Santa Marta, *Linden s.n.* (P); Ocaña, *Schlim* 39 (K, P); *Schlim s.n.* (SEL). ECUADOR: Baeza, *McCook* (BH); Tunguragua: Chimborazo, road from Baños to Riobamba, *Lugo* 1862 (AMES); road between Quito and Santo Domingo, 2000 m, *Luer & Kent* 331 (SEL); *Dodson & Thien* 1224 (SEL); Baños, 16 July 1971, *Ackerman* 1051 (SEL); Tungurahua: 14 Dec. 1880, *Lehmann* 217 (G[3 sheets); lava flows of Tunguragua, *McCook* 1150 (BH); lava fields above Baños, *Luer & Dent* 201 (SEL); Baños, *Harling, Storm & Strom* 9865 (AMES); Baños, 1877, *Lehmann s.n.* (W); Tunguragua, 1877, *Lehmann s.n.* (W); Baños, *Pearce* 429 (W); Cultivated, ex Baños, *Ackerman* 1051 (SEL)); road from Chiriboga to Santo Domingo, *Asplund* 17357 (AMES, S), *Asplund 17433* (AMES, NY, S) & *Dodson, Williams & Adams* 3800 (SEL); Pichincha, Quito to Baños rd, 1400-1450 m, 1 Nov. 1941, *Dodson & Thien* 1224 (SEL); above Baños, 2000 m, 8 Oct. 1961, *Dodson & Thien* 857 (SEL) & 2100 m, 7 Oct. 1961, *Dodson & Thien* 867 (SEL); Pichincha, road from Nono to Tandayapa, along Rio Alambi, 1850 m, 14 May 1981, *Dodson, Dodson, Clendenin & Clendenin* 10821 (SEL); Napo, road from Quito to Lago Agrio, *Gentry* 12610 (M0); Napo, km 29 road from Baeza to Lago Agrio, 24 Nov. 1983, *Besse, Halton & Besse* 1901 (MO, SEL); Carchi, nr Selva Alegra, 2100 m, 1 March 1992, *Dalström* 1591 (SEL).

P. warszewiczianum

VENEZUELA: Alrededores de Merida, *Linden s.n.* (W); COLOMBIA: Palmira, 11 July 1894, *White in Lehmann* HK1299 (K); cult. Sander (K). ECUADOR: Carchi Prov., Maldonado to Tulcan, 1605 m, 1 May 1992, *Dodson, Hagsater & Embree* 19072 (MO, QCNE); Zamora-Chinchipe Prov., Jan. 1993, *Müller* 1291 (RENZ); Zamora, *Lehmann* 6868 (K, W) & *Lehmann* 10027 (K); Choco, 7 July 1869, *Elwes s.n.* (BM); Morona-Santiago Prov., Rio Upano, 14 km N of bridge at Macas, 23 Nov. 1988, *Whitten s.n.* (QCNE) & 29 Aug. 1987, *Whitten* 87309 (QCNE). Cult. Lawrence, June 1891 (K); cult. Glasnevin, 8 April 1885 (K); cult. Sander, 21 March 1921 (K); cult. Swansea Borough, 18 June 1924 (K); cult. Veitch, Feb. 1888 (K).

SELENIPEDIUM

S. aequinoctiale

ECUADOR: Esmeraldas, Lita, *Javita & Epling* 860 (S, SEL!) & *Harling* 4304 (K, S); border area between Carchi and Esmeraldas provinces, about 290 km past Lita on road to Alto Tambo, 550 m, 24 June 1991, *van der Werff* 11970 (MO, QCNE); Lita to San Lorenzo, km 4.5, 500 m, 3 Sept. 1993, *Dodson et al.* 19090 (RPSC); Carchi, Reserva Awa, Sabalera, 650 m, 19 June 1992, *G. Tipaz et al.* 1512 (QCNE); Imbabura, Lita, 600 m, 20 July 1964, *Javita & Epling* 860 (S, SEL, UCLA). COLOMBIA: photograph from Malaga Bay (P. Quintana, pers. comm.).

S. chica

PANAMA: without prov., *Duchassaing s.n.* (P!, W!); 20 miles NE of Panama City, *Powell 258* (MO) & *Powell* 3007 (MO); 4 km NE of city, Nov 1922, *Powell* 258 (K).
Ancon Hill, *Killip* 3043 (MO); Canyon of Rio Chagres, 70-80 m, *Steyermark & Allen* 17506 (MO).

S. dodsonii

ECUADOR: Succumbios Province, Rio Quiwado and Rio Tiwaeno, 200 m, 28 April 1981, *E.W. Davis & Yost* 1048 (AMES); Pastaza Province, Pozo Petrolero, 'Namoyacu', de Unocal, 30 km S of Pueblo de Curaray, 290 m, 13-30 Nov. 1990, *S. Espinoza & T. Coba* 670 (MO, CR. OCNE); 10 km S of Sarayacu, trail to Copataza, 19 Aug. 1979, *H. Lugo* 5538 (AMES).

S. isabelianum

BRAZIL: Para State, Antonio Lemos iguapos, Aug. 1927, *P. von Luetzelburg* 21229 (W); Para, Ariboquinha, 9 Aug. 1999, *Guedes* 1673 (SEL); Para, April 1889, *Rand in L. Linden s.n.* (K); Bahia, Van den Berg, Briggs, Lucas (all sight and photographic records).

S. palmifolium

GUIANA: Barama, 1896, *Im Thurn* C233 (K); Kulaboon, Nov 1886, *Jenman* 2519 (K); Upper Demerara R, Sept. 1887, *Jenman* 4007 (K); Demerara, *Jenman* 6680 (K); Marabilli creek, Essequibo R, 29 Sept 1938, *Forest Dept s.n.* (K) & *Forest Dept* F119 (2728); Bartica to Potaro road, nr 14th milepost, 19 Aug 1937, *Sandwith* 1146 (K); 50th mile[post, 13 Sept 1937, *Sandwith* 1519 (K); 1-6 km W of Kamarang, Mazarumi R, 450-550 m, 18 Aug 1977, *Maas & Boyan* 2534 (K); without exact prov., *Appon 832* (K) & *Schomburgk* 778 (1341B0; Pakaraima Mts, Kamarang, 450 m, 8 Sept 1979, *Maas & Westra* 3962 (K); *Baumrucker & Calzardilla* 22A; *Rutkis & Dressler* 11 & 28; *Rusby & Squires s.n.*; *Schomburgk s.n.* (K) & 33 (P!); *Rand s.n.* (K); *Martius* s.n.; *im Thurn C2* (K); *Maas & Westra 3962* (K, U); *Maas, Maas & Boyan 2534* (K); *British Guyana Forest Dept. 2728* (K); *Poiteau s.n.*(K); *Rothery* 44 & 49 (K); *Sandwith* 1146 & 1519 (K); *Appun 832* (K); *Jenman* 2519, 4007 & 6680 (K); *Schomburgk* 778 (K). FRENCH GUYANA: Cayenne, 18 March 1961, *Aubreville* 27 (P); 1838 & 1850, *Leprieur s.n.* (P); *Delongchamps* 94 (P); Cayenne, 1859, *Sagot s.n.* (P); Mt Cabasson, *Nichuri s.n.* (P); Saul, Mts Belvedere, 3.5 km SE of village, 27 April 1973, *de Granville* 1576 (P); Cayenne, 1853, *Rothery 44* (K, P); without exact provenance, *Le Prieur s.n.* (P!); *Duchaissang* 94 (P!). TRINIDAD: Long Stretch Savanna, 26 July, 1977, *Philcox, Andrews & Kelly* 8055 (K! P!); Valencia, 15 Oct. 1862, Herb. Trin. 1819 (AMES 67803a!) & (AMES 67172!); between Tamana rd and next mile post, 6 Nov. 1924, *Williams & Freeman s.n.*, Herb. Trinidad 67803 (AMES, K); Valencia road (Old Matura road), 12 May 1929, *Broadway* 7186 (K). VENEZUELA: Lower Orinoco, *Rusby & Squires* 393 (K).

S. steyermarkii
VENEZUELA: Venezuela, Estado Bolivar, Fila de la Danta, between Luepa and Cerro Venamo, 132 km from El Dorado on way to Santa Elena de Uairén, *Steyermark & Nilsson* 273 (holo. VEN, iso. NY); Cerro Venamo, about 130 km S of El Dorado, 800-1200 m, *Dunsterville* 777A (AMES, K); *Steyermark* 60854 (AMES, F), 94117; *Steyermark, Dunsterville & Dunsterville* 92324 & 92911 (AMES); *Steyermark & Aristeguieta* 23 (AMES) & 88 (AMES).

GENERAL INDEX

SCIENTIFIC INDEX